STUDY GUIDE AND SELECTED SOLUTIONS

Bette Kreuz

University of Michigan

GENERAL, ORGANIC, AND BIOLOGICAL CHEMISTRY

AN INTEGRATED APPROACH

FROST • DEAL • TIMBERLAKE

Prentice Hall

Boston Columbus Indianapolis New York San Francisco Upper Saddle River
Amsterdam Cape Town Dubai London Madrid Milan Munich Paris Montréal Toronto
Delhi Mexico City São Paulo Sydney Hong Kong Seoul Singapore Taipei Tokyo

Editor in Chief: Adam Jaworski
Assistant Editor: Carol G. DuPont
Marketing Manager: Erin Gardner
Managing Editor, Chemistry and Geosciences: Gina M. Cheselka
Operation Specialist: Maura Zaldivar
Supplement Cover Designer: Paul Gourhan
Cover Image: *Grapes and Pasta:* Jupiter Unlimited; *Dancer:* Shutterstock; *Molecular Art:* Precision

Pearson Education, Inc.
Pearson Prentice Hall
Upper Saddle River, NJ 07458

Printed in the United States of America

10 9 8 7 6 5 4 3 2 1

ISBN-13: 978-0-805-38186-3
ISBN-10: 0-805-38186-4

Prentice Hall
is an imprint of

PEARSON

www.pearsonhighered.com

Contents

Chapter 1

Chemistry: It's All About "Stuff"

1.1 Stuff *is* Matter

Learning Objectives

Upon completion of this material, a student should be able to do the following:

A. define the following key terms:

matter	gram	state of matter
volume	liter	solid
mass	meter	liquid
metric system	conversion factor	gas

B. state the metric units of measurement for length, volume, and mass
C. list the five prefixes commonly used in the metric system along with their symbols and values
D. use the units on numbers to create conversion factors
E. convert from metric to English units and vice versa
F. perform word problems and simple dosage calculations using unit conversions.
G. classify the three states of matter (solid, liquid, and gas) based on shape and volume characteristics.

1.2 A Closer Look at Gases – An Introduction to the Behavior of Gases

Learning Objectives

Upon completion of this material, a student should be able to do the following:

A. define the following key terms:

pressure	significant figures
pounds per square inch	temperature
millimeters of mercury	

B. state two common units used for measuring pressure
C. explain the basic relationships of gas pressure, temperature, volume, and amount.
D. apply Boyle's law
E. given appropriate information about the initial and final states, use the Boyle's law equation to calculate the initial or final pressure or to calculate the initial or final volume
F. identify the number of significant figures in a measurement
G. report the answers to addition, subtraction, multiplication, and division calculations using the proper number of significant figures
H. tell how temperature and motion of gas particles are related
I. list the three temperature scales in common use
J. write out the equations which convert temperatures from °C to °F, from °F to °C, and from °C to K; use these equations to perform temperature conversions from one scale to another

K. apply Charles's law
L. given appropriate information about the initial and final states, use the Charles's law equation to calculate the initial or final temperature or to calculate the initial or final volume

1.3 Classifying Matter: Mixture or Pure Substance

Learning Objectives

Upon completion of this material, a student should be able to do the following:

A. define the following key terms:

mixture	element
homogeneous mixture	atom
heterogeneous mixture	compound
pure substance	

B. distinguish between a pure substance and a mixture
C. classify a mixture as a homogeneous mixture or a heterogeneous mixture

1.4 Elements, Compounds, and the Periodic Table

Learning Objectives

Upon completion of this material, a student should be able to do the following:

A. define the following key terms:

group	period	chemical formula

B. classify a pure substance as an element or a compound
C. explain how the elements are arranged in the periodic table
D. tell where the main-group elements, transition elements, and inner transition elements are located in the periodic table
E. classify an element as a metal, nonmetal, or metalloid based upon its location in the periodic table (see Figure 1.11)
F. classify an element as an alkali metal, alkaline earth metal, halogen, or noble gas based upon its location in the periodic table (see Figure 1.11)
G. if given the chemical formula, identify and provide the number of atoms of each element present in a compound

1.5 How Matter Changes

Learning Objectives

Upon completion of this material, a student should be able to do the following:

A. define the following key terms:

reactant product	chemical reaction
product	chemical equation

B. distinguish physical changes and chemical reactions.
C. identify the components of a chemical equation (reactants, reaction arrow, products)
D. identify the physical state of a reactant or product from its state labels in a chemical equation

Practice Test for Chapter 1

1. Length is measured in which of the following units?
   A. cc
   B. mg
   C. km
   D. cal

2. Mass is measured in which of the following units?
   A. kL
   B. ns
   C. J
   D. mg

3. Which of the following is larger than 10 mg?
   A. 0.0000001 g
   B. 0.0001 kg
   C. 1 cg
   D. 10,000 μg

4. How many mL are in 0.0452 L?
   A. 45.2 mL
   B. 0.0000452 mL
   C. 4.52 mL
   D. 22.6 mL

5. In one second, light travels 29,979,245,800 cm. How many kilometers is this?
   A. 299,792.458 km
   B. 2,997,924,580 km
   C. 2,997,924,580,000,000 km
   D. 299,792,458,000 km

6. A liquid has __________________________.
   A. neither definite shape nor volume
   B. both definite shape and volume
   C. a definite shape but no definite volume
   D. a definite volume but no definite shape

7. Pressure is the force exerted against a given __________.
   A. volume
   B. distance
   C. area
   D. time

8. The temperature is held constant for a gas trapped in a cylinder with a piston. If the piston doubles the pressure, how has the volume changed?
   A. expanded to double the original value
   B. decreased to one fourth the original value
   C. decreased to one half the original value

D. a conclusion cannot be reached without more data

9. At a constant temperature of 32 °C, the volume of a steel cylinder is decreased by a piston from 5.00 L to 3.00 L. If the cylinder initially contained hydrogen gas at 14.0 psi, what is the final pressure of the gas?
   A. 23.3 psi
   B. 0.0429 psi
   C. 8.40 psi
   D. 183 psi

10. Which of the following has more than two significant figures?
    A. 20
    B. 1.20
    C. 430000
    D. 0.015

11. Perform the addition 53,511 + 4,200 and report your answer to the correct number of significant figures.
    A. 58,000
    B. 57,700
    C. 57,711
    D. 49,300

12. Do the following calculation and report your answer to the correct number of significant figures.
    0.00040 / 0.000000602 = ?
    A. 664.5
    B. 665
    C. 660
    D. 670

13. As the temperature of a gas increases ______________________________.
    A. the gas particles lose kinetic energy
    B. the potential energy of the gas particles increases
    C. collisions between particles no longer occur
    D. the gas particles move faster

14. The boiling point of methanol is 65 °C. What is this temperature in degrees Fahrenheit?
    A. 18 °F
    B. 149 °F
    C. 175 °F
    D. 4 °F

15. A Sahara desert summer could be as hot as 56 °C. What is this temperature on the Kelvin scale?
    A. 329 K
    B. 133 K
    C. −217 K
    D. 13 K

16. Initially at 298 K, a 2.0 L balloon is heated at constant pressure to 360 K. Determine the final volume.
   A. 2.4 L
   B. 1.8 L
   C. 2.0 L
   D. 808 mmHg

17. Which of the following is a pure substance?
   A. copper wire
   B. a wooden baseball bat
   C. iced - tea
   D. clean air

18. Which of the following statements about heterogeneous mixtures is **false**?
   A. The individual components are distinguishable.
   B. Components of the mixture can be separated by their physical properties.
   C. They are composed of two or more elements that are chemically joined together.
   D. No two samples contain the same substances in the same amount.

19. Which of the following is a mixture?
   A. muddy water
   B. table salt, $NaCl$
   C. oxygen, O_2
   D. aluminum foil

20. Which of the following is a compound?
   A. CO
   B. Cu
   C. Al
   D. Sn

21. What are the atomic symbols of silver and iron?
   A. Ag and Fe
   B. Au and Pb
   C. Si and I
   D. S and Ir

22. Which of the following elements is a transition element?
   A. magnesium
   B. manganese
   C. helium
   D. sodium

23. Which of the following elements is a metalloid?
   A. Ba
   B. Be
   C. B
   D. Br

24. Which of the following elements is an alkaline earth metal?
    A. potassium
    B. strontium
    C. neon
    D. iodine

25. What main group element is located in Group 4A of Period 3 on the Periodic Table?
    A. Si
    B. Ga
    C. Ge
    D. In

26. Which of the following compounds contains six oxygen atoms?
    A. $C_6H_{12}O_6$
    B. $Mg(NO_3)_2$
    C. $Ca_3(PO_3)_2$
    D. all of the above

27. Which of the following compounds contains the element sodium?
    A. SO_2
    B. $NaNO_3$
    C. K_3PO_4
    D. Sm_2O_3

28. Which of the following is a physical change?
    A. rusting iron
    B. digesting food
    C. freezing water
    D. burning propane

29. Which of the following is a chemical change?
    A. boiling water
    B. making ammonia from nitrogen and hydrogen
    C. melting lead
    D. mixing salt with sand

30. Identify the gaseous product(s) in the following chemical equation:

$$CaCO_3(s) \rightarrow CaO(s) + CO_2(g)$$

i. $CaCO_3(s)$ ii. $CaO(s)$ iii. $CO_2(g)$

    A. i. only
    B. iii. only
    C. ii. and iii.
    D. i. and ii.

Answers

1. C	**2.** D	**3.** B	**4.** A	**5.** A	**6.** D	**7.** C	**8.** C	**9.** A	**10.** B
11. B	**12.** C	**13.** D	**14.** B	**15.** A	**16.** A	**17.** A	**18.** C	**19.** A	**20.** A
21. A	**22.** B	**23.** C	**24.** B	**25.** A	**26.** D	**27.** B	**28.** C	**29.** B	**30.** B

Chapter 1 – Solutions to Odd-Numbered Problems

Practice Problems

1.1 Contrast the movement of particles in a liquid with that of a gas.

The particles of a gas are far apart from each other and are free to move about their container. The particles of a liquid are free to move relative to one another; however, they are in close proximity to one another, giving the liquid a definite volume.

1.3 Indicate whether each of the following describes a solid, liquid, or gas:
a. This substance has no definite volume or shape.
b. The particles in this substance are rigidly held in place.

a. Gases have no definite volume or shape since the particles that make up the gas are not in close proximity and are, therefore, free to move throughout the entire volume of the container.

b. Solids have particles that are held rigidly in place, giving the solid both definite volume and shape.

1.5 A balloon is filled with helium gas, which is lighter than air. When the following changes are made at a constant temperature, which of the diagrams (A, B, or C) show the new volume of the balloon?

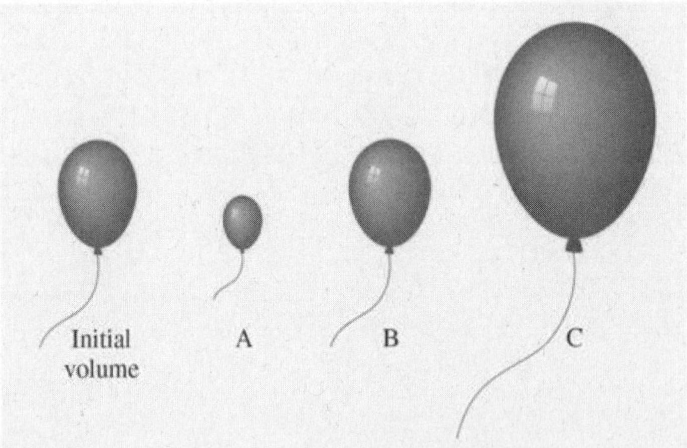

a. The balloon floats to a higher altitude where the outside pressure is lower.
b. The balloon is taken inside a house, but the atmospheric pressure remains the same.
c. The balloon is placed in a hyperbaric chamber where the pressure is increased.

a. Diagram C best describes the result. If the pressure is lowered at constant temperature, the volume of the gas will increase.

b. Diagram B best describes the result. If the pressure remains unchanged at constant temperature, the volume will not change.

c. Diagram A best describes the result. If the pressure is increased at constant temperature, the volume of the gas will decrease.

1.7 A child's balloon containing 8.5 L of helium gas at sea level is released and floats away into the atmosphere. What will the volume of the helium be when the balloon rises to the point where the atmospheric pressure is 380 mmHg?

If the temperature does not change, the Boyle's law equation can be used to solve this problem:

$$P_i \times V_i = P_f \times V_f$$

Identify what is known and what is unknown:

P_i = the initial pressure: atmospheric pressure or 760 mmHg

V_i = the initial volume: 8.5 L

P_f = the final pressure: 380 mmHg

V_f = the final volume: unknown

The final volume is unknown. Solve the Boyle's law equation for the unknown variable to get:

$$\frac{P_i \times V_i}{P_f} = V_f$$

Substituting the values into the formula and canceling units gives:

$$\frac{760\,\cancel{\text{mmHg}} \times 8.5\ \text{L}}{380\,\cancel{\text{mmHg}}} = V_f \qquad \frac{6460\ \text{L}}{380} = 17\ \text{L} = V_f$$

The volume of the helium balloon will be 17 L.

1.9 To keep a room comfortable, the air is heated or cooled to remain at approximately 75 °F. This is often referred to as ambient or room temperature. What is room temperature in °C?

Substitute the value into the conversion formula:

$$^\circ\text{C} = \frac{(^\circ\text{F} - 32)}{1.8} = \frac{(75^\circ\text{F} - 32)}{1.8} = \frac{43}{1.8} = 24$$

Room temperature is 24 °C.

1.11 Select the diagram that shows the new volume of a balloon when the following changes are made at constant pressure.

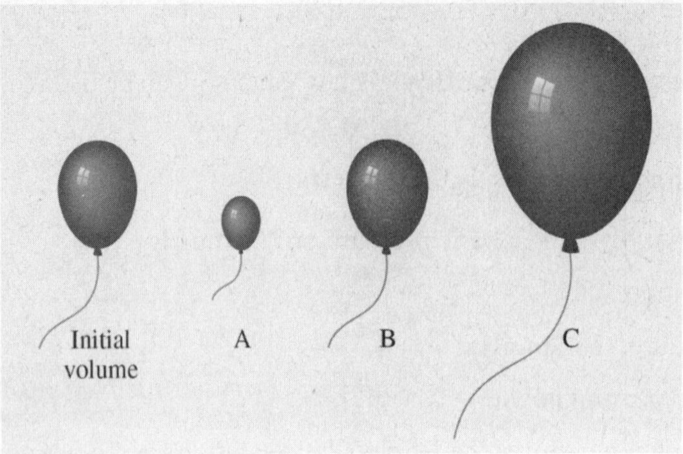

a. If the temperature is changed from 150 K to 300 K.
b. If the balloon is placed in a freezer.
c. If the balloon is warmed and then cooled to its initial temperature.

a. Diagram C best describes the result. If the temperature is increased at constant pressure, the volume of the gas will increase.

b. Diagram A best describes the result. If the temperature is decreased at constant pressure, the volume of the gas will decrease.

c. Diagram B best describes the result. If the temperature is returned to its initial temperature, the volume of the gas will be at its initial volume.

1.13 A gas has a volume of 10 L at 0 °C. What is the final temperature of the gas (in °C) if its volume increases to 25 L?

If the pressure does not change, the Charles's law equation can be used to solve this problem:

$$\frac{V_i}{T_i} = \frac{V_f}{T_f}$$

Identify what is known and what is unknown:

V_i = the initial volume: 10 L

T_i = the initial temperature: 0 °C

V_f = the final volume: 25 L

T_f = the final temperature: unknown

The final temperature is unknown. Solve the Charles's law equation for the unknown variable to get:

$$\frac{V_f \times T_i}{V_i} = T_f$$

To use the Charles's law equation, the initial temperature must be in Kelvin units, not degrees Celsius. To convert to Kelvin, substitute 0 °C into the conversion formula:

$$K = °C + 273 \qquad K = 0\ °C + 273 = 273$$

T_i = 273 K. Substituting the values into the formula and canceling units gives:

$$\frac{25\cancel{L} \times 273\ K}{10\cancel{L}} = T_f \qquad \frac{6825\ K}{10} = 683\ K = T_f$$

The final temperature is 683 K. In °C, this becomes 410 °C as shown below:

$$°C = K - 273 \qquad °C = 683 - 273 = 410\ °C.$$

1.15 Classify each of the following mixtures as homogeneous or heterogeneous.

a. a bottle of sports drink **b.** a blueberry pancake
c. gasoline **d.** tapioca pudding

One way to distinguish between homogeneous and heterogeneous mixtures is by their appearance. Homogeneous mixtures appear to be pure substances because their composition is consistent throughout. If you can see the components of the mixture, it is heterogeneous.

a. A sports drink appears to be a pure substance; therefore, it is a mixture whose composition is the same throughout. It is a homogeneous mixture.

b. You can see the components of a blueberry pancake—the berries stand out from the pancake; therefore, the mixture does not have a uniform composition. It is a heterogeneous mixture.

c. Gasoline appears to be a pure substance; therefore, it is a mixture whose composition is the same throughout. It is a homogeneous mixture.

d. You can see the components of tapioca pudding—the rice stands out from the pudding; therefore, the mixture does not have a uniform composition. It is a heterogeneous mixture.

1.17 Classify each of the following substances as a pure substance or a mixture.
a. salt water **b.** distilled water **c.** concrete

a. Salt water contains at least two substances: water and salt. Therefore, it is a mixture.

b. Distilled water consists of a single substance: water. Therefore, it is a pure substance.

c. Concrete does not have a uniform composition; therefore, it must be a mixture.

1.19 Classify each of the following as an element or a compound.
a. Fe **b.** $CaCl_2$ **c.** Si **d.** KI

a. Fe is the element iron.

b. $CaCl_2$ contains two elements: calcium and chlorine. Therefore, $CaCl_2$ is a compound.

c. Si is the element silicon.

d. KI contains two elements: potassium and iodine. Therefore, KI is a compound.

1.21 Identify and give the number of particles of each element in the following compounds.
a. $NaNO_2$, sodium nitrite, a preservative **b.** $C_6H_{22}O_{11}$, table sugar
c. $CaCO_3$, calcium carbonate

a. Sodium nitrite, $NaNO_2$, contains 1 atom of sodium, 1 atom of nitrogen, and 2 atoms of oxygen.

b. Table sugar, $C_6H_{22}O_{11}$, contains 6 atoms of carbon, 22 atoms of hydrogen, and 11 atoms of oxygen.

c. Calcium carbonate, $CaCO_3$, contains 1 atom of calcium, 1 atom of carbon, and 3 atoms of oxygen.

1.23 Determine whether each of the following is a physical change or a chemical reaction.
a. A copper penny turns green.
b. Sugar is melted to make candy.
c. An antacid tablet is dropped into water and bubbles of carbon dioxide form.

Some indications that a chemical reaction has taken place are a change in color, the production of a flame or light, the formation of a gas, or the formation of a solid.

a. A copper penny turning green represents a chemical change because there is a change in color.

b. Melting sugar is an example of a physical change. Melting is a change in phase from solid to liquid. The sugar does not change its identity during this phase change; thus, it is a physical change. All phase changes are physical changes.

c. The formation of a gas (bubbles of carbon dioxide) indicates that a chemical change has occurred.

Additional Problems

1.25 When completing the following conversions, would you expect the number to get larger or smaller?

a. 50 mg to g
b. 0.23 g to μg
c. 5.2 m to km
d. 800 kL to dL

a. The number should decrease because g is a larger unit than mg. 1000 mg = 1 g

b. The number should increase because μg is a smaller unit than g. 1,000,000 μg = 1 g

c. The number should decrease because km is a larger unit than m. 1 km = 1000 m

d. The number should increase because dL is a smaller unit than kL. 1 kL = 10,000 dL

$$\left(\frac{1000\cancel{\text{L}}}{1\,\text{kL}}\right)\times\left(\frac{10\,\text{dL}}{1\cancel{\text{L}}}\right)=\left(\frac{10{,}000\,\text{dL}}{1\,\text{kL}}\right)$$

1.27 Supply the missing information in each of the following conversion factors:

a. $\frac{?\,\text{mL}}{1\,\text{L}}$ **b.** $\frac{?\,\text{kg}}{1000\,\text{g}}$ **c.** $\frac{100\,\text{cm}}{?\,\text{m}}$

d. $\frac{1\,\text{L}}{?\,\text{dL}}$ **e.** $\frac{?\,\mu\text{g}}{1\,\text{g}}$

a. The missing value is 1000. There are 1000 mL in 1 L.

b. The missing value is 1. There are 1000 g in 1 kg.

c. The missing value is 1. There are 100 cm in 1 m.

d. The missing value is 10. There are 10 dL in 1 L.

e. The missing value is 1,000,000. There are 1,000,000 μg in 1 g.

1.29 The typical runner's stride is about 1 meter. If a runner completes a 5K (5-kilometer) race, how many strides did he take? (Assume all his strides were the same length.)

To solve the problem, we need to convert kilometers to meters then meters to strides. The two equalities we will use are 1000 m = 1 km and 1 m = 1 stride.

$$5\,\cancel{\text{km}}\times\left(\frac{1000\cancel{\text{m}}}{1\cancel{\text{km}}}\right)\times\left(\frac{1\,\text{stride}}{1\cancel{\text{m}}}\right)=5000\,\text{strides}$$

The runner takes 5000 strides during the race.

1.31 Soft drinks are sold in plastic bottles that will hold 1.0 L or 2.0 L. How many mL of soft drink are contained in each size bottle?

To solve the problem, we need to convert liters to milliliters. The equality needed to solve the problem is 1000 mL = 1 L.

$$1.0\,\cancel{L} \times \left(\frac{1000\text{ mL}}{1\,\cancel{L}}\right) = 1000\text{ mL} \qquad 2.0\,\cancel{L} \times \left(\frac{1000\text{ mL}}{1\,\cancel{L}}\right) = 2000\text{ mL}$$

There are 1000 mL (1.0×10^2 showing two significant figures) in a 1.0 L bottle and 2000 mL (2.0×10^2 showing two significant figures) in a 2.0 L bottle.

1.33 A raisin has a mass of approximately 1 g.
a. Write a conversion factor relating number of raisins and number of grams.
b. If a box contains 4500 cg of raisins, how many 1-g raisins are present in the box?

a. The conversion factors relating the number of raisins and number of grams are

$$\left(\frac{1\text{ g}}{1\text{ raisin}}\right) \qquad \left(\frac{1\text{ raisin}}{1\text{ g}}\right)$$

b. To answer this question, first convert cg to g, then use a conversion factor from part a.

$$4500\,\cancel{cg} \times \left(\frac{1\text{ g}}{100\,\cancel{cg}}\right) = 45\text{ g} \qquad 45\,\cancel{g} \times \left(\frac{1\text{ raisin}}{1\,\cancel{g}}\right) = 45\text{ raisins}$$

There are 45 1-g raisins in a 4500-cg box of raisins.

1.35 In a high-pressure chamber, the pressure can easily rise to 1140 mmHg. Convert this pressure to psi.

To solve the problem, we need to convert mmHg to psi. The equality needed to solve the problem is 760 mmHg = 14.7 psi.

$$1140\,\cancel{mmHg} \times \left(\frac{14.7\text{ psi}}{760\,\cancel{mmHg}}\right) = \frac{16758\text{ psi}}{760} = 22.1\text{ psi}$$

A pressure of 1140 mmHg is equivalent to 22.1 psi.

1.37 In this chapter, you learned that the volume of a gas changes with the applied pressure. The same is not true for a solid. In terms of particle arrangement, explain why solids are not compressible.

The particles of a gas are very far apart with a lot of empty space between them. Increasing the pressure forces the particles of a gas to move closer together. The particles of a solid are held rigidly in place without much empty space between them. Increasing the pressure cannot force the particles of a solid any closer together; thus, solids are incompressible.

1.39 A sealed syringe contains 5 cc of oxygen gas under 14.7 psi of pressure. If the pressure on the syringe is increased to 36.8 psi, but the temperature remains constant, what is the volume of the oxygen gas in the syringe?

If the temperature does not change, the Boyle's law equation can be used to solve this problem:

$$P_i \times V_i = P_f \times V_f$$

Identify what is known and what is unknown:

P_i = the initial pressure: 14.7 psi

V_i = the initial volume: 5 cc (1 cc = 1 cubic centimeter = 1 mL)

P_f = the final pressure: 36.8 psi

V_f = the final volume: unknown

The final volume is unknown. Solve the Boyle's law equation for the unknown variable to get:

$$\frac{P_i \times V_i}{P_f} = V_f$$

Substituting the values into the formula and canceling units gives:

$$\frac{14.7\,\cancel{\text{psi}} \times 5\text{ cc}}{36.8\,\cancel{\text{psi}}} = V_f \qquad \frac{73.5\text{ cc}}{36.8} = 2\text{ cc} = V_f$$

The volume of the oxygen gas in the syringe will be 2 cc following the increase in pressure.

1.41 The lowest recorded atmospheric pressure occurred during a typhoon in the Pacific Ocean in 1979 and measured approximately 652 mmHg. The highest atmospheric pressure on record occurred in Mongolia in 2001 and measured approximately 814 mmHg. If a balloon were inflated with air to a volume of 10. L under the conditions in Mongolia and then rapidly transported to the conditions described for the typhoon in the Pacific, what would the volume of the balloon be under these new conditions, assuming no change in temperature?

If the temperature does not change, the Boyle's law equation can be used to solve this problem:

$$P_i \times V_i = P_f \times V_f$$

Identify what is known and what is unknown:

P_i = the initial pressure: 814 mmHg

V_i = the initial volume: 10. L

P_f = the final pressure: 652 mmHg

V_f = the final volume: unknown

The final volume is unknown. Solve the Boyle's law equation for the unknown variable to get:

$$\frac{P_i \times V_i}{P_f} = V_f$$

Substituting the values into the formula and canceling units gives:

$$\frac{814\,\cancel{\text{mmHg}} \times 10.\,\text{L}}{652\,\cancel{\text{mmHg}}} = V_f \qquad \frac{8140\ \text{L}}{652} = 12\ \text{L} = V_f$$

The volume of the balloon would become 12 L after the change in conditions.

1.43 Give the number of significant figures in each of the following measurements:

a. 25 minutes **b.** 44.30 °C
c. 0.037010 m **d.** 800. L

All numbers in a value are significant except zeroes at the beginning of a decimal number or at the end of whole numbers. The significant figures in the following measurements are underlined.

a. There are 2 significant figures in the measurement 25 minutes. All non-zero numbers in a value are significant.

b. There are 4 significant figures in the measurement 44.30 °C. 44.30 is not a whole number; it is a decimal number. Thus, the zero at the end is significant.

c. There are 5 significant figures in the measurement 0.037010 m. 0.037010 is a decimal number; therefore, the zeroes at the beginning are not significant. The zero at the end of the value is.

d. There are 3 significant figures in the measurement 800. L. The inclusion of the decimal point indicates that this measurement is known to the ones place. If the decimal point were not present (i.e., 800 L), there would only be 1 significant figure.

1.45 Give the number of significant figures in each of the following measurements:

a. 200 °C **b.** 0.080 m
c. 10,581,000 kg **d.** 1.40 L

All numbers in a value are significant except zeroes at the beginning of a decimal number or at the end of whole numbers. The significant figures in the following measurements are underlined.

a. There is 1 significant figure in the measurement 200 °C. 200 is a whole number; therefore the zeroes at the end are not significant.

b. There are 2 significant figures in the measurement 0.080 m. 0.080 is a decimal number; therefore, the zeroes at the beginning of the value are not significant. The zero at the end is.

c. There are 5 significant figures in the measurement 10,581,000 kg. 10,581,000 is a whole number; therefore, the zeroes at the end are not significant.

d. There are 3 significant figures in the measurement 1.40 L. 1.40 is a decimal number; therefore, the zero at the end of the value is significant.

1.47 Complete each of the following calculations and give your answer to the correct number of significant figures.

a. 100 × 23 = _______
b. 97.5 – 43.02 = _______
c. 8,064/0.0360 = _______
d. 54.00 + 78 = _______

a. $100 \times 23 = 2300$. When multiplying or dividing, the resulting answer can be no more certain than the least certain number you started with. The value 100 has 1 significant figure. The value 23 has 2 significant figures. The least certain value is 100 with 1 significant figure; thus, the answer should be reported to 1 significant figure: 2000.

b. $97.5 - 43.02 = 54.48$. When adding or subtracting, the answer is rounded to give the same number of decimal places as the starting number with the fewest decimal places. The value 97.5 has 1 decimal place. The value 43.02 has 2 decimal places. The answer should be rounded to the tenths place: 54.5.

c. $8{,}064/0.0360 = 224{,}000$. When multiplying or dividing, the resulting answer can be no more certain than the least certain number you started with. The value 8,064 has 4 significant figures. The value 0.0360 has 3 significant figures. The least certain value is 0.0360 with 3 significant figures; thus, the answer should be reported to 3 significant figures: 224,000.

d. $54.00 + 78 = 132.00$. When adding or subtracting, the answer is rounded to give the same number of decimal places as the starting number with the fewest decimal places. The value 54.00 has 2 decimal places. The value 78 has 0 decimal places. The answer should be rounded to the ones place: 132.

1.49 A temperature of – 40 °F is dangerously cold. What is this temperature on the Celsius scale? On the Kelvin scale?

To convert to degrees Celsius, substitute the value into the conversion formula:

$$^{\circ}\mathrm{C} = \frac{(^{\circ}\mathrm{F} - 32)}{1.8} = \frac{(-40^{\circ}\mathrm{F} - 32)}{1.8} = \frac{-72}{1.8} = -40$$

A temperature of –40 °F is equivalent to –40 °C. To convert to Kelvin, substitute –40 °C into the conversion formula:

$$\mathrm{K} = {}^{\circ}\mathrm{C} + 273 \qquad \mathrm{K} = -40\ {}^{\circ}\mathrm{C} + 273 = 233$$

A temperature of –40 °F is equivalent to 233 K (or 230 K if –40 °F is only known to the tens place).

1.51 The hottest temperature ever recorded in the U.S. was 56.7 °C in Death Valley, California. Convert this temperature to Kelvin. What is this temperature on the Fahrenheit scale?

To convert to Kelvin, substitute 56.7 °C into the conversion formula:

$$\mathrm{K} = {}^{\circ}\mathrm{C} + 273 \qquad \mathrm{K} = 56.7\ {}^{\circ}\mathrm{C} + 273 = 329.7$$

A temperature of 56.7 °C is equivalent to 330. K To convert to degrees Fahrenheit, substitute the value into the conversion formula:

$$^{\circ}\mathrm{F} = 1.8^{\circ}\mathrm{C} + 32 \qquad ^{\circ}\mathrm{F} = 1.8 \times 56.7 + 32 = 102.06 + 32 = 134.06$$

A temperature of 56.7 °C is equivalent to 134 °F.

1.53 When working with Charles's law, what variable must be assumed to remain constant?

Charles investigated how changing the temperature of a gas affected its volume. He held the pressure and amount of gas constant in his experiments.

1.55 A beach ball is filled with 10.0 L of air in a hotel room where the temperature is 65 °F. What is the volume of the air in the beach ball out on the beach where the temperature is 38 °C?

If the pressure does not change, the Charles's law equation can be used to solve this problem:

$$\frac{V_i}{T_i} = \frac{V_f}{T_f}$$

Identify what is known and what is unknown:

V_i = the initial volume: 10.0 L

T_i = the initial temperature: 65 °F

V_f = the final volume: unknown

T_f = the final temperature: 38 °C

The final volume is unknown. Solve the Charles's law equation for the unknown variable to get:

$$\frac{V_i \times T_f}{T_i} = V_f$$

To use the Charles's law equation, the initial and final temperatures must be in Kelvin units, not degrees Fahrenheit or degrees Celsius. To convert the initial temperature into Kelvin, degrees Fahrenheit must be converted to degrees Celsius first. To do so, substitute 65 °F into the conversion formula:

$$^\circ\text{C} = \frac{(^\circ\text{F} - 32)}{1.8} = \frac{(65^\circ\text{F} - 32)}{1.8} = \frac{33}{1.8} = 18$$

To convert to Kelvin, substitute 18 °C into the conversion formula:

$$\text{K} = {}^\circ\text{C} + 273 \qquad \text{K} = 18\ {}^\circ\text{C} + 273 = 291$$

T_i = 291 K. To convert the final temperature into Kelvin, substitute 38 °C into the conversion formula:

$$\text{K} = {}^\circ\text{C} + 273 \qquad \text{K} = 38\ {}^\circ\text{C} + 273 = 311$$

T_f = 311 K. Substituting the values into the formula and canceling units gives:

$$\frac{10.0\ \text{L} \times 311\ \cancel{\text{K}}}{291\ \cancel{\text{K}}} = V_f \qquad \frac{3110\ \text{L}}{291} = 10.7\ \text{L} = V_f$$

The final volume of the beach ball would be 10.7 L. (There should be three significant figures in the answer. When converting temperatures, 65 °F became 291 K; and 38 °C became 311 K. Both 291 K and 311 K have three significant figures, as does the initial volume 10.0 L. Thus, the final answer can be reported to three significant figures.)

1.57 If a hot air balloon has an initial volume of 1000 L at 50 °C, what is the temperature (in °C) of the air inside the balloon if the volume expands to 2500 L?

If the pressure does not change, the Charles's law equation can be used to solve this problem:

$$\frac{V_i}{T_i} = \frac{V_f}{T_f}$$

Identify what is known and what is unknown:

V_i = the initial volume: 1000 L

T_i = the initial temperature: 50 °C

V_f = the final volume: 2500 L

T_f = the final temperature: unknown

The final temperature is unknown. Solve the Charles's law equation for the unknown variable to get:

$$\frac{V_f \times T_i}{V_i} = T_f$$

To use the Charles's law equation, the initial temperature must be in Kelvin units, not degrees Celsius. To convert the initial temperature into Kelvin, substitute 50 °C into the conversion formula:

$$K = °C + 273 \qquad K = 50\ °C + 273 = 323$$

T_i = 323 K. Substituting the values into the formula and canceling units gives:

$$\frac{2500\cancel{L} \times 323\ K}{1000\cancel{L}} = T_f \qquad \frac{807500\ K}{1000} = 807.5\ K = T_f$$

The final temperature of the air inside the balloon is 808 K. In °C, this becomes 535 °C as shown below:

$$°C = K - 273 \qquad °C = 808 - 273 = 535\ °C.$$

1.59 Classify each of the following as a mixture or a pure substance:

a. ink in a pen **b.** wood
c. blood **d.** ammonia (NH_3)
e. silicon

Pure substances are elements or compounds. If a substance has more than one component, it is a mixture. There are very few examples of pure substances that can be found in nature.

a. The ink in a pen is a mixture.

b. Wood is a mixture. You can see different colors in the grain and throughout the wood.

c. Blood is a mixture of many components.

d. Ammonia is a pure substance, as indicated by the chemical formula.

e. Silicon is an element, making it a pure substance.

1.61 For each of the substances that you classified as a pure substance in Problem 1.59, determine whether it is an element or a compound.

d. Ammonia is a compound containing two elements: nitrogen and hydrogen, as indicated by the chemical formula NH_3.

e. Silicon is an element which has the atomic symbol Si.

1.63 For each of the substances that you classified as a mixture in Problem 1.62, determine whether it is a homogeneous or heterogeneous mixture.

One way to distinguish between homogeneous and heterogeneous mixtures is by their appearance. Homogeneous mixtures appear to be pure substances because their composition is consistent throughout. If you can see the components of the mixture, it is heterogeneous.

a. Paint is a homogeneous mixture; it appears to have constant composition.

d. Most milk available in stores has been homogenized to produce a homogenous mixture. However, some high-fat milk is a heterogeneous mixture that will separate upon standing for 12 to 24 hours. Homogenization is a process that prevents the high-fat cream layer from separating out from the low-fat milk layer.

1.65 Use the periodic table to supply the missing information in the following chart.

Name	Elemental Symbol	Group	Period	Metal or Nonmetal
Lithium				
		3B	4	
	S			
		4A		Nonmetal

Lithium has the elemental symbol Li. It can be found in Group 1A and Period 2. Lithium is a metal.

The element found in Group 3B of Period 4 is scandium. Scandium has the elemental symbol Sc and is a metal.

The element which has the symbol S is sulfur. Sulfur is found in Group 6A of Period 3. Sulfur is a nonmetal.

The only nonmetal in Group 4A is carbon. Carbon has the elemental symbol C and can be found in Period 2.

The complete table would be:

Name	Elemental Symbol	Group	Period	Metal or Nonmetal
Lithium	**Li**	**1A**	**2**	**Metal**
Scandium	**Sc**	3B	4	**Metal**
Sulfur	S	**6A**	**3**	**Nonmetal**
Carbon	**C**	4A	**2**	Nonmetal

1.67 Give the name and number of particles of each element in the following compounds:

a. TiO_2, titanium dioxide, white pigment in paint
b. NH_3, ammonia, a household cleaner
c. Na_2CO_3, sodium carbonate, baking powder
d. N_2O, nitrous oxide, laughing gas
e. KOH, potassium hydroxide, used in some drain cleaners

a. Titanium dioxide, TiO_2, contains 1 atom of titanium and 2 atoms of oxygen.

b. Ammonia, NH_3, contains 1 atom of nitrogen and 3 atoms of hydrogen.

c. Sodium carbonate, Na_2CO_3, contains 2 atoms of sodium, 1 atom of carbon, and 3 atoms of oxygen.

d. Nitrous oxide, N_2O, contains 2 atoms of nitrogen and 1 atom of oxygen.

e. Potassium hydroxide, KOH, contains 1 atom of potassium, 1 atom of oxygen, and 1 atom of hydrogen.

1.69 Identify each of the following as a physical change or a chemical reaction:

a. bleaching your hair
b. cutting your finger
c. a cut healing
d. nitric acid dissolving copper metal to form copper nitrate
e. striking a match

Some indications that a chemical reaction has taken place are a change in color, the production of a flame or light, the formation of a gas, or the formation of a solid.

a. Bleaching your hair is an example of a chemical reaction. Color change is an indication that a chemical change has taken place.

b. Cutting your finger is an example of a physical change. Cutting may change the shape or appearance of an object, but it does not change its composition.

c. A cut healing would be an example of a chemical reaction—actually, many chemical reactions. Enzymes in your body work to repair the cellular damage.

d. This is an example of a chemical reaction. Although the term "dissolving" is used, which usually indicates a physical change, the production of a new compound from the starting materials, nitric acid and copper metal, indicates that a chemical reaction has occurred.

e. Striking a match is an example of a chemical reaction between phosphorus present on the tip of the match and oxygen in the air. The resulting flame is an indication that a chemical reaction is taking place.

Challenge Problems

1.71 The brakes in your car work on a hydraulic system. When you depress the brake pedal, fluid in your brake line is compressed. In turn, this fluid presses against the brake pads, which then press against the wheels causing them to stop. If air gets into your brake lines, the brakes will work but require you to step on the brake pedal harder to make your car stop. Using your knowledge of the difference in the molecular arrangement in liquids and gases and your understanding of gas behavior, explain why air in your brake lines causes your brakes to work less efficiently.

The particles of a liquid are loosely held together without much empty space between them. Increasing the pressure cannot force the particles of a liquid any closer together; thus, when force is applied at one end, the fluid presses against the brake pad on the other end. The particles of a gas are very far apart with a lot of empty space between them. When force is applied by stepping on the brake pedal, the particles of a gas simply move closer together, preventing the brakes from working efficiently.

1.73 Table sugar ($C_{12}H_{22}O_{11}$) is dissolved in water to make sugar water.

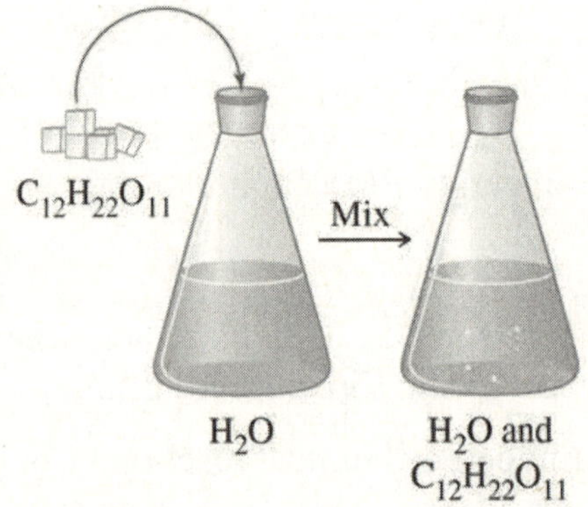

a. Is water an element or compound? How do you know?
b. Is sugar water a homogeneous or heterogeneous mixture? How do you know?
c. When sugar is dissolved in water, is it a physical change or a chemical reaction?

a. Water is a compound. There are two elements present in its chemical formula, H_2O.

b. Sugar water is a homogeneous mixture. Once dissolved completely, the resulting mixture would have a uniform composition.

c. Sugar dissolving in water is an example of a physical change. The sugar is still present, as indicated by the sweet taste of the mixture, and can be collected by evaporating the water away.

Chapter 2

Atoms and Radioactivity

2.1 Atoms and Their Components

Learning Objectives

Upon completion of this material, a student should be able to do the following:

A. define the following key terms:

electron	proton	neutron
nucleus	atomic mass unit (amu)	subatomic particle
electron cloud		

B. state the charge on the common subatomic particles (Table 2.1)
C. compare the masses of the common subatomic particles (Table 2.1)
D. describe the structure of the atom in terms of subatomic particles
E. describe the nucleus of an atom
F. specify the unit chemists use to measure the mass of an atom

2.2 Atomic Number and Mass Number

Learning Objectives

Upon completion of this material, a student should be able to do the following:

A. define the following key terms:

mass number	atomic number

B. distinguish isotopes in terms of subatomic particles present
C. locate the atomic number of an element on the periodic table
D. state the relationship between the number of electrons and protons in a neutral atom
E. provide the number of protons, neutrons, and electrons in an atom based on its atomic number and mass number
F. represent an atom and its subatomic composition using symbolic notation

2.3 Isotopes and Atomic Mass

Learning Objectives

Upon completion of this material, a student should be able to do the following:

A. define the following key terms:

isotopes	atomic mass

B. distinguish mass number and atomic mass
C. determine the composition of an isotope of an element from its symbolic notation
D. determine the composition of an isotope of an element from is atomic mass and atomic number
E. find the atomic mass of an element on the periodic table

2.4 Counting Atoms: The Mole

Learning Objectives

Upon completion of this material, a student should be able to do the following:

A. define the following key terms:

mole	molar mass
Avogadro's number (N)	equivalent units

B. express a very small or very large decimal number in scientific notation
C. convert a decimal number to scientific notation
D. give the value of Avogadro's number
E. define a mole
F. describe the relationship between Avogadro's number and the mole
G. describe how Avogadro's number and the mole are related to atomic mass
H. express the equivalent unit relating Avogadro's number and the mole
I. express the atomic mass of an element as a conversion factor relating mass and moles
J. calculate the number of moles of an element given its mass and vice versa.
K. calculate the number of atoms present from the given moles of an element.
L. calculate the number of atoms present from the given mass of an element.

2.5 Electron Arrangements

Learning Objectives

Upon completion of this material, a student should be able to do the following:

A. define the following key terms:

electron cloud	valence electrons
valence shell (level)	
	n

B. determine the number of electrons in each energy level of an atom (Table 2.4)
C. determine the value of *n* for the valence shell of a given atom
D. determine the number of valence electrons present in the main-group elements in the first three periods of the periodic table.
E. recognize the relationship between the group (column) number on the periodic table and the number of valence electrons in an atom
F. recognize the relationship between the period (row) number on the periodic table and the outermost shell (level) containing electrons in a given atom

2.6 Radioactivity and Radioisotopes

Learning Objectives

Upon completion of this material, a student should be able to do the following:

A. define the following key terms:

nuclear radiation	radioactive
radioisotope	radioactive decay
alpha (α) particle	beta (β) particle
gamma (γ) ray	ionizing radiation

B. describe radiation as a form of energy
C. state how nuclear radiation and radioactive decay are related

D. give examples of how radioisotopes are used in medicine (Table 2.5)
E. give the symbols and charges for the common types of nuclear radiation (Table 2.6)
F. describe why radioactive emissions can be dangerous to living organisms
G. list the common types of radiation in order of increasing penetrating power
H. state how X-rays and gamma rays are similar and how they differ

2.7 Nuclear Equations and Radioactive Decay

Learning Objectives

Upon completion of this material, a student should be able to do the following:

A. define the following key terms:

nuclear decay equation	alpha (α)particle
beta (β) particle	gamma (γ) ray

B. write nuclear decay equations for the common forms of ionizing radiation (α, β, and γ)
C. state how radioisotopes can be prepared in a chemical laboratory
D. write a nuclear equation for the bombardment of a given isotope with neutrons

2.8 Radiation Units and Half-Lives

Learning Objectives

Upon completion of this material, a student should be able to do the following:

A. define the following key terms:

curie	Curie
half-life	

B. give the symbol for the common units for radiation activity (Table 2.8)
C. calculate the remaining radioisotope after a certain time period given the half-life of the radioactive compound
D. state how the half-lives of naturally occurring radioisotopes differ from those used in medicine

2.9 Medical Applications for Radioisotopes

Learning Objectives

Upon completion of this material, a student should be able to do the following:

A. define the following key term:
   tracer
B. state the two main uses of radioisotopes in medicine
C. describe how the two main uses of radioisotopes in medicine differ in terms of
   1. dose
   2. penetrating power
D. distinguish "hot" spot and "cold" spot
E. use half-lives to determine the amount of radioactivity left after a given amount of time in a medical application

Practice Test for Chapter 2

1. A certain subatomic particle is located in the nucleus and has a relative mass of 1. This particle could have:
   A. no charge.
   B. a positive charge.
   C. a negative charge.
   D. no charge or a positive charge.

2. A subatomic particle with a large mass and a charge opposite that of an electron is represented by which of the following symbols?
   A. e^-
   B. p^+
   C. n^0
   D. n

3. What does a chemist use an amu to measure?
   A. the charge of an electron
   B. the mass of an atom
   C. radiation penetrating power
   D. a mole of atoms

4. Where is most of the mass of an atom located?
   A. electron cloud
   B. in the neutrons
   C. nucleus
   D. in the protons

5. The atomic number of an atom is equal to the number of:
   A. protons
   B. neutrons
   C. neutrons plus protons
   D. electrons plus protons

6. Consider the following symbolic notation.

$$^{27}_{13}\text{Al}$$

   This isotope has:
   A. 13 protons, 13 electrons, and 27 neutrons.
   B. 14 protons, 13 electrons, and 13 neutrons.
   C. 27 protons, 13 electrons, and 13 neutrons.
   D. 13 protons, 13 electrons, and 14 neutrons.

7. An isotope has 29 protons, 29 electrons, and 36 neutrons. Which of the following is the correct symbolic representation?
   A. $^{36}_{29}Cu$
   B. $^{29}_{36}Cu$
   C. $^{65}_{36}Cu$
   D. $^{65}_{29}Cu$

8. Gallium (Ga) has two naturally occurring isotopes, gallium-69 and gallium-71. If the atomic mass of gallium is 69.72, this indicates that the abundance of gallium-69
   A. is greater than that of gallium-71.
   B. is less than that of gallium-71.
   C. is about the same as that of gallium-71.
   D. A conclusion cannot be reached without more data.

9. The following represents one of the element blocks on a periodic table.

26 Fe 55.85

   The atomic mass of this element is:
   A. 26 amu
   B. 55.85 amu
   C. 81.85 amu
   D. 56 amu

10. Convert the following number to standard scientific notation.

$$0.0000000789 \text{ g}$$

   A. 7.89×10^{-8}
   B. 7.89×10^{8}
   C. 7.89×10^{-9}
   D. 7.89×10^{7}

11. Express the following number in decimal format.

$$4.4 \times 10^{6}$$

   A. 440,000,000
   B. 0.00000044
   C. 4,400,000
   D. 0.0000044

12. Which of the following is a valid conversion factor based on the atomic mass of gold (Au)?

$$197.0 \text{ g Au} = 1 \text{ mole Au}$$

A. $\dfrac{197.0 \text{ mole Au}}{1 \text{ g Au}}$

B. $\dfrac{197.0 \text{ g Au}}{1 \text{ mole Au}}$

C. $\dfrac{1 \text{ mole Au}}{197.0 \text{ g Au}}$

D. either B or C

13. What is the mass of 3.50 moles of sodium?
   A. 80.5 g
   B. 3.50 g
   C. 22.99 g
   D. 6.57 g

14. How many moles of platinum atoms are in 10.0 g of Pt?
   A. 195 mole of Pt
   B. 0.0513 mole of Pt
   C. 3.08×10^{22} mole of Pt
   D. 1950 mole of Pt

15. How many calcium atoms are in a beaker containing 22.5 g of Ca?
   A. 0.561 atoms
   B. 5.42×10^{26} atoms
   C. 1.07×10^{24} atoms
   D. 3.38×10^{23} atoms

16. Which of the following contains the largest number of atoms?
   A. 1.00 mole of K
   B. 1.00 mole of Li
   C. 1.00 mole of Na
   D. All contain the same number of atoms.

17. Which of the following contains the largest number of atoms?
   A. 5.00 g of copper
   B. 5.00 g of carbon
   C. 5.00 g of cesium

D. All contain the same number of atoms.

18. Which of the following represent the number of electrons in each energy level of magnesium?

A.

N	Number of Electrons
1	8
2	2
3	

B.

N	Number of Electrons
1	2
2	8
3	

C.

N	Number of Electrons
1	2
2	8
3	2

D.

N	Number of Electrons
1	2
2	8
3	12

19. How many electrons constitute a full energy level if $n = 2$?
    A. 2
    B. 6
    C. 8
    D. 10

20. How many valence electrons are there in an atom of phosphorus (P)?
    A. 3
    B. 5
    C. 15
    D. 16

21. Which of the following types of radiation has the largest penetrating power?
    A. α
    B. β
    C. ${}^{0}_{-1}e$
    D. γ

22. Which of the following types of radiation has a positive charge?
    A. α

B. β
C. ${}^{0}_{-1}e$
D. γ

23. Which of the following most closely resembles an X-ray?
A. α
B. β
C. ${}^{0}_{-1}e$
D. γ

24. An isotope containing 9 protons and 9 neutrons is used in PET scanning. Which of the following correctly characterizes this isotope?
A. fluorine-18
B. emits ${}^{0}_{1}e$
C. mass number = 18
D. A and C
E. A, B, and C

25. Consider the following nuclear equation.

$${}^{66}_{29}\text{Cu} + {}^{0}_{-1}\text{e} \longrightarrow \text{X}$$

In the equation X is:
A. ${}^{66}_{30}\text{Zn}$
B. ${}^{66}_{28}\text{Ni}$
C. ${}^{66}_{28}\text{Cu}$
D. ${}^{65}_{29}\text{Cu}$

26. Consider the following nuclear equation.

$${}^{212}_{84}\text{Po} + \text{X} \longrightarrow {}^{216}_{86}\text{Rn}$$

In the equation X is:
A. ${}^{4}_{2}\text{He}$
B. ${}^{0}_{1}\text{n}$
C. ${}^{0}_{1}\text{e}$
D. ${}^{0}_{-1}\text{e}$

27. Chromium-51 has a half-life of 28 days. If a 222 g sample is received in a clinical lab, how long will it take until only 27.8 g of radioactive chromium-51 remains?
A. 28 days
B. 3 days
C. 56 days
D. 84 days

28. Technicium-99m is a useful imaging tool for medical diagnosis. This isotope has a short half-life of 6 hours. If a patient receives a dose of 28 mCi, how much radioactivity will remain when the patient goes home 24 hours later?

A. 4.0 mCi
B. 7.0 mCi
C. 3.5 mCi
D. 1.8 mCi

29. A patient receives a 50 μCi does of a gamma emitter. Which of the following is most likely correct?
A. The gamma emitter is a tracer.
B. The procedure is diagnostic in nature.
C. The procedure is therapeutic in nature.
D. A and B
E. A and C

30. Consider the following data.

Radioisotope	Half-Life
Hydrogen-3	12.3 years
Iron-59	46 days
Iodine-131	8 days
Iodine-123	13.3 hours

Which isotope would be the **least** likely to be used in a medical application?
A. hydrogen-3
B. iron-59
C. iodine-131
D. iodine-123
E. All would be likely candidates.

Answers

1. D	**2.** B	**3.** B	**4.** C	**5.** A	**6.** D	**7.** D	**8.** A	**9.** B	**10.** A
11. C	**12.** D	**13.** A	**14.** B	**15.** D	**16.** D	**17.** B	**18.** C	**19.** C	**20.** B
21. D	**22.** A	**23.** D	**24.** E	**25.** B	**26.** A	**27.** D	**28.** D	**29.** D	**30.** A

Chapter 2 – Solutions to Odd-Numbered Problems

Practice Problems

2.1 How can you determine the following?

a. the number of protons present in an atom
b. the number of neutrons present in an atom
c. the number of electrons present in an atom

a. The atomic number of an element gives the number of protons. Look up the atomic number on the periodic table.
b. The mass number is the number of protons plus the number of neutrons and the atomic number is the number or protons. The number of neutrons is the mass number minus the atomic number.
c. In a neutral atom the number of positive charges (from the protons) must be the same as the number of negative charges (from the electrons). The number of electrons is the same as the number of protons in a neutral atom or is equal to the atomic number.

2.3 Provide the name and atomic symbol of the element that has the following atomic number:

a. 8 **b.** 12 **c.** 10 **d.** 29 **e.** 47

Find the atomic number on the periodic table. This will provide the symbol of the element. A listing of the elements will provide the name and symbol.
a. oxygen, O
b. magnesium, Mg
c. neon, Ne
d. copper, Cu
e. silver, Ag

2.5 What is the mass number for a nitrogen atom that contains 7 neutrons?

The mass number is the number of neutrons plus protons. From the periodic table the atomic number (number of protons) for a nitrogen atom is 7. Therefore, the mass number is 14 or 7 protons plus 7 neutrons equals 14.

2.7 Determine the number of protons, neutrons, and electrons in the following:

a. a bromine atom that has a mass number of 80
b. a sodium atom that has a mass number of 23

The atomic number of an element gives the number of protons.
The number of neutrons is the mass number minus the atomic number.
The number of electrons is the same as the number of protons in a neutral atom or is equal to the atomic number.
a. bromine-80
1. 35 protons (the atomic number)
2. 45 neutrons (mass number - atomic number = $80 - 35 = 45$)
3. 35 electrons (same as the atomic number in a neutral atom)
b. sodium-23

1. 11 protons (the atomic number)
2. 12 neutrons (mass number - atomic number = 23 − 11 = 12)
3. 11 electrons (same as the atomic number in a neutral atom)

2.9 Provide the symbolic notation for atoms with the following:

a. 2 protons, 2 neutrons
b. 17 protons, 18 neutrons
c. 16 protons, 16 neutrons
d. 55 protons, 78 neutrons

The symbolic notation gives the mass number (neutrons + protons) of the isotope as the superscript that precedes the symbol for the elements. The atomic number (number of protons) is the subscript that precedes the symbol of the element.

a. ${}^{4}_{2}He$
b. ${}^{35}_{17}Cl$
c. ${}^{32}_{16}S$
d. ${}^{133}_{55}Cs$

2.11 Provide the number of protons, neutrons, and electrons for each of the following atoms:

a. ${}^{18}_{8}O$ **b.** ${}^{40}_{20}Ca$ **c.** ${}^{108}_{47}Ag$ d. ${}^{207}_{82}Pb$

The symbolic notation gives the mass number (neutrons + protons) of the isotope as the superscript that precedes the symbol for the elements. The atomic number (number of protons) is the subscript that precedes the symbol of the element.

a. 8 protons, 10 neutrons (18 − 8 = 10), 8 electrons
b. 20 protons, 20 neutrons (40 − 20 = 20), 20 electrons
c. 47 protons, 61 neutrons (108 − 47 = 61), 47 electrons
d. 82 protons, 125 neutrons (207 − 82 = 125), 82 electrons

2.13 There are three naturally occurring isotopes of magnesium, magnesium-24, magnesium-25, and magnesium-26.

a. How many neutrons are present in each isotope?
b. Write complete symbolic notation for each isotope.
c. Based on the average atomic mass given in the periodic table, which isotope of magnesium is the most abundant?

The number of neutrons is the mass number minus the atomic number.

a. magnesium-24: 12 neutrons (24 −12 = 12)
magnesium-25: 13 neutrons (25 −12 = 13)
magnesium-26: 12 neutrons (26 −12 = 14)

The symbolic notation gives the mass number (neutrons + protons) of the isotope as the superscript that precedes the symbol for the elements. The atomic number (number of protons) is the subscript that precedes the symbol of the element.

b. magnesium-24: $^{24}_{12}Mg$

magnesium-25: $^{25}_{12}Mg$
magnesium-26: $^{26}_{12}Mg$

c. Since average atomic mass (from the periodic table) is 24.31, the isotope with the mass number closest to this value must be the most abundant or magnesium-24.

2.15 Compare (a) the number of atoms and (b) the number of grams present in 1.00 mol of silver (Ag) and 1.00 mol of gold (Au).

a. 1.00 mol contains Avogadro's number. 1.00 mol of silver and 1.00 mol gold contain the same number of atoms, Avogadro's number.
b. The atomic mass of an element contains one mole or Avogadro's number of atoms. The atomic mass of silver is 107.9 g and that of gold is 197.0 g. One mole of gold has a larger mass than one mole of silver.

2.17 Calculate the following:
a. the number of Na atoms in 0.250 mol of Na
b. the number of moles in 4.0 g of Pb
c. the number of atoms in 12.0 g of Si

Avogadro's number (6.02×10^{23} Na atoms = 1 mole Na) provides the conversion factor to convert from atoms to moles. Set up the calculation so that the units cancel appropriately to give the desired units on the answer.

a. $0.250 \text{ mol Na} \times \dfrac{6.02\times10^{23} \text{ Na atoms}}{1 \text{ mole Na atoms}} = 1.51\times10^{23} \text{ Na atoms}$

The atomic mass (207.2 g Pb = 1 mole Pb) provides the conversion factor to convert from moles to grams.

b. $4.0 \text{ g Pb} \times \dfrac{1 \text{ mole Pb}}{207.2 \text{ g Pb}} = 0.019 \text{ mole Pb}$

Avogadro's number (6.02×10^{23} Si atoms = 1 mole Si) and the atomic mass (28.09 g Si = 1 mole Si) must be used to convert atoms to grams. The calculation involves two conversions shown in the calculation below with the numbered boxes.

c. $12.0 \text{ g Si} \times \dfrac{1 \text{ mole Si}}{28.09 \text{ g Si}} \times \dfrac{6.02\times10^{23} \text{ Si atoms}}{1 \text{ mole Si atoms}} = 2.57\times10^{23} \text{ Si atoms}$

$\boxed{1}$ $\boxed{2}$

2.19 How many electrons constitute a full energy level n=1? Energy level n=2?

See Table 2.4: n=1 has a capacity of 2 electrons
n=2 has a capacity of 8 electrons

2.21 How many electrons are in each energy level of the following elements?

a. He **b.** C **c.** Na **d.** Ne

Find the number electrons for each atom using the atomic number. The number of electrons equal the atomic number for a neutral atom. Use Table 2.4 to place the electrons in the energy levels.

a. He: atomic number 2, 2 electrons

n	Number of Electrons
1	2

b. C: atomic number 6, 6 electrons

n	Number of Electrons
1	2
2	4

c. Na: atomic number 11, 11 electrons

n	Number of Electrons
1	2
2	8
3	1

d. Ne: atomic number 10, 10 electrons

n	Number of Electrons
1	2
2	8

2.23 How many valence electrons are present in the following atoms?

a. O **b.** C **c.** P **d.** Na

The number of valence electrons is the number contained in the highest (outermost) energy level. Use Table 2.4 to determine the electron arrangement.

a. O: atomic number 8, 8 electrons

n	Number of Electrons
1	2
2	6

6 valence electrons

b. C: atomic number 6, 6 electrons

n	Number of Electrons
1	2
2	**4**

4 valence electrons

c. P: atomic number 15, 15 electrons

n	Number of Electrons
1	2
2	8
3	**5**

5 valence electrons

d. Na: atomic number 11, 11 electrons

n	Number of Electrons
1	2
2	8
3	**1**

1 valence electron

2.25 If the symbol for an alpha particle is ${}^{4}_{2}\text{He}$, how is an alpha particle different from a helium atom?

An alpha particle is a helium nucleus. A helium atom consists of both a nucleus and the electron cloud. For helium, there are two electrons in the electron cloud that surrounds the nucleus.

2.27 Which type of nuclear radiation has the highest energy?

Gamma (γ) rays are the type of radiation that has the highest energy.

2.29 Complete the table below:

Isotope Name	Symbolic Notation	Number of Protons	Number of Neutrons	Mass Number	Medical Use
Thallium-201					Tumor imaging
		53		123	Thyroid imaging
		54		133	Pulmonary ventilation imaging
Fluorine-18					Positron emission tomography (PET) imaging

The atomic number of an element gives the number of protons.

The number of neutrons is the mass number minus the atomic number.

The symbolic notation gives the mass number (neutrons + protons) of the isotope as the superscript that precedes the symbol for the elements. The atomic number (number of protons) is the subscript that precedes the symbol of the element.

Isotope Name	Symbolic Notation	Number of Protons	Number of Neutrons	Mass Number	Medical Use
Thallium-201	${}^{201}_{81}Tl$	81	120	201	Tumor imaging
Iodine-123	${}^{123}_{53}I$	53	70	123	Thyroid imaging
Xenon-133	${}^{133}_{54}Xe$	54	79	133	Pulmonary ventilation imaging
Fluorine-18	${}^{18}_{9}F$	9	9	18	Positron emission tomography (PET) imaging

2.31 Provide a balanced nuclear decay equation for the decay of each of the following:

a. carbon-14 undergoing beta decay
b. polonium-212 undergoing alpha decay
c. copper-66 undergoing beta decay
d. americium-241 undergoing alpha decay

The mass and atomic numbers must be the same on both sides of the equation. An alpha particle is represented as ${}^{4}_{2}He$ and a beta particle is represented as ${}^{0}_{-1}e$. The alpha or beta particle is one of the products of the reaction, the other product must be determined based on the new atomic and mass number of the isotope produced.

a. carbon-14, $^{14}_{6}C$, beta decay, $^{0}_{-1}e$

The mass number of the missing product must be 14 to maintain the balance of 14 on either side of the equation. The atomic number of the missing element must be 7 to maintain the balance of 6 (7 +(−1)=6) on either side of the equation. Nitrogen (N) has an atomic number of 7.

$$^{14}_{6}C \rightarrow ^{14}_{7}N + ^{0}_{-1}e$$

b. polonium-212, $^{212}_{84}Po$, alpha decay, $^{4}_{2}He$

The mass number of the missing product must be 208 to maintain the balance of 212 (208 + 4 =212) on either side of the equation. The atomic number of the missing element must be 82 to maintain the balance of 84 (82 +2=84) on either side of the equation. Lead (Pb) has an atomic number of 82.

$$^{212}_{84}Po \rightarrow ^{208}_{82}Pb + ^{4}_{2}He$$

c. copper-66, $^{66}_{29}Cu$, beta decay, $^{0}_{-1}e$

The mass number of the missing product must be 30 to maintain the balance of 29 (30 + (-1) =30) on either side of the equation. The atomic number of the missing element must be 66 to maintain the balance of 66 (66 +0=66) on either side of the equation. Zinc (Zn) has an atomic number of 30.

$$^{66}_{29}Cu \rightarrow ^{66}_{30}Zn + ^{0}_{-1}e$$

d. americium-241, $^{241}_{95}Am$, alpha decay, $^{4}_{2}He$

The mass number of the missing product must be 237 to maintain the balance of 241 (237 + 4 =241) on either side of the equation. The atomic number of the missing element must be 93 to maintain the balance of 95 (93 +2=95) on either side of the equation. Neptunium (Np) has an atomic number of 93.

$$^{241}_{95}Am \rightarrow ^{237}_{93}Np + ^{4}_{2}He$$

2.33 How do the radioisotopes used in medical imaging differ from naturally occurring radioisotopes?

The half-lives of naturally occurring isotopes are much longer than those used in medical applications. The short half-life means that the radioactivity will more quickly dissipate which is better for the patient.

2.35 Radioactive iodine-131 has a half-life of 8 days. If a dose with an activity of 400 μCi of ^{131}I is given to a patient on October 1, how much of the ^{131}I will still be active on November 1 (32 days later)?

The half-life is how long it takes one-half of the original sample to decay. Determine the number of half-lives that have passed between October 1 and November 1. This illustrates how many times the activity will be cut in half sequentially.

$$32 \text{ days} \times \frac{1 \text{ half-life}}{8 \text{ days}} = 4 \text{ half-lives}$$

The activity will be reduced sequentially four times by a factor of two.

$$400\ \mu\text{Ci} \times \frac{1}{2} \times \frac{1}{2} \times \frac{1}{2} \times \frac{1}{2} = 25\ \mu\text{Ci}$$

2.37 Because calcium is an element in bone, why do you think it might be useful to use the radioisotope calcium-47 in the diagnosis and treatment of bone diseases?

Calcium-47 will be taken up by the bones and an image of the bone can be recorded for diagnostic purposes when administered in tracer amounts. Higher levels might be used for therapeutic purposes.

2.39 Chromium-31 is used in imaging red blood cells and has a half-life of 28 days. If a dose with an activity of 50 μCi is given to a patient, how long will it take for the patient to have less than 5 μCi (one-tenth the original dose) present?

One-tenth (0.100) the original dose would be between 3 $(\frac{1}{2} \times \frac{1}{2} \times \frac{1}{2} = 0.125)$ and 4 $(\frac{1}{2} \times \frac{1}{2} \times \frac{1}{2} \times \frac{1}{2} = 0.0625)$ half lives. Three half-lives would be 84 days (3 ×28 days). Four half-lives would be 112 days (4 ×28 days). 0.100 is **about** half way between 0.125 and 0.0625, so an estimate would be about half way between 84 and 112 days$(\frac{84 \text{ days} + 112 \text{ days}}{2} = 98 \text{ days})$ or 98 days.

Additional Problems

2.41 Complete the following statements:

a. A ________ is a subatomic particle that has a neutral charge.
b. The atomic number on the periodic table equals the number of ________ in the atom.
c. Two atoms that have the same number of protons and electrons but a different number of neutrons are called ___________.

Refer to Table 2.1 for the properties of the common subatomic particles.
a. neutron
Use the definition of atomic number.
b. protons
The number of protons is the same so the atoms are the same element. Use the definition of an isotope.
c. isotopes

2.43 Provide the number of protons and electrons in the following atoms:
a. gold
b. Zn
c. element number 29

The atomic number, found on the periodic table, gives the number of protons in the nucleus of the atom. In a neutral atom, the number of protons and electrons are the same.
a. gold (Au), atomic number 79: 79 protons and 79 electrons
a. Zn (zinc), atomic number 30: 30 protons and 30 electrons
a. element 29, atomic number 29: 29 protons and 29 electrons. Note this element is Cu (copper) but this isn't really needed to answer the question.

2.45 Determine the number of protons, neutrons, and electrons in the following atoms:
a. $^{55}_{26}Fe$ **b.** $^{15}_{7}N$ c. $^{52}_{24}Cr$ **d.** $^{137}_{56}Ba$

The symbolic notation gives the mass number (neutrons + protons) of the isotope as the superscript that precedes the symbol for the element. The atomic number (number of protons) is the subscript that precedes the symbol of the element.

In a neutral atom, the number of protons and electrons are the same.

a. $^{55}_{26}Fe$ 26 protons, 29 neutrons (55 − 26 = 29), 26 electrons

b. $^{15}_{7}N$ 7 protons, 8 neutrons (15 − 7 = 8), 7 electrons

c. $^{52}_{24}Cr$ 24 protons, 28 neutrons (52 − 24 = 28), 24 electrons

d. $^{137}_{56}Ba$ 56 protons, 81 neutrons (137 − 56 = 81), 56 electrons

2.47 Complete the following table:

Symbol	Number of Protons	Number of Neutrons	Number of Electrons	Mass Number	Name
$^{1}_{1}H$					
	12	12	12		
$^{9}_{4}Be$		5			

The symbolic notation gives the mass number (neutrons + protons) of the isotope as the superscript that precedes the symbol for the element. The atomic number (number of protons) is the subscript that precedes the symbol of the element.

The atomic number, found on the periodic table, gives the number of protons in the nucleus of the atom. In a neutral atom, the number of protons and electrons are the same.

The mass number is the number of neutrons plus the number of protons.

Symbol	Number of Protons	Number of Neutrons	Number of Electrons	Mass Number	Name
$^{1}_{1}H$	1	0	1	1	Hydrogen-1
$^{24}_{12}Mg$	12	12	12	24	Magnesium-12
$^{9}_{4}Be$	4	5	4	9	Beryllium-9

2.49 What is the mass of 4.00 mol of the following elements?
a. H **b.** Li **c.** Na **d.** Cl

The atomic mass (obtained from the periodic table) provides the conversion factor to convert from moles to grams. Set up the calculation so that the units cancel appropriately to give the desired units on the answer.

a. Atomic mass H: 1.008 g H = 1 mole H

$$4.00 \text{ mol H} \times \frac{1.008 \text{ g H}}{1 \text{ mole H}} = 4.03 \text{ g H}$$

b. Atomic mass Li: 6.941 g Li = 1 mole Li

$$4.00 \text{ mol Li} \times \frac{6.941 \text{ g Li}}{1 \text{ mole Li}} = 27.8 \text{ g Li}$$

c. Atomic mass Na: 22.99 g Na = 1 mole Na

$$4.00 \text{ mol Na} \times \frac{22.99 \text{ g Na}}{1 \text{ mole Na}} = 92.0 \text{g Na}$$

d. Atomic mass Cl: 35.45 g Cl = 1 mole Cl

$$4.00 \text{ mol Cl} \times \frac{35.45 \text{ g Cl}}{1 \text{ mole Cl}} = 142 \text{ g Cl}$$

2.51 How many atoms are present in a 1.00 kg block of gold?

Several conversions are involved:
1.) Convert the mass in kg to g (See Chapter 1).

2.) Convert gram to moles. The molar mass (197.0 g Au = 1 mole Au, from the periodic table) provides the conversion factor to convert from grams to moles.
3.) Convert from moles to atoms. Avogadro's number (6.02×10^{23} Au atoms = 1 mole Au) provides the conversion factor to convert from atoms to moles.

Set up the calculation so that the units cancel appropriately to give the desired units on the answer. All conversions are combined in the following set up and are indicated by the numbered boxes.

$$1.00\text{ kg Au} \times \underbrace{\frac{1000\text{ g Au}}{1\text{ kg Au}}}_{\boxed{1}} \times \underbrace{\frac{1\text{ mole Au}}{197.0\text{ g Au}}}_{\boxed{2}} \times \underbrace{\frac{6.02\times10^{23}\text{ Au atoms}}{1\text{ mole Au atoms}}}_{\boxed{3}} = 3.06\times10^{24}\text{ Au atoms}$$

The unit on the answer is atoms which is the desired unit and the numerical answer is large which is expected for a number of atoms.

2.53 Provide scientific notation for the following numbers:
a. 203,000,000 **b.** 12.4 **c.** 0.0000000278

The general form for scientific notation is

$$C \text{ x } 10^n$$

where C is called the coefficient and is a number between 1 and 10 and n is the exponent telling us the number of tens places applying to the coefficient. A positive exponent tells us the actual number is greater than 1, and a negative exponent tells us the number is less than 1. In scientific notation, only significant figures are shown in the coefficient (all place-holding zeros are not part of the notation).

a. 203,000,000
This number is larger than one so the exponent on the 10 (n) will be a positive number. Count the number of tens places to the right of the first number. In this number there are 8 tens places to the right of the 2, so the exponent (n) is 8. Place the decimal after the first number, here the 2. Any other significant numbers go after the decimal place. So, C is 2.03

$$2.03 \times 10^8$$

b. 12.4
This number is larger than one so the exponent on the 10 (n) will be a positive number. Count the number of tens places to the right of the first number. In this number there is 1 tens places to the right of the 1, so the exponent (n) is 1. Place the decimal after the first number, here the 1. Any other significant figures go after the decimal place. So, C is 1.24

$$1.24 \times 10^1$$

c. 0.0000000278
This number is smaller than one so the exponent on the 10 (n) will be a negative number. Count the number of tens places between the decimal point and the first number and add one. In this number there are 7 tens places between the decimal and the 2, adding 1 gives 8 (7+1). The

exponent (n) is the negative of this number or -8. Place the decimal after the first nonzero number, here the 2. Any other significant figures go after the decimal place. So, C is 2.78.

$$2.78 \times 10^{-8}$$

2.55 Express the following numbers in a decimal format:
a. 1.56×10^{-5} **b.** 2.8×10^{5} **c.** 9.0×10^{-2}

a. 1.56×10^{-5}
Negative exponents indicate that this number is smaller than 1, but greater than zero. The number in the exponent tells you how many tens places should appear to the left of the decimal place in the decimal notation. The number 1 will occupy one of these places, so 4 more zeros should appear in the final answer to the right of the decimal place. All digits in the scientific notation are significant figures and should appear in the final answer.
0.0000156

b. 2.8×10^{5}
The positive exponent indicates that this number is larger than one. The number in the exponent tells you how many tens places belong behind the decimal place in decimal notation. In this problem, the number 8 will occupy one of the tens places, so four more zeros after the 8 are necessary. All digits in the scientific notation are significant and should appear in the final answer.

280,000

c. 9.0×10^{-2}
Negative exponents indicate that this number is smaller than 1, but greater than zero. The number in the exponent tells you how many tens places should appear to the left of the decimal place in the decimal notation. The number 9 will occupy one of these places, so 1 more zero should appear in the final answer to the right of the decimal place. All digits in the scientific notation are significant and should appear in the final answer.
0.090

2.57 What is the number of electrons in the outer energy level and the group number for each of the following elements?
Example: Fluorine – 7 electrons, Group 7A

a. magnesium **b.** chlorine **c.** oxygen **d.** nitrogen
e. barium **f.** bromine

Find the location of the element on the periodic table. The number of outer energy level (valence) electrons corresponds to the element group number on the periodic table. The group (column) number represents the number of valence electrons for the main-group elements.

a. magnesium– 2 electrons, Group 2A
b. chlorine – 7 electrons, Group 7A
c. oxygen – 6 electrons, Group 6A
d. nitrogen – 5 electrons, Group 5A
e. barium – 2 electrons, Group 2A

f. bromine – 7 electrons, Group 7A

2.59 How many valence electrons are present in the following atoms?
a. C **b.** I **c.** K **d.** B

Find the location of the element on the periodic table. The number of outer energy level (valence) electrons corresponds to the element group number on the periodic table. The group (column) number represents the number of valence electrons for the main-group elements.

a. C (Group 4A): 4 valence electrons
b. I (Group 7A): 7 valence electrons
c. K (Group 1A): 1 valence electron
d. B (Group 3A): 3 valence electrons

2.61 Identify the ionizing radiation associated with the following:
a. particle that has the same number of protons and neutrons as a helium nucleus
b. an electron that is emitted from an atom's nucleus
c. has no mass

Refer to Table 2.6 that summarizes the type and properties of nuclear radiation.
a. α particle, also represented as ${}^{4}_{2}He$
b. β particle, also represented as ${}^{0}_{-1}e$
c. γ ray, also represented as ${}^{0}_{0}\gamma$

2.63 Provide the symbolic notation for the following radioactive isotopes.
a. phosphorus-32
b. cobalt with 33 neutrons
c. the element number 24 containing 27 neutrons.

The symbolic notation gives the mass number (neutrons + protons) of the isotope as the superscript that precedes the symbol for the element. The atomic number (number of protons) is the subscript that precedes the symbol of the element.

The atomic number can be found on the periodic table and is the same of the number of protons.

The mass number is equal to the number of protons plus neutrons and is often shown as a number following the name of the element.

a. phosphorus-32: 32 is the mass number of the isotope and the atomic number of phosphorus is 15.
${}^{32}_{15}P$

b. The atomic number of cobalt is 27. The mass number of this isotope is 60 (33 + 27 = 60)
${}^{60}_{27}Co$

c. The atomic number of the element is 24 and is the element chromium (Cr). The mass number of this isotope is 51 (24 + 27 = 51)

$^{51}_{24}Cr$

2.65 Complete the following nuclear decay equations:

a. $^{15}_{8}O \rightarrow \ ^{15}_{9}F + ?$ **b.** $^{46}_{23}V \rightarrow \ ? + ^{0}_{-1}e$

c. $^{234}_{92}U \rightarrow \ ? + ^{4}_{2}He$ **d.** $^{8}_{4}Be \rightarrow \ ? + ^{4}_{2}He$

The mass and atomic numbers must be the same on both sides of the equation. Determine the mass number and atomic number of the missing substance. If it is an element, the symbol can be found on the periodic table. If it is a type of nuclear radiation, the symbol can be found in Table 2.6.

a. The mass number of the missing substance is 0, to balance the mass numbers on both sides of the equation as 15. The atomic number of the missing particle must be −1 to balance the atomic numbers on both sides of the equation to be 8 (9 + (−1) =8). This combination corresponds to a beta particle or $^{0}_{-1}e$

$$^{15}_{8}O \rightarrow ^{15}_{9}F + ^{0}_{-1}e$$

b. The mass number of the missing substance is 46, to balance the mass numbers on both sides of the equation as 46. The atomic number of the missing particle must be 24 to balance the atomic numbers on both sides of the equation to be 23 (24 + (−1) =23). This combination corresponds to the element chromium, Cr.

$$^{46}_{23}V \rightarrow ^{46}_{24}Cr + ^{0}_{-1}e$$

c. The mass number of the missing substance is 230, to balance the mass numbers on both sides of the equation as 234. The atomic number of the missing particle must be 90 to balance the atomic numbers on both sides of the equation to be 92 (90 + 2=92). This combination corresponds to the element thorium, Th

$$^{234}_{92}U \rightarrow ^{230}_{90}Th + ^{4}_{2}He$$

d. The mass number of the missing substance is 4, to balance the mass numbers on both sides of the equation as 8. The atomic number of the missing particle must be 2 to balance the atomic numbers on both sides of the equation to be 4 (2 + 2=4). This combination corresponds to the element helium, He (or a second α particle).

$$^{8}_{4}Be \rightarrow ^{4}_{2}He + ^{4}_{2}He$$

2.67 Write balanced nuclear decay equations for each of the following emitters:

a. copper-66 (β) **b.** platinum-192 (α)

c. tin-126 (β) **d.** gallium-72 produces germanium-72

The mass and atomic numbers must be the same on both sides of the equation. An alpha particle is represented as ${}^{4}_{2}\text{He}$ and a beta particle is represented as ${}^{0}_{-1}\text{e}$ and a gamma ray as ${}^{0}_{0}\gamma$. The alpha or beta particle or gamma ray is one of the products of the reaction, the other product must be determined based on the new atomic and mass number of the isotope produced.

a. copper-66 is ${}^{66}_{29}\text{Cu}$, β decay emits ${}^{0}_{-1}\text{e}$

The mass number of the missing product must be 66 to maintain the balance of 66 (66 + 0 =66) on either side of the equation. The atomic number of the missing element must be 30 to maintain the balance of 30 (30 +(−1)=29) on either side the equation. Zinc (Zn) has an atomic number of 30.

$${}^{66}_{29}\text{Cu} \rightarrow {}^{66}_{30}\text{Zn} + {}^{0}_{-1}\text{e}$$

b. platinum-192 is ${}^{192}_{78}\text{Pt}$, α decay emits ${}^{4}_{2}\text{He}$

The mass number of the missing product must be 188 to maintain the balance of 192 (188 + 4 =192) on either side of the equation. The atomic number of the missing element must be 76 to maintain the balance of 78 (76 +2=78) on either side the equation. Osmium (Os) has an atomic number of 76.

$${}^{192}_{78}\text{Pt} \rightarrow {}^{188}_{76}\text{Os} + {}^{4}_{2}\text{He}$$

c. tin-126 is ${}^{126}_{50}\text{Sn}$, β decay emits ${}^{0}_{-1}\text{e}$

The mass number of the missing product must be 126 to maintain the balance of 126 (126 + 0 =126) on either side of the equation. The atomic number of the missing element must be 51 to maintain the balance of 50 (51 +(−1)=50) on either side the equation. Antimony (Sb) has an atomic number of 51.

$${}^{126}_{50}\text{Sn} \rightarrow {}^{126}_{51}\text{Sb} + {}^{0}_{-1}\text{e}$$

d. gallium-72 is ${}^{72}_{31}\text{Ga}$, germanium-72 is ${}^{72}_{32}\text{Ge}$

The mass number of the missing product must be 0 to maintain the balance of 72 (72 + 0 =72) on either side of the equation. The atomic number of the missing element must be −1 to maintain the balance of 31 (32 +(−1)=31) on either side the equation. This combination represents a beta particle, ${}^{0}_{-1}\text{e}$.

$${}^{72}_{31}\text{Ga} \rightarrow {}^{72}_{32}\text{Sb} + {}^{0}_{-1}\text{e}$$

2.69 When boron-10 is bombarded with alpha particles, nitrogen-12 and neutrons are produced. Provide a nuclear equation for this reaction.

Boron-10 is ${}^{10}_{5}\text{B}$ and an alpha particle is ${}^{4}_{2}\text{He}$. The product, nitrogen -12 is, ${}^{12}_{7}\text{N}$. Neutrons (${}^{1}_{0}\text{n}$) are also a product. These symbols give the equation for the alpha decay of boron-10.

$${}^{10}_{5}\text{B} + {}^{4}_{2}\text{He} \rightarrow {}^{12}_{7}\text{N} + ?{}^{1}_{0}\text{n}$$

Determine the number of neutrons needed by applying the principle that the mass numbers and atomic numbers on either side of the equation must balance. The atomic numbers are already balanced at 7 on either side of the equation. There must be 2 neutrons to maintain the mass number balance at 14 (12 +2=14) on either side of the equation.

$$^{10}_{5}B + ^{4}_{2}He \rightarrow ^{12}_{7}N + 2^{1}_{0}n$$

2.71 Fluorine-18, which has a half-life of 110 minutes, is used in tomography scans. If 100 mg of fluorine-18 is shipped at 8:00 AM, how many milligrams of the radioisotope are still active if the sample arrives at the nuclear medicine laboratory at 1:30 PM the same day?

The time elapsed from shipment to delivery is 5 hours and 30 minutes. This must be converted to minutes to have the same time units as the given half-life.

$$5\text{ hours} \times \frac{60\text{ minutes}}{1\text{ hour}} = 300\text{ minutes}$$

$$\text{Total Time}: 300\text{ minutes} + 30\text{ minutes} = 330\text{ minutes}$$

Find the number of half-lives that 330 minutes represents.

$$330\text{ minutes} \times \frac{1\text{ half-life}}{110\text{ minutes}} = 3\text{ half-lives}$$

The amount of radioactive material will be reduced sequentially three times by a factor of two.

$$100\text{ mg} \times \frac{1}{2} \times \frac{1}{2} \times \frac{1}{2} = 12.5\text{ mg} \approx 10\text{ mg}$$

Answer above is shown to 1 significant figure as allowed by the data, 100 mg.

2.73 If the amount of radioactive iodine-123 in a sample decreases from 0.4 g to 0.1 g in 26.2 hours, what is the half-life of iodine-123?

Use the initial mass (0.4 g) and the final mass (0.1 g) to determine how many half-lives have passed. To determine this take the initial amount and divide by 2 sequentially until the final amount is obtained.

$$0.4\text{ g} \times \frac{1}{2} = 0.2\text{ g}$$

$$0.2\text{ g} \times \frac{1}{2} = 0.1\text{ g}$$

2 half-lives have passed in 26.2 hours. The half-life is then calculated as:

$$26.2\text{ hours} \times \frac{1}{2\text{ half-lives}} = \frac{13.1\text{ hours}}{1\text{ half-life}}$$

Challenge Problems

2.75 Precious metals are commonly measured out in troy ounces. One troy ounce of platinum is equivalent to 31.10 g of platinum. How many atoms of platinum are in one troy ounce of platinum?

Two conversions are needed to determine the final answer. 1.) The mass in grams must be converted to moles using the molar mass of Pt (195.1 g Pt = 1 mole Pt). 2.) the number of moles must be converted to atoms by using Avogadro's number (1 mole Pt = 6.02×10^{23} atoms Pt). Set up the conversions so that units cancel to give the desired units on the final answer. The conversions are indicated by the boxed numbers.

$$\frac{31.10\text{ g Pt}}{1\text{ troy ounce}} \times \underbrace{\frac{1\text{ mole Pt}}{195.1\text{ g Pt}}}_{\boxed{1}} \times \underbrace{\frac{6.02\times10^{23}\text{ atoms Pt}}{1\text{ mole Pt}}}_{\boxed{2}} = \frac{9.60\times10^{22}\text{ atoms Pt}}{1\text{ troy ounce}} = 9.60\times10^{22}\text{ atoms Pt/ troy oz}$$

The units are correct and the size of the number makes sense since a large number of Pt atoms would be present in the sample.

2.77 Iron-59 has a half-life of 46 days. If 168 g of a radioactive iron (^{59}Fe) is received in the lab today, what percentage of the original is left after 276 days?
First determine the number of half-lives that have passed in 276 days.

$$276\text{ days} \times \frac{1\text{ half-life}}{46\text{ days}} = 6.0\text{ half-lives}$$

Determine the amount of radioactive iron left after half-lives by sequentially dividing the original sample by 2 six times.

$$168\text{ g} \times \frac{1}{2} \times \frac{1}{2} \times \frac{1}{2} \times \frac{1}{2} \times \frac{1}{2} \times \frac{1}{2} = 2.6\text{ g}$$

Last determine what percent this mass represents of the original sample.

$$\frac{2.6\text{ g}}{168\text{ g}} \times 100\% = 1.6\%$$

Chapter 3

Compounds: Putting Particles Together

3.1 The Octet Rule

Learning Objectives

Upon completion of this material, a student should be able to do the following:

A. define the following key term:
   octet rule
B. describe the relationship between reactivity and stability of an element
C. recognize that the electron arrangement of the noble gases is important in determining reactivity and stability
D. recall how the group number on the periodic table relates to the number of valence electrons
E. realize that some atoms give away electrons, some gain electrons, and others share electrons to achieve a valence octet

3.2 In Search of an Octet Part 1: Ion Formation

Learning Objectives

Upon completion of this material, a student should be able to do the following:

A. define the following key terms:

ion	anion	isoelectronic
cation	polyatomic ion	

B. describe the process of anion formation in terms of electron arrangement
C. describe the process of cation formation in terms of electron arrangement
D. determine the number of protons and electrons in an ion
E. understand that ion formation produces an electron arrangement usually isolectronic with that of a noble gas
F. understand that ionic charges are related to the position of an element on the periodic table
G. predict the charge on an ion formed by main-group elements based on their group number on the periodic table
H. formulate the systematic name for metal cations and nonmetal anions
I. know how to determine the name of a polyatomic ion (Table 3.1)
J. recognize some of the biologically important ions (Table 3.2)

3.3 Ionic Compounds − Electron Give and Take

Learning Objectives

Upon completion of this material, a student should be able to do the following:

A. define the following key terms:

ionic bond	ionic compound

B. understand the electrical charge nature of an ionic bond
C. understand that an ionic compound must be electrically neutral and that the charge of the ions determines the subscripts in the formula
D. use the crossover approach as a straightforward method of getting the correct formula for an ionic compound
E. predict the formula for an ionic compound given the ions present
F. name ionic compounds

3.4 In Search of an Octet Part 2: Covalent Bonding

Learning Objectives

Upon completion of this material, a student should be able to do the following:

A. define the following key terms:

covalent bond	covalent compound	molecule
bonding pair	lone pair	single bond
double bond	triple bond	Lewis structure
binary compound	molecular formula	

B. contrast the formation of a covalent bond to that of an ionic bond
C. relate the number of bonds an element forms to the number of electrons that it needs to complete its octet
D. use the number of valence electrons to determine the number and type of covalent bonds an element forms
E. represent the valence electrons for an element using an electron dot symbol
F. use electron dots or lines to represent covalent bonds
G. recognize the preferred covalent bonding patterns for main-group elements (Table 3.3)
H. draw the Lewis structure for a covalent compound based on the molecular formula
I. name binary covalent compounds based on the molecular formula

3.5 Getting Covalent Compounds into Shape

Learning Objectives

Upon completion of this material, a student should be able to do the following:

A. define the following key terms:

tetrahedral	valence-shell electron-pair repulsion	VSEPR
charge clouds	nonbonded pairs	

B. realize that a chemical formula and the Lewis structure do not specify the actual shape (geometry) of a molecule
C. understand that molecules have a three-dimensional shape that affects their properties
D. describe how valence-shell electron-pair repulsion (VSEPR) predicts the shape of a molecule
E. identify basic molecular shapes for carbon, oxygen, and nitrogen (Table 3.5)
F. identify bonding and nonbonding clouds in a Lewis structure
G. describe the effect of nonbonding clouds on molecular shape

H. use VSEPR to determine the shape of a simple molecule
I. apply VSEPR to determine the shapes within larger molecules

3.6 Electronegativity and Molecular Polarity

Learning Objectives

Upon completion of this material, a student should be able to do the following:

A. define the following key terms:

electronegativity	polar covalent bond	nonpolar covalent bond
polar molecule	nonpolar molecule	

B. use electronegativities to predict when a covalent bond is polar
C. use the dipole moment arrow (arrow with a hash mark) to represent a polar covalent bond
D. determine the polarity of a covalent bond
E. relate an element's electronegativity to the position on the periodic table
F. determine the direction of the polarity of a molecule and indicate this using the dipole moment arrow.

Practice Test for Chapter 3

1. How many protons and electrons, respectively, are in the following ion?

$$Co^{3+}$$

A. 30, 27
B. 27, 24
C. 27, 27
D. 27, 30

2. What is the correct name for the following species?

$$P^{3-}$$

A. phosphorous ion
B. phosphorous(III) ion
C. phosphate
D. phosphide

3. An ion containing 38 protons and 36 electrons could be represented and named as:
A. Sr^{2+}, strontium ion.
B. Kr^{2-}, kryptonide.
C. Sr^{2-}, strontide ion.
D. Kr^{2+}, krypton ion.

4. Which of the following is a polyatomic cation?
A. NH_4^+
B. SO_4^{2-}
C. OH^-
D. all of the above

5. Determine the correct formula of the ionic compound formed from the following combination of ions: potassium ions and nitride ions

A. KN
B. K_2N
C. KN_3
D. K_3N

6. Determine the number and type of ions present in the following formula.

$$Ca_3(PO_4)_2$$

A. 3 Ca^{3+} and 2 PO_4^{2-}
B. 3 Ca^{2+} and 2 PO_4^{3-}
C. 1 Ca_3^{2+} and 2 PO_4^{3-}
D. 3 Ca^{2+} and 8 PO_4^{3-}

7. The name for the ionic compound given below is:

$$Fe(NO_3)_2$$

A. iron nitride.
B. iron(II) nitride.
C. iron nitrate.
D. iron(II) nitrate.

8. The formula for barium carbonate is:
A. BaC.
B. $BaCO_3$.
C. $Ba(HCO_3)_2$.
D. Ba_2C.

9. Which of the following is, respectively, the correct name and composition of the ionic compound formed between ions of nickel(III) and sulfide?
A. nickel(II) sulfate, 1 Ni^{2+} and S^{2-}
B. nickel(III) sulfate, 3 Ni^{3+} and 2 S^{2-}
C. nickel(III) sulfide, 2 Ni^{3+} and 3 S^{2-}
D. nickel(II) sulfate, 1 Ni^{2+} and S^{2-}

10. The correct Lewis structure for OF_2 would contain:

A. 1 O—F bond and 1 F—F bond.
B. 1 O=F bond and 1 F—F bond.
C. 2 O—F bonds.
D. 2 O=F bonds.

11. If Cl atoms replaced the Br atoms in the molecule shown, which of the following would describe the new Lewis structure?

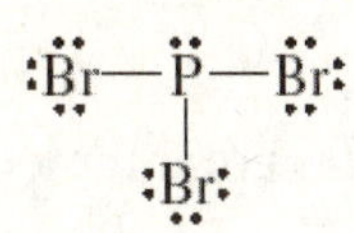

A. There would be the same number of lone pairs.
B. One of the bonds between P and Br would be double.
C. P would no longer have a lone pair.
D. all of the above

12. Which of the following pairs contains an ionic compound and a covalent compound?
    A. NH_3 and NH_4Cl
    B. PCl_3 and ClBr
    C. AlI_3 and SF_2
    D. both A and C

13. Based on Lewis structures, which of the following compounds is **not** likely to exist?
    A. CCl_4
    B. H_3S
    C. OCl_2
    D. PI_3

14. The correct name for P_2O_5 would be:
    A. diphosphorous pentoxide.
    B. phosphorous oxide.
    C. phosphorous(II) oxide.
    D. pentaphosphorous dioxide.

15. The formula for dinitrogen monoxide is
    A. NO.
    B. N_2O.
    C. NO_2.
    D. N_2O_3.

16. What is the VSEPR form of the following molecule?

    SO_3

    A. AB_2
    B. AB_3
    C. AB_2N_2
    D. AB_4

17. What is the molecular shape (geometry) of a molecule of NBr_3?
    A. tetrahedral
    B. pyramidal
    C. trigonal planar
    D. bent

18. In which of the following molecules is the molecular shape (geometry) around each carbon atom tetrahedral?

A.

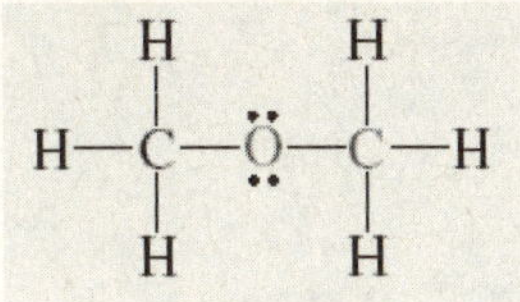

B.

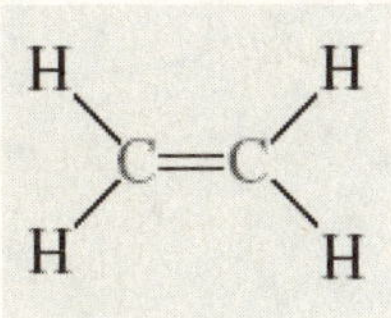

C.

D.

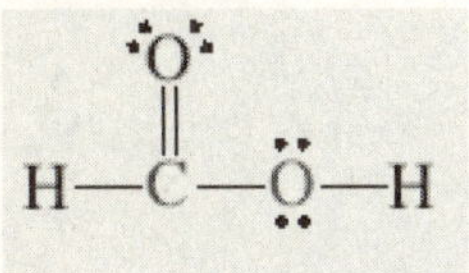

19. Which of the following atoms is the most electronegative?
   A. N
   B. Br
   C. P
   D. As

20. In which of the following bonds will the Br atom exist with a δ−?
   A. Br—O
   B. Br—F
   C. Br—N
   D. Br—Te

21. A student wrote the following as the Lewis structure for C_2H_4.

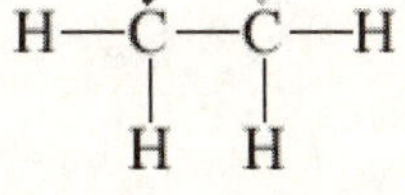

What error(s), if any, was(were) made?
   A. The number of atoms of C and H are incorrect.
   B. The H atoms do not form the correct number of bonds.
   C. The C atoms do not have an octet of electrons.
   D. All of the above are errors present in the structure.
   E. The structure is correct as written.

22. In which of the following bonds will the dipole moment arrow ($+\!\!\!\rightarrow$) point toward N?
   A. N–S
   B. N–O
   C. N–Cl
   D. N–F

23. Which of the following molecules contains polar bonds?
   A. CH_4
   B. NCl_3
   C. OF_2
   D. All contain polar bonds.

24. The most polar bond among those listed would be:
   A. F–F.
   B. O–F.
   C. C–F.
   D. N–F.

25. Of the compounds listed, which would consist of nonpolar molecules?

   A.

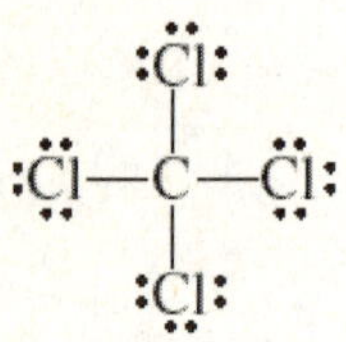

   B.

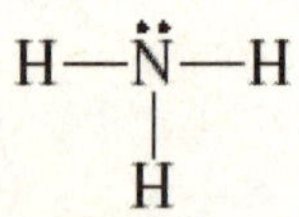

   C.

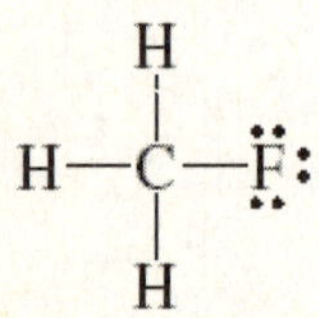

   D.

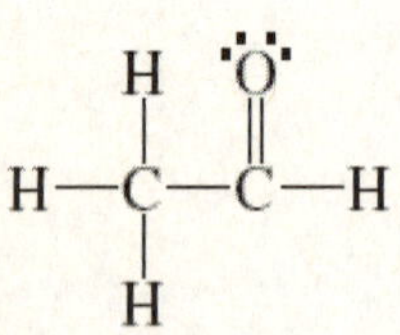

26. In the molecule shown below, how many of the carbon atoms have trigonal planar shape?

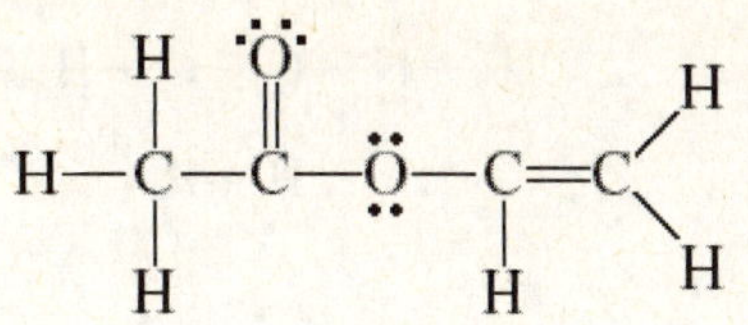

A. 1
B. 2
C. 3
D. 4
E. none

27. For the following molecule, the arrow for the molecular dipole should be drawn as:

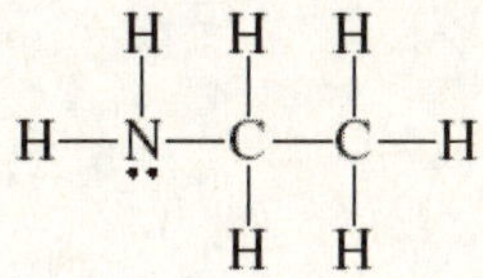

A. 
B. 
C. 
D. The molecule does not have a dipole.

28. In the following structure, the geometry around the N atom is:

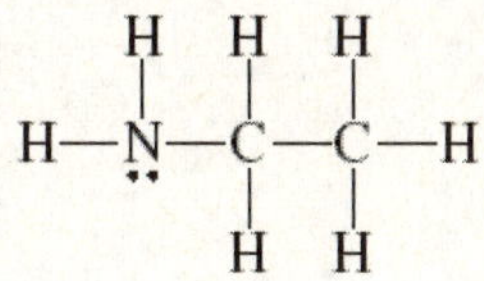

A. tetrahedral.
B. pyramidal.
C. trigonal planar.
D. linear.

29. In the following structure, the most polar bond would be:

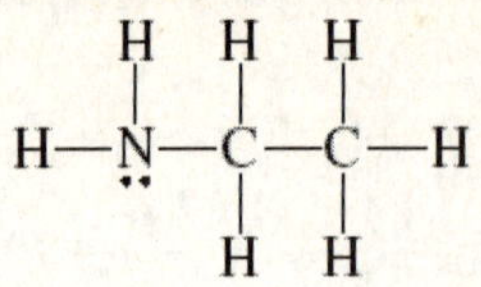

A. C—H.
B. C—N.
C. N—H.
D. All are about the same polarity.

30. An ionic compound consists of two elements, X and Z. If the formula for the compound is X_3Z_2, which of the following is true?

A. X is a metal.
B. Z is a nonmetal.
C. X might be found in Group 2A
D. Z might be found in Group 5A
E. all of the above

Answers

1. B	**2.** D	**3.** A	**4.** A	**5.** D	**6.** B	**7.** D	**8.** B	**9.** C	**10.** C
11. A	**12.** D	**13.** B	**14.** A	**15.** B	**16.** B	**17.** B	**18.** A	**19.** A	**20.** D
21. C	**22.** A	**23.** C	**24.** C	**25.** A	**26.** C	**27.** C	**28.** B	**29.** C	**30.** E

Chapter 3 – Solutions to Odd-Numbered Problems

Practice Problems

3.1 What is the difference between an atom and an ion of the same element?

An ion is an atom that has either lost or gained one or more electrons. The ion has the same number of protons and neutrons as the neutral atom, but differs in the number of electrons and electrical charge. An ion can be considered an electrically charged atom.

3.3 How many protons and electrons are present in the following ions?
a. Ca^{2+} **b.** I^- **c.** S^{2-} **d.** Zn^{2+}

The number of protons in the ions is the same as the atom which is the atomic number.

In a cation (positive ion), the number of electrons **lost** is equal to ion charge. The number of electrons remaining in the ion is then equal to the atomic number **minus** the number of electrons lost.

In an anion (negative ion), the number of electrons **gained** is equal to ion charge. The number of electrons present in the ion is then equal to the atomic number **plus** the number of electrons gained.

a. Ca^{2+}, atomic number 20: 20 protons, 18 electrons $(20 - 2 = 18)$
b. I^-, atomic number 53: 53 protons, 54 electrons $(53 + 1 = 54)$
c. S^{2-}, atomic number 16: 16 protons, 18 electrons $(16 + 2 = 18)$
d. Zn^{2+}, atomic number 30: 30 protons, 28 electrons $(30 - 2 = 28)$

3.5 Name the ions in problem 3.3 above
a. Ca^{2+} **b.** I^- **c.** S^{2-} **d.** Zn^{2+}

Metal ions (cations) are named by simply using the name of the elements and adding the word **ion**.

Transition metal ions (cations) typically form more than one ion. A Roman numeral equal to the charge is placed in parentheses directly after the name of the metal.

Nonmetal ions (anions) are named by replacing the last few letters of the name of the element with the suffix **ide**.

a. Ca^{2+}, the cation of calcium: calcium ion
b. I^-, the anion of iodine: iodide
c. S^{2-}, the anion of sulfur: sulfide ion
d. Zn^{2+}, the cation of the transition metal, zinc: zinc ion*

*Zinc only forms one ion, so the Roman numeral is omitted.

3.7 Give the name and symbol of the ion with the following number of protons and electrons:
a. 9 protons, 10 electrons
b. 24 protons, 21 electrons

Names:
Metal ions (cations) are named by simply using the name of the elements and adding the word **ion**.

Transition metal ions (cations) typically form more than one ion. A Roman numeral equal to the charge is placed in parentheses directly after the name of the metal.

Nonmetal ions (anions) are named by replacing the last few letters of the name of the element with the suffix **ide**.

Symbols:
The number of protons in the ions is the same as the atom which is the atomic number.

In a cation (positive ion), the number of electrons is **less** than the number of protons. The charge on the ion is equal to the number of protons **minus** the number of electrons. The charge will have a **positive** sign.

In an anion (negative ion), the number of electrons is **greater** than the number of protons. The charge on the ion is equal to the number of electrons **minus** the number of protons. The charge will have a **negative** sign.

a. 9 protons, 10 electrons: 1 more electron than protons, a negative ion with a charge of 1−
fluoride, F^-

b. 24 protons, 21 electrons: three more protons than electrons, a positive ion with a charge of 3+
chromium(III), Cr^{3+}

3.9 Name the following ions:
a. NH_4^+ **b.** $C_2H_3O_2^-$ **c.** CN^-

The names of polyatomic ions must be memorized. (See Table 3.1.) As in the case of simple cations, the names of positive polyatomic ions end with the word **ion**.

a. ammonium ion
b. acetate
c. cyanide

3.11 Give the formula and name for the ionic compounds formed from the following pairs of elements:
a. gold(II) and chloride, an antimicrobial
b. calcium and carbonate, the mineral limestone
c. magnesium and hydroxide, the main compound in milk of magnesia

Formulas:
The compound is written with no charges shown. When writing formulas for chemical compounds, we represent the number of each particle (ions in this case) in the compound with a subscript. A "1" is understood in the formula for the compound.

The number of ions that combine to form an ionic compound is determined by the charge of the cation and the charge of the anion. Cations and anions will combine so that the charges cancel each other out when combined, providing a formula with no net charge.

The *crossover approach* is a straightforward way of getting the correct formula for an ionic compound. The number representing the magnitude of the charge on the cation becomes the subscript on the anion and vice – versa. In the case where the charges on the ions are equal and opposite, for example +2 and –2, the formula is reduced to the smallest common ratio.

a. gold(II) and chloride: Au^{2+} and Cl^{-}, two Cl^{-} for each Au^{2+} gives a net charge of 0, $AuCl_2$
b. calcium and carbonate: Ca^{2+} and CO_3^{2-}, one CO_3^{2-} for each Ca^{2+} gives a net charge of 0, $CaCO_3$
c. magnesium and hydroxide: Mg^{2+} and OH^{-}, two OH^{-} for each Mg^{2+} gives a net charge of 0, $Mg(OH)_2$ Remember that polyatomic ions must be enclosed in parentheses with the subscript outside the parenthesis when the subscript is greater than 1.

Names:
To name an ionic compound, put the names of the two ions together, always remembering to list the cation first.
When naming the cation of a transition metal, a Roman numeral is used to designate the charge on the transition metal.
If the compound contains one of the polyatomic ions, the name of the ion remains unchanged in the name of the compound.

a. gold(II) and chloride: gold(II) chloride
b. calcium and carbonate: calcium carbonate
c. magnesium and hydroxide: magnesium hydroxide

3.13 Give the number of each type of ion present in each of the ionic compounds in problem 3.11 above.
a. gold(II) and chloride, an antimicrobial
b. calcium and carbonate, the mineral limestone
c. magnesium and hydroxide, the main compound in milk of magnesia

a. gold(II) and chloride: $AuCl_2$ one Au^{2+} and two Cl^{-}
b. calcium and carbonate: $CaCO_3$ one Ca^{2+} and one CO_3^{2-}
c. magnesium and hydroxide: $Mg(OH)_2$ one Mg^{2+} and two OH^{-}

3.15 Repeat problem 3.14 (Give the formula and name for the ionic compound formed from the combination of the indicated metal ion with the given anion) using the acetate, ($C_2H_3O_2^{-}$), polyatomic ion
a. Na
b. Ba
c. Al

Formulas:

Predict the charge on the cation using the rule: Charge = Group Number

The compound is written with no charges shown. When writing formulas for chemical compounds, we represent the number of each particle (ions in this case) in the compound with a subscript. A "1" is understood in the formula for the compound.
The number of ions that combine to form an ionic compound is determined by the charge of the cation and the charge of the anion. Cations and anions will combine so that the charges cancel each other out when combined, providing a formula with no net charge.

The *crossover approach* is a straightforward way of getting the correct formula for an ionic compound. The number representing the magnitude of the charge on the cation becomes the subscript on the anion and vice – versa. In the case where the charges on the ions are equal and opposite, for example +2 and –2, the formula is reduced to the smallest common ratio.

a. Na: Na is in Group 1A: Na^+ and $C_2H_3O_2^-$, one $C_2H_3O_2^-$ for each Na^+ gives a net charge of 0, $NaC_2H_3O_2$
b. Ba: Ba is in Group 2A: Ba^{2+} and $C_2H_3O_2^-$, two $C_2H_3O_2^-$ for each Ba^{2+} gives a net charge of 0, $Ba(C_2H_3O_2)_2$ Remember that polyatomic ions must be enclosed in parentheses with the subscript outside the parenthesis when the subscript is greater than 1.
c. Al: Al is in Group 3A: Al^{3+} and $C_2H_3O_2^-$, three $C_2H_3O_2^-$ for each Al^{3+} gives a net charge of 0, $Al(C_2H_3O_2)_3$ Remember that polyatomic ions must be enclosed in parentheses with the subscript outside the parenthesis when the subscript is greater than 1.

Names:
To name an ionic compound, put the names of the two ions together, always remembering to list the cation first.
When naming the cation of a transition metal, a Roman numeral is used to designate the charge on the transition metal.
If the compound contains one of the polyatomic ions, the name of the ion remains unchanged in the name of the compound. $C_2H_3O_2^-$ is acetate.

a. Na: $NaC_2H_3O_2$ sodium acetate
b. Ba: $Ba(C_2H_3O_2)_2$ barium acetate
c. Al: $Al(C_2H_3O_2)_3$ aluminum acetate

3.17 Give the number of each type of ion present in each of the ionic compounds in problem 3.15 above.
a. Na: $NaC_2H_3O_2$ one Na^+ and one $C_2H_3O_2^-$
b. Ba: $Ba(C_2H_3O_2)_2$ one Ba^{2+} and two $C_2H_3O_2^-$
c. Al: $Al(C_2H_3O_2)_3$ one Al^{3+} and three $C_2H_3O_2^-$

3.19 Determine whether each of the following is a covalent or ionic compound:
a. NCl_3
b. Fe_2O_3
c. Cs_2CO_3
f. NH_4OH
g. SeO_2
h. BaS

d. PBr_3 **i.** CBr_4
e. C_5H_{12} **j.** OF_2

Compounds formed from the combination of a metal and a nonmetal are **ionic** compounds.
Compounds containing polyatomic ions are **ionic** compounds.
Compounds formed from the combination of nonmetals are **covalent** compounds.

a. NCl_3: two nonmetals, covalent
b. Fe_2O_3: a metal and a nonmetal, ionic
c. Cs_2CO_3: a metal and a polyatomic ion, ionic
d. PBr_3: two nonmetals, covalent
e. C_5H_{12}: two nonmetals, covalent
f. NH_4OH: two polyatomic ions, ionic
g. SeO_2: two nonmetals, covalent
h. BaS: a metal and a nonmetal, ionic
i. CBr_4: two nonmetals, covalent
j. OF_2: two nonmetals, covalent

3.21 Draw the correct Lewis structure for each of the following covalent compounds:
a. CH_3Cl, chloroform
b. H_2S, smell of rotting eggs
c. CO_2, dry ice
d. C_2H_2, acetylene, a fuel used in welding

Start by drawing the correct electron dot symbol for each of the atoms involved in the formula. The number of dots is equal to the number of valence electrons which is the same as the group number on the periodic table for the main-group elements.

Next, arrange the atoms so that unpaired electrons are next to each other, much like you line up the pieces of a puzzle in order to fit them together.

To complete the Lewis structure, replace the pairs of dots between atoms with lines representing bonds connecting the two atoms.

Finally, compare the structure to those in Table 3.3, which shows the preferred bonding patterns of the main-group elements.

a. $CHCl_3$:
C (4A) has 4 dots each unpaired, H(1A) has one dot which is unpaired and each Cl (7A) has 7 dots, only one is unpaired
C will form one bond to the H and one to each Cl.
All of the electrons on C are involved in single bonds as are the electrons of H. Each Cl has 3 nonbonding pairs.
Replace the bonding electrons pairs with lines.

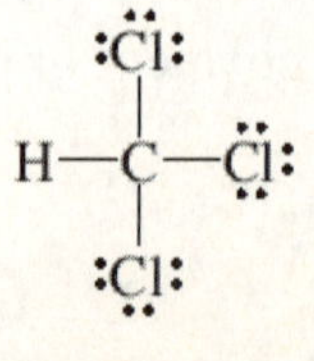

b. H_2S:

Each H(1A) has one dot which is unpaired and S (6A) has 6 dots, only two unpaired
S will form one bond to each H.
All of the electrons of each H are involved in single bonds. S has two nonbonding pairs.
Replace the bonding electron pairs with lines.

$$H-\underset{\cdot\cdot}{\ddot{S}}-H$$

c. CO_2:

C (4A) has 4 dots each unpaired and each O (6A) has 6 dots, only two unpaired
C will form a double bond with each oxygen.
All of the electrons of C are involved in double bonds. Each O has two nonbonding pairs.
Replace the bonding electron pairs with lines.

$$\underset{\cdot\cdot}{\ddot{O}}=C=\underset{\cdot\cdot}{\ddot{O}}$$

d. C_2H_2:

Each C (4A) has 4 dots each unpaired and each H (1A) has one dot which is unpaired
One C will form a triple bond with the other C.
Each C will form a single bond with one H.
All of the electrons of both elements are involved in bonding.
Replace the bonding electron pairs with lines.

$$H-C\equiv C-H$$

3.23 Using Table 3.3, decide whether molecules with the formulas shown are likely to exist (form Lewis structures that obey the octet rule). Briefly explain your answer.

a. CH_2
b. NH_2
c. BrCl
d. H_2S

Use the preferred covalent bonding pattern for each element shown in Figure 3.3 as the criteria to decide if the molecules exist.

a. CH_2: C forms 4 bonds, H forms 1 bond. With only 2 H present, C cannot form 4 bonds. This molecule likely does not exist.
b. NH_2: N forms 3 bonds, H forms 1 bond. With only 2 H present, N cannot form 3 bonds. This molecule likely does not exist.
c. BrCl: Cl forms 1 bond, Br forms 1 bond. Cl can form one bond to Br (or vice – versa). This molecule is likely to exist.
d. H_2S: S forms 2 bonds, H forms 1 bond. With 2 H present, S can form 2 bonds, one with each H. This molecule is likely to exist.

3.25 Complete the table below by supplying the missing name or formula (as indicated) for each covalent compound.

Formula	Name
?	Sulfur trioxide
P_2O_5	?
?	Selenium tetrafluoride
CO	?
?	Dinitrogen trioxide

To name binary compounds:

Step 1 - Name the first element in the formula.
Step 2 - Name the second element in the formula and change the ending to "-ide."
Step 3 - Designate the number of each element present using one of the Greek prefixes shown in Table 3.4.

To determine the formula from the name:

Step 1: Write the symbol for each element given in the name in the order of the name.
Step 2: Use the number that corresponds to Greek prefix in the name as the subscript in the formula. Subscripts of 1 are not written in the formula.

Remember if there is no Greek prefix on the first element in the name, the subscript is 1.

Formula	Name
SO_3	Sulfur trioxide
P_2O_5	Diphosphorus pentoxide
SeF_4	Selenium tetrafluoride
CO	Carbon monoxide
N_2O_3	Dinitrogen trioxide

3.27 Determine the shape of the molecule around the red-colored (shown as lighter gray below) atom (or atoms) in each of the Lewis structures shown below.

a. Dimethylamine, an insect pheromone

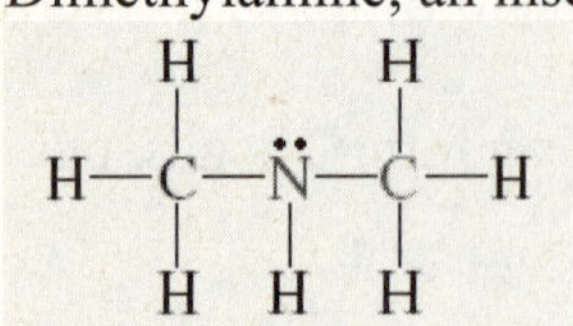

b. Acrylonitrile found in plastics

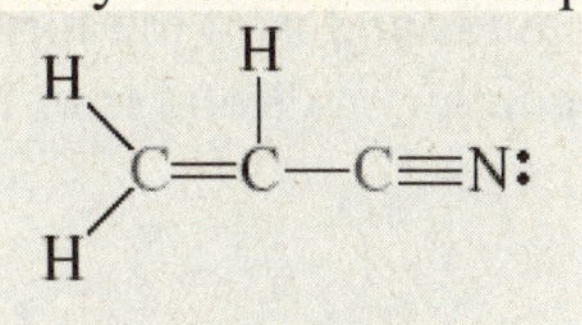

c. Acetic acid, vinegar

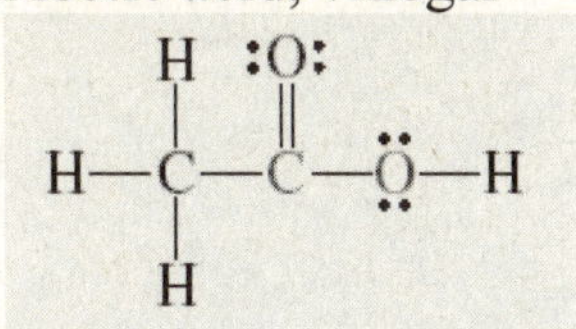

d. Acrylamide, a by-product of cooking starchy foods a high temperatures

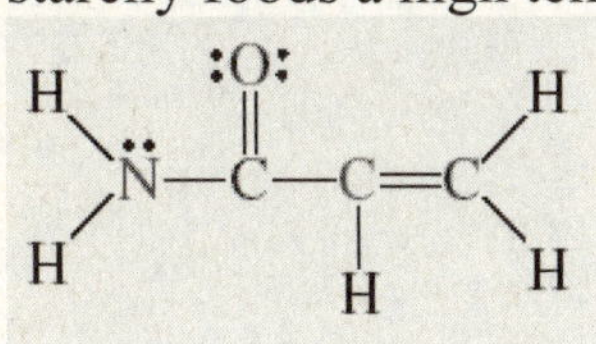

Use VSEPR to determine the shape around a given atom. Examine the atom to determine the number of bonding clouds (B) and the number of nonbonding clouds (N). Find the basic shape that corresponds to the combination in Table 3.5

a. Dimethylamine: N has 3 bonding clouds and one nonbonding cloud and is of the form AB_3N and is pyramidal. C has 4 bonding clouds and zero nonbonding clouds and is of the form AB_4 and is tetrahedral.

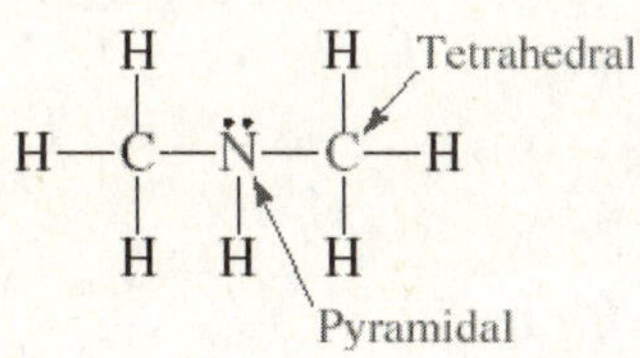

b. Acrylonitrile: The left C has 3 bonding clouds and zero nonbonding clouds and is of the form AB_3 and is trigonal planar. The right C has 2 bonding clouds and zero nonbonding clouds and is of the form AB_2 and is linear.

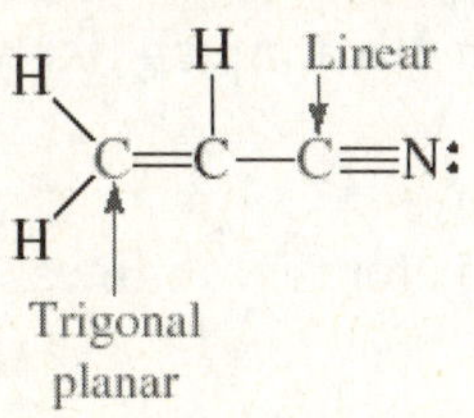

c. Acetic acid: C has 3 bonding clouds and zero nonbonding clouds and is of the form AB_3 and is trigonal planar. O has 2 bonding clouds and 2 nonbonding clouds and is of the form AB_2N_2 and is bent.

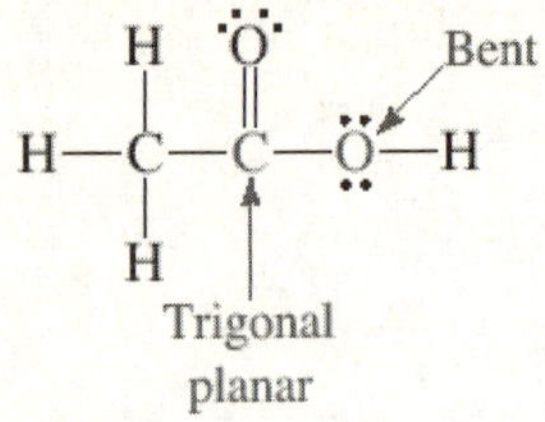

d. Acrylamide: C has 3 bonding clouds and zero nonbonding clouds and is of the form AB_3 and is trigonal planar. N has 3 bonding clouds and 1 nonbonding cloud and is of the form AB_3N and is pyramidal.

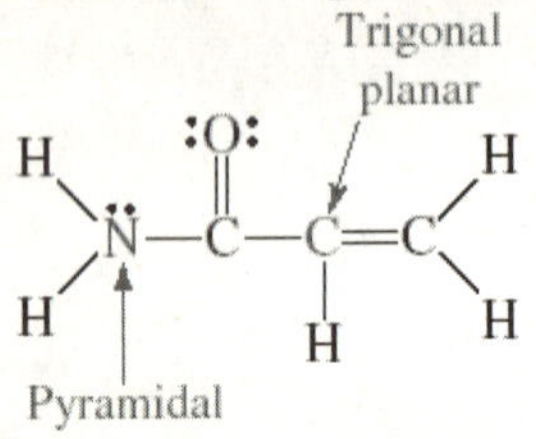

3.29 For each of the following molecules, (1) Draw the correct Lewis structure; (2) Label each polar covalent bond with a dipole moment arrow; (3) Determine if the molecule is polar or nonpolar and draw the dipole moment arrow for the molecule.

a. HCl
b. CH_4
c. NCl_3
d. CSO (C is the central atom)
e. H_2S
f. CH_2Br_2 (C is the central atom.)

(1) Lewis structure
Start by drawing the correct electron dot symbol for each of the atoms involved in the formula. The number of dots is equal to the number of valence electrons which is the same as the group number on the periodic table for the main-group elements.

Next, arrange the atoms so that unpaired electrons are next to each other, much like you line up the pieces of a puzzle in order to fit them together.

To complete the Lewis structure, replace the pairs of dots between atoms with lines representing bonds connecting the two atoms.

Finally, compare the structure to those in Table 3.3, which shows the preferred bonding patterns of the main-group elements.

(2) Use electronegativities to determine the bond polarity and the direction of the dipole. The dipole moment arrow ($\mapsto$) always points toward the more electronegative element.

(3) Use the following rules to determine polarity of the molecules. The dipole moment arrow

(+→) for the molecule points toward the more electronegative element(s).

- If the central atom (or one of the main atoms) has one or more pairs of nonbonding electrons, the molecule is polar.
- If the central atom has no pairs of nonbonding electrons and all of the atoms connected to the central atom are identical, the molecule is nonpolar.
- If the central atom has no pairs of nonbonding electrons and at least one of the atoms connected to the central atom is different from the others that are attached, the molecule is polar.

a. HCl: Both Cl and H form one bond. Since Cl is more electronegative than H, the bond will be polar with the arrow pointing toward Cl. The molecule only involves one bond, the molecular dipole and the bond dipole are the same.

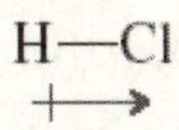

Polar

b. CH_4: C—H bonds are considered to be nonpolar. The molecule only involves C—H bonds so the molecule is nonpolar.

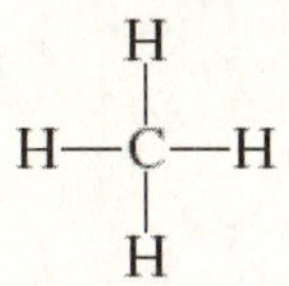

Nonpolar

c. NCl_3: N forms 3 bonds and each Cl forms one bond. The three Cl will be bonded to N with single bonds. N and Cl have the same electronegativity, making the bonds nonpolar. The N has three bonding electron clouds and 1 nonbonding electron cloud and is of the form AB_3N and is pyramidal which makes the molecule polar.

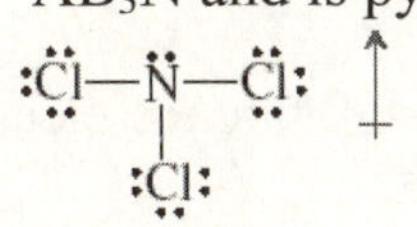

Polar

d. CSO (C is the central atom) : C forms 4 bonds and S and O each two. The C will form one double bond with S and one double bond with O. S and C have the same electronegativity so the C=S bond is nonpolar. O is more electronegative than C, making the C=O bond polar with the arrow pointing toward O. The C has two bonding electron clouds and zero nonbonding electron clouds and is of the form AB_2 and is linear. The molecule is polar because the bond dipole of C=O is not canceled by that of C=S.

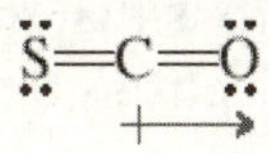

Polar

e. H_2S: S forms two bonds and H forms one bond. S will form a single bond with each of the H atoms. Since S is more electronegative than H, the bonds will be polar with the arrow pointing toward S. The S has two bonding electron clouds and two nonbonding electron clouds and is of the form AB_2N_2 and is bent making the molecule polar with the molecular dipole arrow pointing in the direction of S.

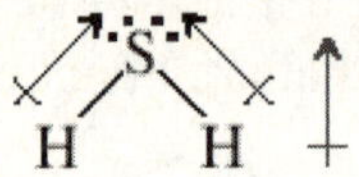

Polar

f. CH_2Br_2 (C is the central atom.): C forms 4 bonds and both Br and H form one bond. The C will form a single bond with each H and each Br. The C—H bond is considered to be nonpolar. Since Br is more electronegative than C, the C—Br bonds will be polar with the arrow pointing toward Br. The C has four bonding electron clouds and zero nonbonding electron clouds and is of the form AB_4 and is tetrahedral. The molecule is polar because the bond dipoles of the C—H bonds do not cancel that of the C—Br bonds. The molecular polarity arrow will point in the direction of the Brs.

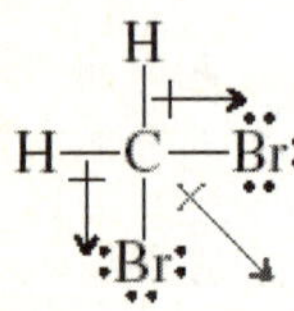

Polar

Additional Problems

3.31 How many valence electrons are present in the following atoms?

a. C **b.** Cl **c.** K **d.** Al

The group number of the main-group elements is equal to the number of valence electrons.
a. C (Group 4A) 4 valence electrons
b. Cl (Group 7A) 7 valence electrons
c. K (Group 1A) 1 valence electron
d. Al (Group 3A) 3 valence electrons

3.33 For each of the atoms in problem 3.31 above, determine the minimum number of electrons the atom needs to gain, lose, or share to become stable (isoelectronic with a noble gas).

Use these quick rules used for formation of ions:

1. Cationic Charge = Group Number (This gives the number of electrons that must be lost.)
2. Anionic Charge = Group Number − 8 (This gives the number of electrons that must be gained.)
3. Number of Electrons Shared = 8 − Group Number (Or use Table 3.3 to determine the number of bonds formed, this is the same as the number of electrons shared.)

a. C (Group 4A) shared 4 electrons
b. Cl (Group 7A) share or gain 1 electron
c. K (Group 1A) lose 1 electron
d. Al (Group 3A) lose 3 electrons

3.35 Complete the following statements:
a. An anion has more/fewer (*circle one*) protons than electrons.
b. The electrons found in the outermost shell of an atom are called ________ electrons.
c. Ions formed from metals have a positive/negative (*circle one*) charge.

An atom loses electrons to form an anion.
a. more

The outermost electrons determine the charge on an ion, historically referred to as an atom's valence.
b. valence
Metal ion valence electrons are loosely held and are easily lost.
c. positive

3.37 Determine the number of protons and electrons in the following ions:

a. chloride
b. Fe^{3+}
c. Cr^{6+}
d. N^{3-}
e. sodium ion
f. H^{+}

The number of protons in the ions is the same as the atom which is the atomic number.

In a cation (positive ion), the number of electrons **lost** is equal to ion charge. The number of electrons remaining in the ion is then equal to the atomic number **minus** the number of electrons lost. Cationic charge = Group Number

In an anion (negative ion), the number of electrons **gained** is equal to ion charge. The number of electrons present in the ion is then equal to the atomic number **plus** the number of electrons gained. Anionic charge = Group Number − 8

a. chloride, Cl atomic number 17 (Group 7A); 17 protons and 18 electrons $(7 - 8 = -1)$
b. Fe^{3+}, Fe atomic number 26, 26 protons, charge 3+, 3 electrons lost, 23 electrons $(26 - 3 = 23)$
c. Cr^{6+}, Cr atomic number 24, 24 protons, charge 6+, 6 electrons lost, 18 protons $(24 - 6 = 18$
d. N^{3-}, N atomic number 7, 7 protons, charge 3−. 3 electrons gained, 10 electrons $(7 + 3 = 10)$
f. H^{+}, H atomic number 1, 1 proton, charge +1. 1 electron lost, 0 electrons $(1 - 1 = 0)$
e. sodium ion, atomic number 11, 11 protons, charge +1. 1 electron lost, 10 electrons $(11 - 1 = 10)$

3.39 Give the name of each of the unnamed ions in problem 3.37 above.

Metal ions (cations) are named by simply using the name of the elements and adding the word **ion**.

Transition metal ions (cations) typically form more than one ion. A Roman numeral equal to the charge is placed in parentheses directly after the name of the metal.

Nonmetal ions (anions) are named by replacing the last few letters of the name of the element with the suffix **ide**.

b. Fe^{3+}: transition metal ion, iron(III)
d. N^{3-}: nonmetal ion, nitride
c. Cr^{6+}: transition metal ion, chromium(VI)

f. H^+: positive ion, hydrogen ion

3.41 Give the name and symbol of the ion with the following number of protons and electrons:
a. 3 protons, 2 electrons
b. 35 protons, 36 electrons

Names:
Metal ions (cations) are named by simply using the name of the elements and adding the word **ion**.

Transition metal ions (cations) typically form more than one ion. A Roman numeral equal to the charge is placed in parentheses directly after the name of the metal.

Nonmetal ions (anions) are named by replacing the last few letters of the name of the element with the suffix **ide**.

Symbols:
The number of protons in the ion is the same as the atom which is the atomic number.

In a cation (positive ion), the number of electrons is **less** than the number of protons. The charge on the ion is equal to the number of protons **minus** the number of electrons. The charge will have a **positive** sign.

In an anion (negative ion), the number of electrons is **greater** than the number of protons. The charge on the ion is equal to the number of electrons **minus** the number of protons. The charge will have a **negative** sign.

a. 3 protons, 2 electrons: Atomic number 3 is lithium (Li). 1 more proton than electrons, a positive ion with a charge of +1, lithium ion, Li^+
b. 35 protons, 36 electrons: Atomic number 35 is bromine (Br). 1 more electron than protons, a negative ion with a charge of 1−, bromide, Br^-

3.43 Each of the following ions is isoelectronic with a noble gas. Give the symbol of the noble gas.
a. Rb^+
b. F^-
c. S^{2-}
d. Ca^{2+}

Determine the number of electrons in each ion based on the atomic number and the charge on the ion. In the neutral noble gas, this will be the number of protons or the atomic number. Look up the noble gas on the periodic table.

a. Rb^+: Atomic number 37, 1+ charge indicates 36 electrons ($37 - 1 = 36$). Kr is the noble gas with an atomic number of 36.
b. F^-: Atomic number 9, 1− charge indicates 10 electrons ($9 + 1 = 10$). Ne is the noble gas with an atomic number of 10.
c. S^{2-}: Atomic number 16, 2− charge indicates 18 electrons ($16 + 2 = 18$). Ar is the noble gas with an atomic number of 18.
d. Ca^{2+} Atomic number 20, 2+ charge indicates 18 electrons ($20 - 2 = 18$). Ar is the noble gas with an atomic number of 18.

3.45 Complete the table of polyatomic ions below by supplying the missing information.

	Name	Formula
a.	acetate	
b.	bicarbonate	
c.		NO_3^-
d.		CN^-

The names and formulas of polyatomic ions must be memorized. (See Table 3.1.) As in the case of simple cations, the names of positive polyatomic ions end with the word **ion**.

	Name	Formula
a.	acetate	$C_2H_3O_2^-$
b.	bicarbonate	HCO_3^-
c.	nitrate	NO_3^-
d.	cyanide	CN^-

3.47 Give the formula for the ionic compound formed by the ammonium ion and each of the following anions:

a. sulfide
b. chloride
c. sulfate
d. hydroxide

The compound is written with no charges shown. When writing formulas for chemical compounds, we represent the number of each particle (ions in this case) in the compound with a subscript. A "1" is understood in the formula for the compound.
The number of ions that combine to form an ionic compound is determined by the charge of the cation and the charge of the anion. Cations and anions will combine so that the charges cancel each other out when combined, providing a formula with no net charge.

The *crossover approach* is a straightforward way of getting the correct formula for an ionic compound. The number representing the magnitude of the charge on the cation becomes the subscript on the anion and vice – versa. In the case where the charges on the ions are equal and opposite, for example +2 and –2, the formula is reduced to the smallest common ratio.

The positive ion in all these compounds will be the ammonium ion, NH_4^+. The formula for the anion can be derived from its name. Nonmetal ions (anions) are named by replacing the last few letters of the name of the element with the suffix **ide**. The formulas of polyatomic ions must be memorized. (See Table 3.1.)

a. sulfide (derived from sulfur, Column 6A, Anionic Charge = Group Number − 8 = 6 − 8 = 2−)
NH_4^+ and S^{2-}: $(NH_4)_2S$
b. chloride (derived from chlorine, Column 7A, Anionic Charge = Group Number − 8 = 7 − 8 = 1−)
NH_4^+ and Cl^-: NH_4Cl
c. sulfate (polyatomic ion, SO_4^{2-}) NH_4^+ and SO_4^{2-}: $(NH_4)_2SO_4$
d. hydroxide (polyatomic ion, OH^-) NH_4^+ and OH^-: NH_4OH

3.49 Give the formula for each of the following ionic compounds:

a. sodium hydroxide, drain cleaner
b. aluminum oxide, dull coating that forms on aluminum metal
c. potassium nitrate, an ingredient in gunpowder

Formulas:
The compound is written with no charges shown. When writing formulas for chemical compounds, we represent the number of each particle (ions in this case) in the compound with a subscript. A "1" is understood in the formula for the compound.
The number of ions that combine to form an ionic compound is determined by the charge of the cation and the charge of the anion. Cations and anions will combine so that the charges cancel each other out when combined, providing a formula with no net charge.

The *crossover approach* is a straightforward way of getting the correct formula for an ionic compound. The number representing the magnitude of the charge on the cation becomes the subscript on the anion and vice – versa. In the case where the charges on the ions are equal and opposite, for example +2 and –2, the formula is reduced to the smallest common ratio.

First determine the ions and their charges.

a. sodium hydroxide: sodium (Na) Group 1A, Cationic Charge = Group Number, Na^+
hydroxide (polyatomic ion) OH^-
$NaOH$

b. aluminum oxide: aluminum (Al) Group 3A, Cationic Charge = Group Number, Al^{3+}
oxide from oxygen (O) Group 6 A, Anionic Charge = Group Number – 8 = 2–, O^{2-}
Al_2O_3

c. potassium nitrate: potassium (K) Group 1A, Cationic Charge = Group Number, K^+
nitrate (polyatomic ion), NO_3^-
KNO_3

3.51 Name the following ionic compounds:

a. Na_2O
b. $BaSO_4$
c. $CuCl_2$
d. $Mg(NO_3)_2$
e. Fe_2O_3
f. KF

Names:
To name an ionic compound, put the names of the two ions together, always remembering to list the cation first.

Metal ions (cations) are named by simply using the name of the element.

Transition metal ions (cations) typically form more than one ion. A Roman numeral equal to the charge is placed in parentheses directly after the name of the metal.

Nonmetal ions (anions) are named by replacing the last few letters of the name of the element with the suffix **ide**.

If the compound contains one of the polyatomic ions, the name of the ion remains unchanged in the name of the compound.

a. Na_2O: sodium oxide (metal ion and nonmetal ion)
b. $BaSO_4$: barium sulfate (metal ion and polyatomic ion)
c. $CuCl_2$: copper(II) chloride (transition metal ion and nonmetal ion)
d. $Mg(NO_3)_2$: magnesium nitrate (metal ion and polyatomic ion)
e. Fe_2O_3: iron(III) oxide (transition metal ion and nonmetal ion)
f. KF: potassium fluoride (metal ion and nonmetal ion)

3.53 Supply the missing information in each of the following formulas:
a. $CaCl_?$
b. $Sr_?(PO_4)_?$
c. $Rb_?CO_3$
d. $K_?S$

The subscripts in a formula for ionic compound are determined by the charges on the ions. The ions must combine in a ratio that results in net charge of 0. Cations and anions will combine so that the charges cancel each other out when combined, providing a formula with no net charge.

The *crossover approach* is a straightforward way of getting the correct formula for an ionic compound. The number representing the magnitude of the charge on the cation becomes the subscript on the anion and vice – versa. In the case where the charges on the ions are equal and opposite, for example +2 and –2, the formula is reduced to the smallest common ratio.

a. $CaCl_?$: Ca Group 2A, Cationic Charge = Group Number, Ca^{2+}
Cl Group 7A, Anionic Charge = Group Number – 8 = 7 – 8 = 1–, Cl^-
$CaCl_2$
b. $Sr_?(PO_4)_?$: Sr Group 2A, Cationic Charge = Group Number, Sr^{2+}
polyatomic ion, PO_4^{3-}
$Sr_3(PO_4)_2$
c. $Rb_?CO_3$: Rb Group 1A, Cationic Charge = Group Number, Rb^+
polyatomic ion, CO_3^{2-}
Rb_2CO_3
d. $K_?S$: K Group 1A, Cationic Charge = Group Number, K^+
S Group 6A, Anionic Charge = Group Number – 8 = 6 – 8 = 2–, S^{2-}
K_2S

3.55 Magnesium salicylate is sold as an over-the-counter drug to treat pain and inflammation. If there are two salicylate ions for each magnesium ion in the compound, what is the charge on the salicylate ion?

Mg, Group 2A, Cationic Charge = Group Number, Mg^{2+}: If there are two salicylate ion for each magnesium, the total positive charge is 2+ and the total negative charge is then 2–. This means that each of the salicylate ions has to be 1–.

3.57 How many covalent bonds does each of the following atoms prefer to form?
a. C
c. P

b. F **d.** O

Based on the number of electrons needed to obtain an octet: Number of Bonds = 8 − Group Number
(Or use Table 3.3 to determine the number of bonds formed, this is the same as the number of electrons shared.)

a. C: Group 4, Number of Bonds = 8 − 4 = 4 bonds
b. F: Group 7, Number of Bonds = 8 − 7 = 1 bonds
c. P: Group 5, Number of Bonds = 8 − 5 = 3 bonds
d. O: Group 6, Number of Bonds = 8 − 6 = 2 bonds

3.59 For each of the following compounds indicate whether it is covalent or ionic and give its correct name:
a. CBr_4
b. SiO_2
c. $MgBr_2$
d. NCl_3
e. $CrCl_3$

To determine the type of compound:
Compounds formed from the combination of a metal and a nonmetal are **ionic** compounds.
Compounds containing polyatomic ions are **ionic** compounds.
Compounds formed from the combination of nonmetals are **covalent** compounds.

To name an ionic compound, put the names of the two ions together, always remembering to list the cation first.
When naming the cation of a transition metal, a Roman numeral is used to designate the charge on the transition metal.
If the compound contains one of the polyatomic ions, the name of the ion remains unchanged in the name of the compound.

To name binary covalent compounds:
Step 1 - Name the first element in the formula.
Step 2 - Name the second element in the formula and change the ending to "-ide."
Step 3 - Designate the number of each element present using one of the Greek prefixes shown in Table 3.4.

a. CBr_4: two nonmetals, covalent compound, carbon tetrabromide
b. SiO_2: two nonmetals, covalent compound, silicon dioxide
c. $MgBr_2$: a metal and a nonmetal, ionic compound, magnesium bromide
d. NCl_3: two nonmetals, covalent compound, nitrogen trichloride
e. $CrCl_3$: a transition metal and a nonmetal, ionic compound, chromium(III) chloride

3.61 Explain the difference between an ionic formula and a molecular formula.

Ionic formulas are written for ionic compounds and indicate the smallest whole number ratio between the ions. Molecular formulas are written for covalent compounds and indicate the exact number and type of each atom present in a molecule of the compound.

3.63 Draw a Lewis structure for each of the following covalent compounds:

a. HCN **c.** PCl_3

b. CS_2 **d.** CH_5N

Start by drawing the correct electron dot symbol for each of the atoms involved in the formula. The number of dots is equal to the number of valence electrons, which is the same as the group number on the periodic table for the main-group elements.

Next, arrange the atoms so that unpaired electrons are next to each other, much like you line up the pieces of a puzzle in order to fit them together.

To complete the Lewis structure, replace the pairs of dots between atoms with lines representing bonds connecting the two atoms.

Finally, compare the structure to those in Table 3.3, which shows the preferred bonding patterns of the main-group elements.

a. HCN: C (4A) has 4 dots each unpaired, H(1A) has one dot which is unpaired and N(5A) has 5 dots, three are unpaired
C will form one bond to the H and three (one triple bond) to the N.
All of the electrons on C are involved in bonds as are the electrons of H. N has 1 nonbonding pair.
Replace the bonding electrons pairs with lines.

H—C≡N:

b. CS_2: C (4A) has 4 dots each unpaired and S (5A) has 6 dots, two are unpaired
C will form two bonds, one double bond to each S.
All of the electrons on C are involved in bonds. Each S has 2 nonbonding pairs.
Replace the bonding electrons pairs with lines.

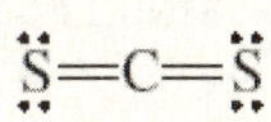

c. PCl_3: P (5A) has 5 dots of which 3 are unpaired, Cl (7A) has seven dots with 1 unpaired.
P will form one bond to each of the three Cl.
The P has 1 nonbonding pair. Each Cl has 3 nonbonding pairs.
Replace the bonding electrons pairs with lines.

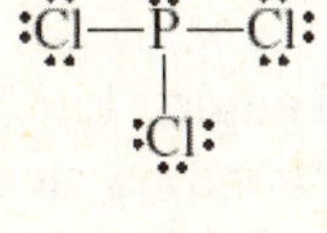

d. CH_5N: C (4A) has 4 dots each unpaired, each H(1A) has one dot which is unpaired and N(5A) has 5 dots, 3 are unpaired.

To be able to connect all of the atoms, C will form one bond to N. Three of the H atoms can then be bonded to the C to complete C's octet. C forms a single bond to each of three of the Hs. Nitrogen will form a single bond to each of the remaining two Hs.

All of the electrons on C are involved in bonds as are the electrons of H. N has 1 nonbonding pair.

Replace the bonding electrons pairs with lines.

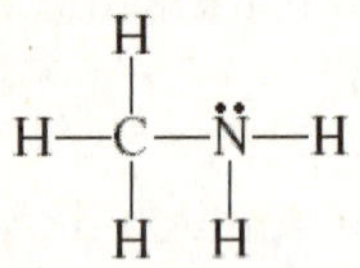

3.65 Give the name of each of the following covalent compounds:

a. SeO_2
b. SiF_4
c. P_4S_3
d. OF_2

To name binary covalent compounds:
Step 1 - Name the first element in the formula.
Step 2 - Name the second element in the formula and change the ending to "-ide."
Step 3 - Designate the number of each element present using one of the Greek prefixes shown in Table 3.4.

a. SeO_2: selenium dioxide
b. SiF_4: silicon tetrafluoride
c. P_4S_3: tetraphosphorus trisulfide
d. OF_2: oxygen difluoride

3.67 Aspartic acid, a naturally occurring amino acid, is a component of the structure of aspartame (Nutrasweet ™), an artificial sweetener widely used in diet soft drinks.

a. Determine the shape of the molecule around each red-colored atom (shown in gray below) in the Lewis structure of aspartic acid, shown below.

b. Give the bond angle of each of the red-colored (shown in gray below) carbon atoms.

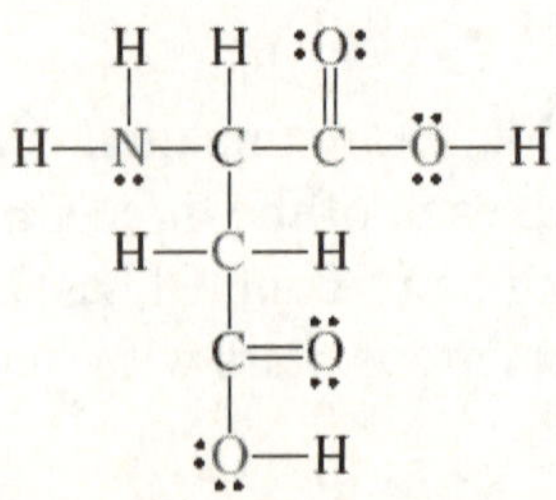

Use VSEPR to determine the shape around a given atom and the bond angle. Examine the atom to determine the number of bonding clouds (B) and the number of nonbonding clouds (N). Find the basic shapes and bond angles that correspond to the combination in Table 3.5.

a. Determine the shape of the molecule around each red-colored atom (shown in gray below) in the Lewis structure of aspartic acid, shown below.

b. Give the bond angle of each of the red-colored (shown in gray below) carbon atoms.

The N has 3 bonding clouds and 1 nonbonding cloud and is of the form AB_3N and is pyramidal with bond angle of 109.5°. The top right C has 3 bonding clouds and zero nonbonding clouds and is of the form AB_3 and is trigonal planar with bond angles of 120°. The middle C (bonded to 2 H) has 4 bonding clouds and zero nonbonding clouds and is of the form AB_4 and is tetrahedral with bond angle of <109.5°. Both the lower O and the O to the right have two bonding clouds and two nonbonding clouds and are of the form AB_2N_2 and are bent with bond angles of about <109.5°.

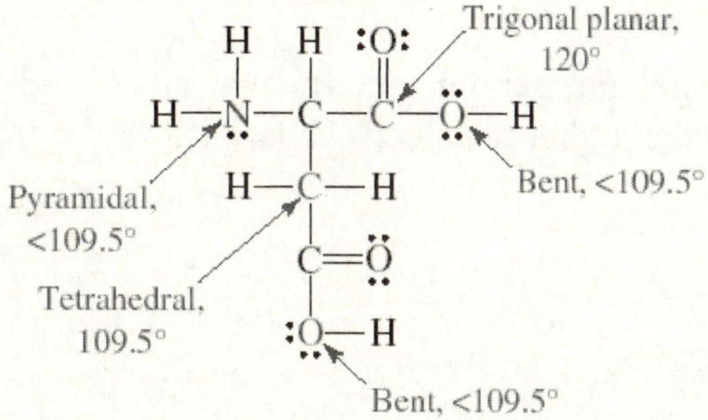

3.69 Explain why the bond angle between H—O—H in water is smaller than the bond angle between H—N—H in ammonia when O and N obey the octet rule.

The O in H—O—H has two nonbonding clouds (lone pairs) while the N in the H—N—H of ammonia (NH_3) has only one nonbonding cloud (lone pair). The two nonbonding clouds repel the adjacent bonding clouds more strongly than the single nonbonding cloud does. This repulsion pushes the bonding clouds away from the nonbonding charge. This results in the bonding clouds being pushed closer together, decreasing the bond angle.

3.71 Methyl isocyanate is used in the manufacturing of pesticides and was the deadly gas released from a chemical plant in the 1984 disaster in Bhopal, India, that killed hundreds. Determine the shape around each of the carbon, nitrogen, and oxygen atoms in the Lewis structure of methyl isocyanate shown below.

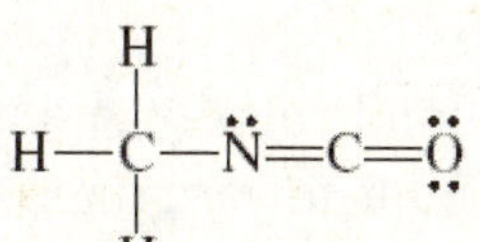

Use VSEPR to determine the shape around a given atom and the bond angle. Examine the atom to determine the number of bonding clouds (B) and the number of nonbonding clouds (N). Find the basic shapes that correspond to the combinations in Table 3.5.

The N has 2 bonding clouds and 1 nonbonding cloud and is of the form AB_2N which is a trigonal planar arrangement of charge clouds. One of the charge clouds is a nonbonded pair so the shape is determined by the remaining two bonding charge clouds which is bent. The right C has 2 bonding clouds and zero nonbonding clouds and is of the form AB_2 and is linear. The left C has 4 bonding clouds and zero nonbonding clouds and is of the form AB_4 and is tetrahedral.

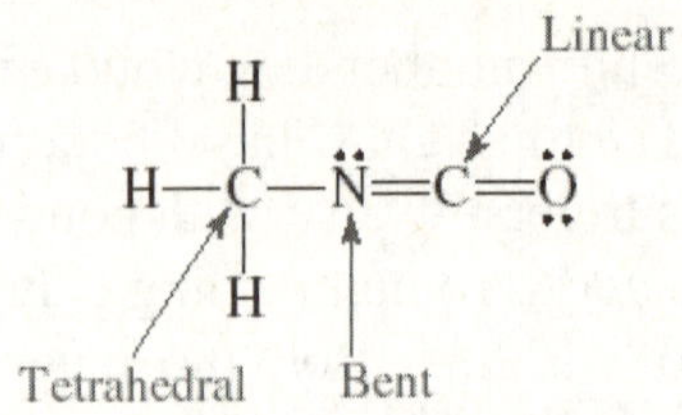

3.73 Identify the more electronegative atom in each of the following pairs:

a. S and Se
b. P and N
c. C and Cl
d. N and I

Use Figure 3.13 or the general trend in electronegativity based on position on the periodic table, that is, the closer to F, the higher the electronegativity.

a. S and Se: S
b. P and N: N
c. C and Cl: Cl
d. N and I: N

3.75 Using the Lewis structures provided below,

a. label each polar covalent bond with a dipole moment arrow.
b. determine if the molecule is polar or nonpolar and draw the molecular dipole,

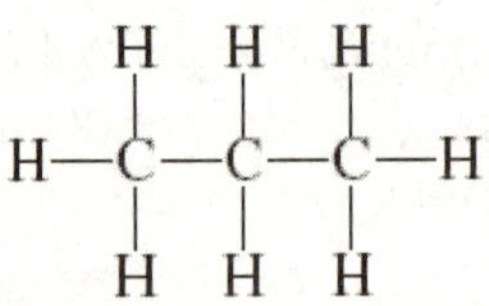

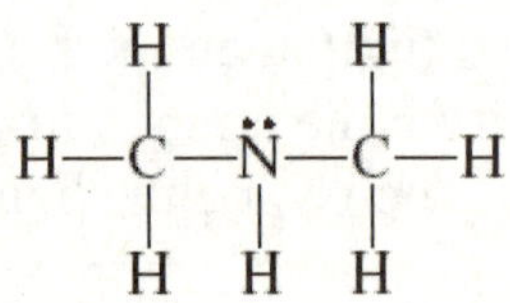

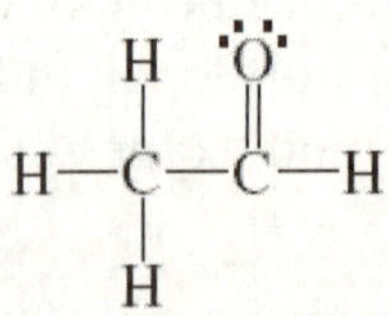

Use electronegativities to determine bond polarity and the direction of the dipole. The dipole moment arrow ($\leftarrow\!\!\!+$ →) always points toward the more electronegative element.

Use the following rules to determine polarity of the molecules. The dipole moment arrow (→) for the molecule points toward the more electronegative element(s).

- If the central atom (or one of the main atoms) has one or more pairs of nonbonding electrons, the molecule is polar.
- If the central atom has no pairs of nonbonding electrons and all of the atoms connected to the central atom are identical, the molecule is nonpolar.

- If the central atom has no pairs of nonbonding electrons and at least one of the atoms connected to the central atom is different from the others that are attached, the molecule is polar.

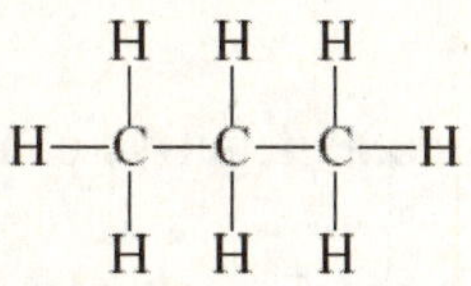

C—H bonds are considered to be nonpolar. The molecule only involves C—H bonds so the molecule is nonpolar.

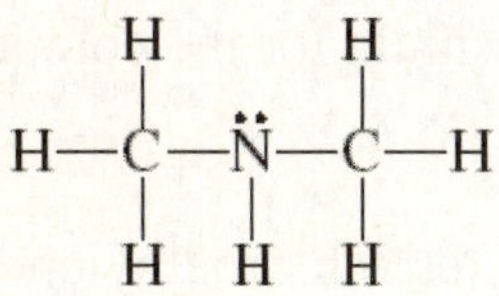

C—H bonds are considered to be nonpolar. N is more electronegative than either C or H, making the two C—N polar and the one N—H bond polar. In both cases the dipole arrows point toward the N. These bond dipoles produce a polar molecule with the dipole arrow in the direction of the N.

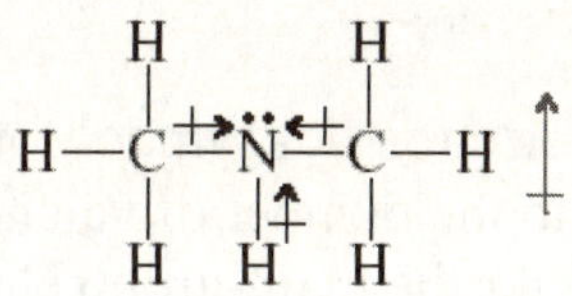

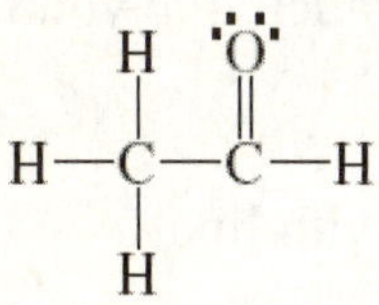

C—H bonds are considered to be nonpolar. O is more electronegative than C, making the C=O bond polar. The dipole arrow will point toward the O. This bond dipoles produce a polar molecule with the dipole arrow in the direction of the O.

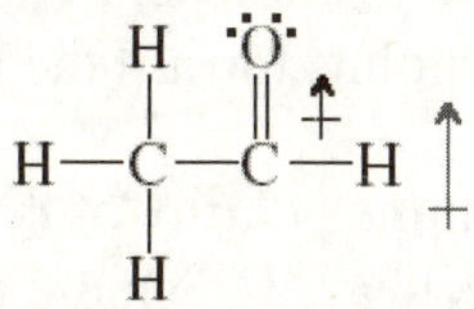

Challenge Problems

3.77 Tooth enamel contains a hard mineral called hydroxyapatite, containing calcium, phosphate, and hydroxide ions. If the formula for the compound contains 6 phosphate ions and 2 hydroxide ions, provide the complete formula for the ionic compound.

Use Table 3.1 and the fact that the total positive charge must equal the total negative charge in an ionic compound.
Each phosphate ion (PO_4^{3-}) has a 3− charge and each hydroxide (OH^-) has a 1− charge. The six phosphate ions carry a total charge of 18− (6 × 3−) and the two hydroxide ions carry a total charge of 2− (2 × 1−). The total negative charge is 20− (18− +(2−)).
The total positive charge must therefore be 20+. Ca is in Group 2A with an ionic charge of 2+ (Ca^{2+}). It requires 10 Ca^{2+} ions to generate a total charge of 20+. The formula for the compound is $Ca_{10}(PO_4)_6(OH)_2$.

3.79 One of the most common compounds used in commercial dry cleaning is tetrachloroethylene, which has the molecular formula C_2Cl_4.
a. Draw the Lewis structure of tetrachlorethylene.
b. Determine the shape of the molecule and the bond angle around each of the carbon atoms.
c. Label each of the polar covalent bonds with a dipole moment arrow.
d. Determine if this compound is polar or nonpolar.

(1) Lewis structure
Start by drawing the correct electron dot symbol for each of the atoms involved in the formula. The number of dots is equal to the number of valence electrons which is the same as the group number on the periodic table for the main-group elements.

Next, arrange the atoms so that unpaired electrons are next to each other, much like you line up the pieces of a puzzle in order to fit them together.

To complete the Lewis structure, replace the pairs of dots between atoms with lines representing bonds connecting the two atoms.

Finally, compare the structure to those in Table 3.3, which shows the preferred bonding patterns of the main-group elements.

(2) Use electronegativities to determine bond polarity and the direction of the dipole. The dipole moment arrow (+→) always points toward the more electronegative element.

(3) Use the following rules to determine polarity of the molecules. The dipole moment arrow (+→) for the molecule points toward the more electronegative element(s).

- If the central atom (or one of the main atoms) has one or more pairs of nonbonding electrons, the molecule is polar.
- If the central atom has no pairs of nonbonding electrons and all of the atoms connected to the central atom are identical, the molecule is nonpolar.
- If the central atom has no pairs of nonbonding electrons and at least one of the atoms connected to the central atom is different from the others that are attached, the molecule is polar.

C forms 4 bonds and each Cl forms one bond. To attain an octet for each atom in the formula, one C will form a double bond to the other C. Two Cls will each form a single bond with one C and the other two Cls will each form a single bond to the other C.

Each C has three bonding electron clouds and zero nonbonding electron clouds and is of the form AB_3 and is trigonal planar with bond angles of 120°. Each Cl will have three nonbonding clouds (lone pairs).

Since Cl is more electronegative than C, the C—Cl bonds will be polar with the arrow pointing toward Cl.

The molecule is nonpolar because the bond dipoles of the C—Cl bonds cancel.

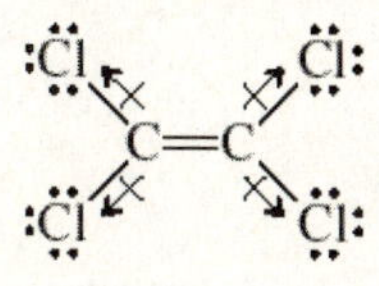

Chapter

4

Introduction to Organic Compounds

4.1 Alkanes: The Simplest Organic Compounds

Learning Objectives

Upon completion of this material, a student should be able to do the following:

A. define the following key terms:

organic compounds	biomolecules	inorganic compounds
organic chemistry	alkanes	saturated hydrocarbons
straight chain alkanes	cycloalkanes	

B. identify molecules as hydrocarbon based on its formula
C. identify a saturated hydrocarbon based on its formula
D. predict the number of carbon atoms in an alkane from its name (Table 4.1)
E. recognize a cycloalkane from either its name or condensed structural formula
F. realize that hydrocarbons are nonpolar in nature

4.2 Representing the Structures of Organic Compounds

Learning Objectives

Upon completion of this material, a student should be able to do the following:

A. define the following key terms:

condensed structural formula	skeletal structure

B. distinguish Lewis structures, condensed structural formulas, ball-and-stick, and skeletal structures
B. convert a Lewis structure to a condensed structural formula
C. convert a condensed structural formula to a Lewis structure
D. draw a skeletal structure from either a Lewis structure or a condensed structural formula

4.3 Fatty Acids–Biological "Hydrocarbons"

Learning Objectives

Upon completion of this material, a student should be able to do the following:

A. define the following key terms:

saturated fatty acids	fatty acids	lipid
functional group	carboxylic acids	amphipathic
unsaturated fatty acids	alkenes	monounsaturated
polyunsaturated	essential fatty acids	

B. determine the type of fatty acid (saturated, monounsaturated, polyunsaturated) from the condensed or skeletal formula
C. label the carboxylic acid functional group in the formula for a fatty acid

D. label the polar and nonpolar portions of an amphipathic molecule
E. identify the alkene functional group in the formula for a fatty acid
F. draw the skeletal structure of a fatty acid from the name and the carbon designation
G. formulate an ω designation for a fatty acid given the structure of a fatty acid
H. appreciate the importance of fatty acids in our diet

4.4 Families of Organic Compounds–Functional Groups

Learning Objectives

Upon completion of this material, a student should be able to do the following:

A. define the following key terms:

heteroatoms	alkynes	aromatic compounds
aromaticity	resonance hybrid	

B. identify common functional groups in a representation of a molecule (Lewis structure, condensed structure, or skeletal structure) (Table 4.6)
C. realize the importance of functional groups in determining the reactivity of molecules
D. identify the families of hydrocarbons from their formulas including:
 1. alkenes
 2. alkynes
 3. aromatics
E. relate aromaticity to resonance hybrids and resulting stability

4.5 Isomerism in Organic Compounds–Part 1

Learning Objectives

Upon completion of this material, a student should be able to do the following:

A. define the following key terms:

structural isomers	branched chain alkanes	conformational isomers
conformers	substituents	alkyl groups
haloalkanes		

B. recognize the difference between structural isomers from conformational isomers
C. recognize structural isomers and conformational isomers
D. formulate an IUPAC name for a branched chain alkane
E. formulate an IUPAC name for a haloalkane
F. formulate an IUPAC name for a cycloalkane with one substituent
G. draw a Lewis structure, condensed structural formula, or skeletal structure from an IUPAC name
H. distinguish structural isomers from conformational isomers by naming
I. draw possible structural isomers based on a molecular formula

4.6 Isomerism in Organic Compounds–Part 2

Learning Objectives

Upon completion of this material, a student should be able to do the following:

A. define the following key terms:

stereoisomers	cis-trans stereoisomers	restricted rotation
enantiomers	chiral	chiral center

B. distinguish stereoisomers and structural isomers
C. be familiar with and understand the use of wedge-shaped bonds and hash-bonds to represent the three-dimensional nature of bonds and molecules

C. explain why restricted rotations gives rise to cis-trans stereoisomers
D. identify cis-trans stereoisomers in cycloalkanes and alkenes
E. distinguish the cis stereoisomer from the trans stereoisomer for a given molecule
F. formulate an IUPAC name for a cycloalkane with two substituents using cis-trans prefixes
G. identify chiral carbon atoms in molecules
H. relate nonsuperimposable images to everyday items as well as to molecular enantiomers
I. recognize that enantiomers have different shapes and the importance of this in biological recognition

Practice Test for Chapter 4

1. Which of the following compounds would be classified as inorganic?

   A. CH_3COOH
   B. CH_2Cl_2
   C. $NaHCO_3$
   D. C_6H_6

2. What is the molecular formula for the structure shown below?

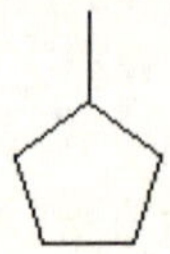

   A. C_6H_{10}
   B. C_6H_{15}
   C. C_6H_8
   D. C_6H_{12}

3. Which of the choices does **not** represent the same molecule as shown below?

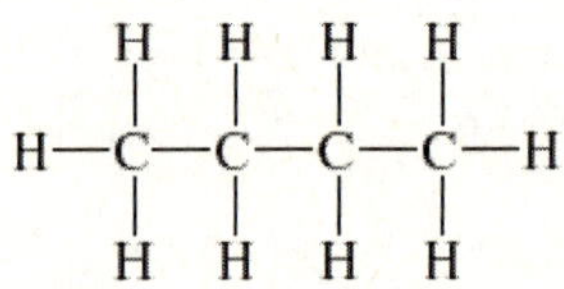

   A.

   B.

C.

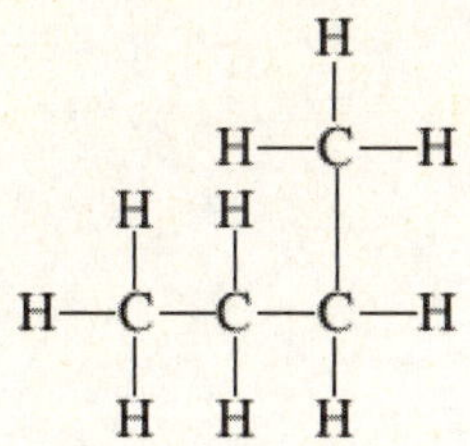

D.

4. Which of the terms does **not** describe the following structure?

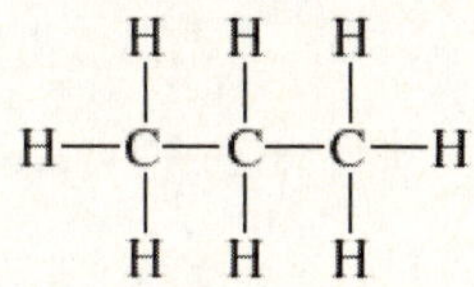

A. hydrocarbon
B. alkane
C. nonpolar
D. unsaturated

5. Which of the given formulas represents a skeletal structure?
A. C_4H_{12}
B.

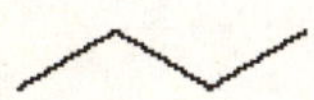

C.

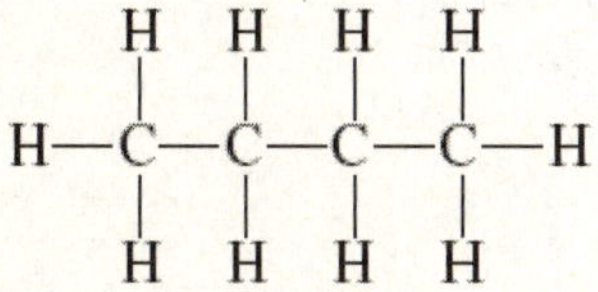

D. $CH_3CH_2CH_2CH_3$

6. Pentane could be represented as:
   A.

   B. $CH_3CH_2CH_2CH_2CH_3$
   C.
   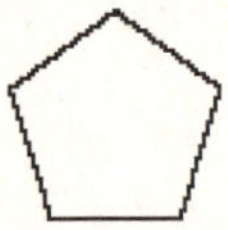
   D. C_6H_{14}

7. Consider the following structure.

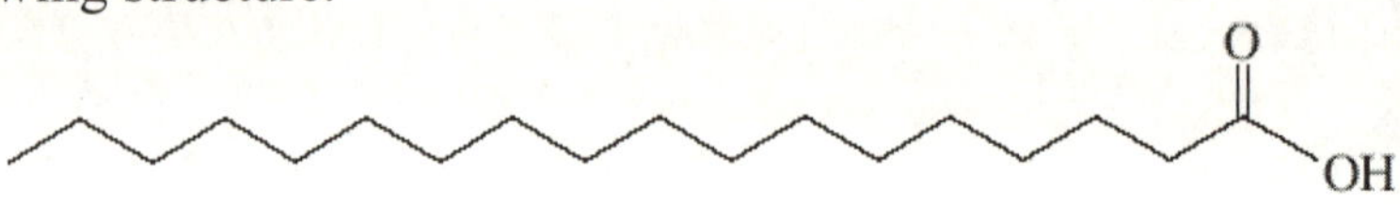

This represents:
A. a fatty acid.
B. stearic acid [18:0].
C. an amphipathic compound.
D. both A and B
E. A, B, and C

8. The compound shown below would be classified as:

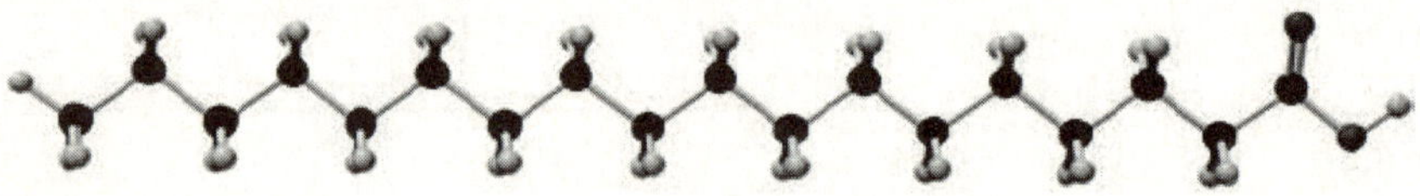

A. saturated.
B. monounsaturated.
C. polyunsaturated.
D. essential.

9. The correct carbon designation for the following fatty acid is:

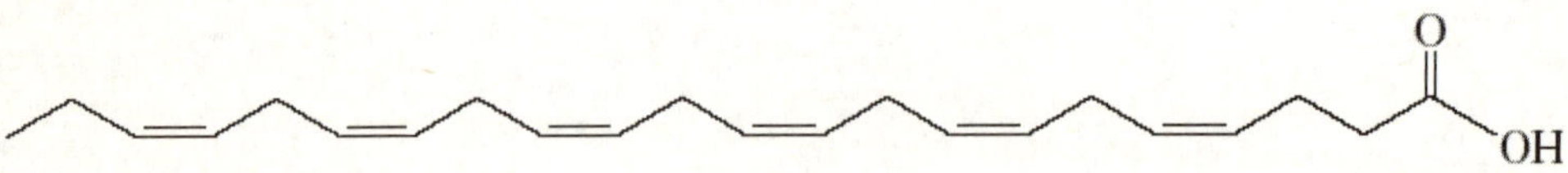

A. [22:6], ω-3.
B. [6:22], ω-3.
C. [22:6], ω-6.
D. [21:6], ω-3.

10. Identify the family of hydrocarbons present in each of the following, respectively.

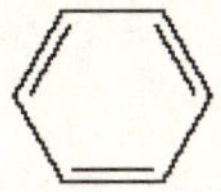
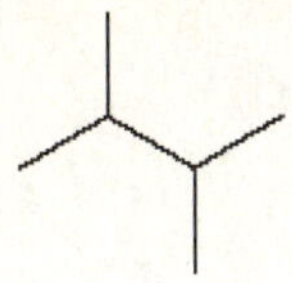

H—C≡C—H

A. aromatic, alkane, alkene
B. alkene, alkane, alkyne
C. aromatic, alkane, alkyne
D. cycloalkene, alkane, alkyne

11. The fundamental feature that distinguishes an alkane from an alkene from an alkyne is
A. type of carbon to carbon bonds.
B. the presence of heteroatoms.
C. the presence of rings.
D. aromaticity.

12. Which of the following functional groups contains a heteroatom?
A. alkyne
B. aldehyde
C. aromatic
D. alkene

13. Identify from left to right the functional groups in the following molecule that contain a heteroatom.

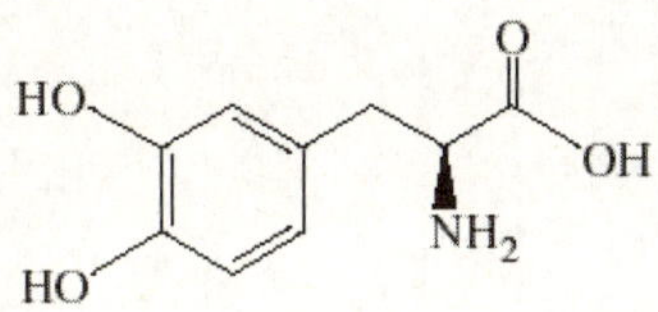

A. alcohol, amine, carboxylic acid
B. alcohol, amine, carboxylate
C. phenol, amine, carboxylic acid
D. phenol, amide, carboxylic acid

14. The members of the following pair are:

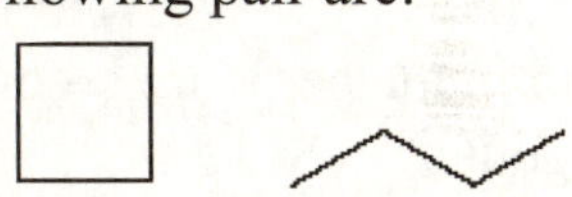

A. not related.
B. structural isomers.
C. conformational isomers.
D. stereoisomers.

15. The members of the following pair are:

$CH_3CH_2CH_2CH_2CH_2CH_3$ 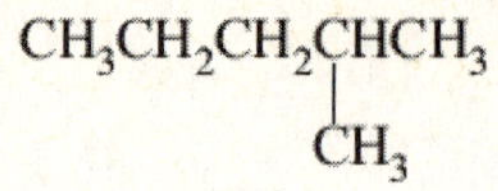

A. not related.
B. structural isomers.
C. conformational isomers.
D. stereoisomers.

16. The members of the following pair are:

 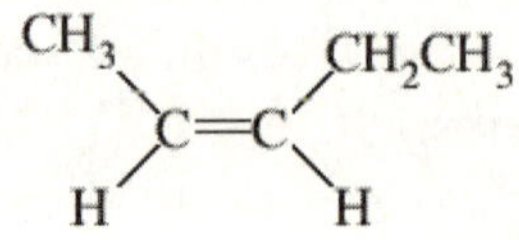

A. not related.
B. structural isomers.
C. conformational isomers.
D. stereoisomers.

17. The members of the following pair are:

$CH_3CH_2CH_2CH_2CH_2CH_3$ 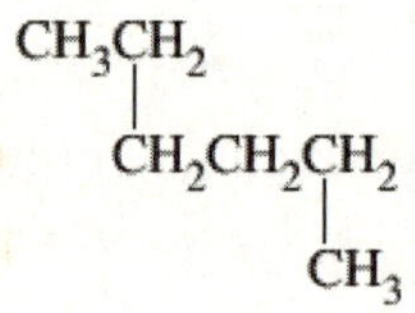

A. not related.
B. structural isomers.
C. conformational isomers.
D. stereoisomers.

18. A propyl alkyl group can be represented as:
A. CH_3CH_2—.
B. CH_3 CH_2CH_2—.
C. CH_3 $CH_2CH_2CH_2$—.
D. $CH_3CH_2CH_2CH_2CH_2$—.

19. If the terms *iodo*, *dimethyl*, and *ene* are found in the name for a compound, which of the following most likely characterizes this compound?
A. It is unsaturated, contains a hetero atom and a CH_3— substituent.
B. It is saturated, contains a hetero atom and two CH_3— substituents.
C. It is unsaturated, contains two hetero atoms and a CH_3— substituent.
D. It is unsaturated, contains a hetero atom and two CH_3— substituents.

20. The IUPAC name for the following compound is:

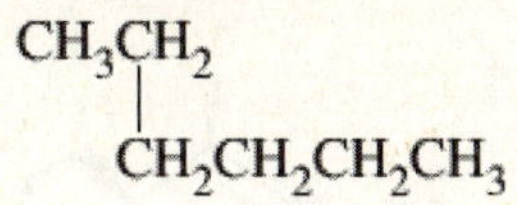

A. 1-ethylbutane.
B. hexane.
C. 1-butylethane.
D. *trans*- 1-ethylbutane.

21. The IUPAC name for the following compound is:

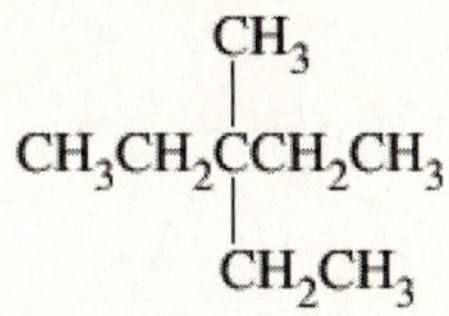

A. 2,2-diethylbutane.
B. 3,3-diethylbutane.
C. 3-ethyl-3-methylpentane.
D. 3-methyl-3-ethylpentane.

22. The IUPAC name for the following compound is:

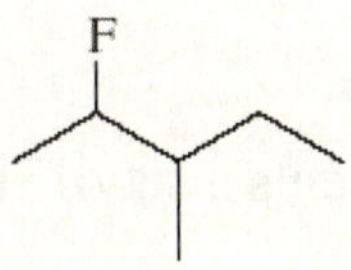

A. 2-fluoro-3-methylpentane.
B. 2-fluoro-3-ethylbutane.
C. 2-fluoroethylbutane.
D. 4-fluoro-3-methylpentane.

23. The IUPAC name for the following compound is:

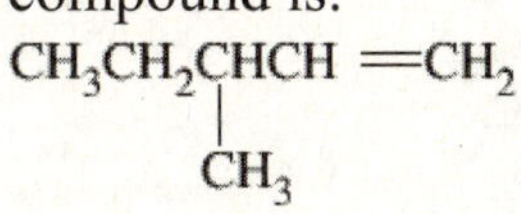

A. 3-methylpentane.
B. 3-methylpentene.
C. 3-methyl-1-pentene.
D. 3-methyl-4-pentene.

24. The IUPAC name for the following compound is:

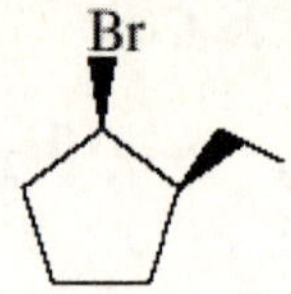

A. *trans*-1-bromo-2-methylpentane.
B. *cis*-1-bromo-2-ethylcyclopentane.
C. *cis*-1-bromo-2-methylcyclopentane.
D. *trans*-1-bromo-2-ethylcyclopentane.

25. A compound with the name 2-chloro-4-methylpentane could be represented as:
A. $CH_3CHClCH_2CH(CH_3)CH_3$.
B.

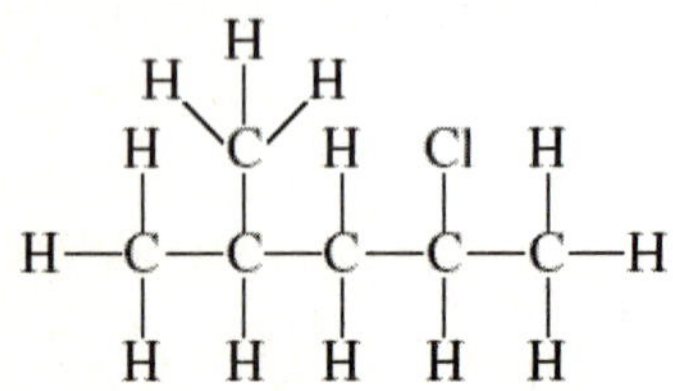

C.

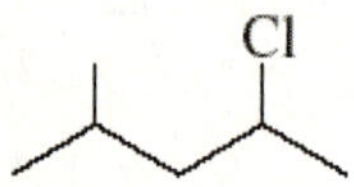

D. B and C
E. A, B, and C

26. Which of the following pairs represents a set of enantiomers?
A.

B.

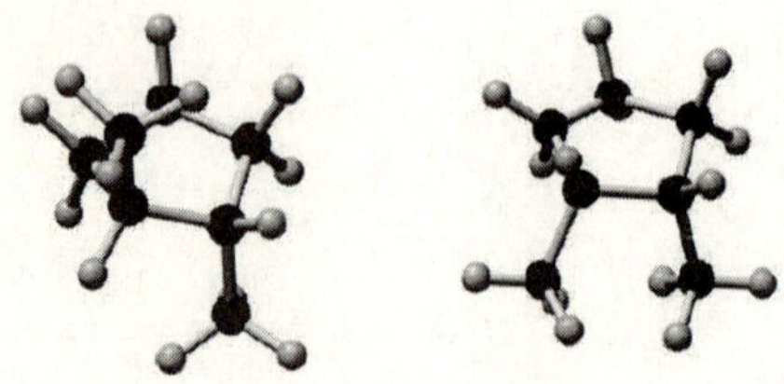

C.

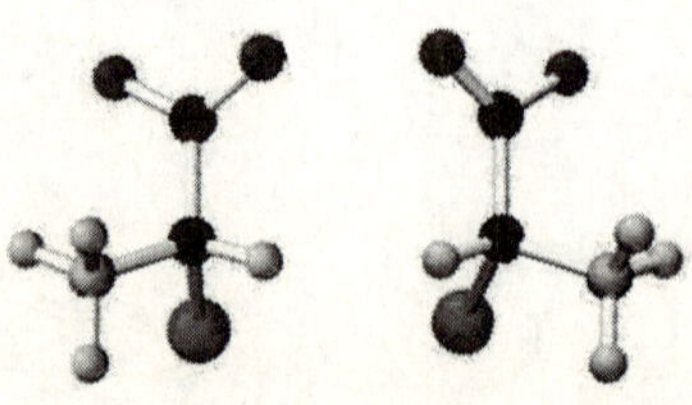

D. B and C

27. In which type of isomer are the molecular formulas for the isomers different?
    A. conformational
    B. structural
    C. stereoisomers
    D. none of the above

28. How many chiral carbon atoms are found in the following molecule?

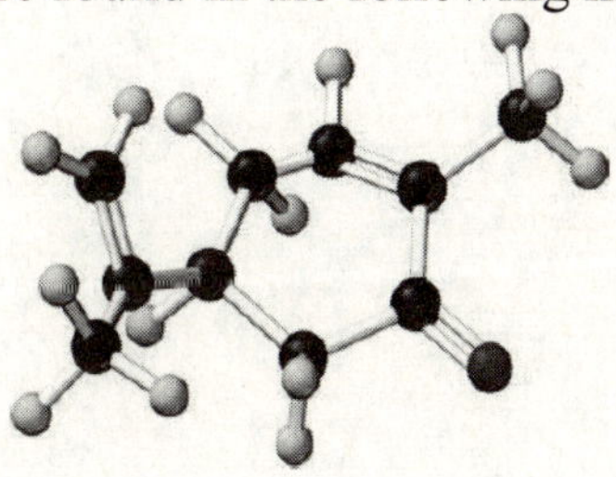

    A. 1
    B. 2
    C. 3
    D. 4
    E. 5

29. Which of the following molecules contains **more than one** chiral carbon atom?
    A.

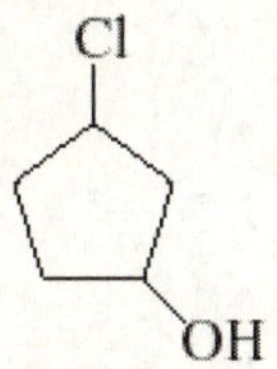

    B.

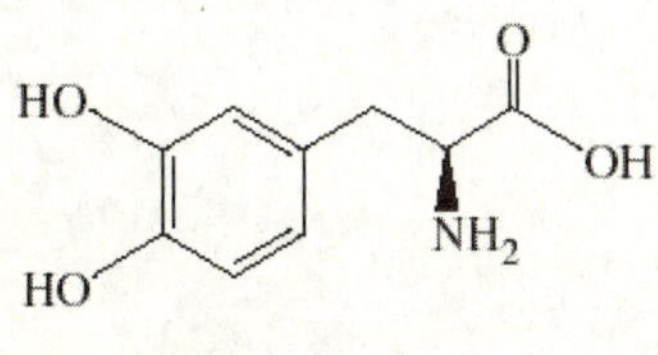

    C.

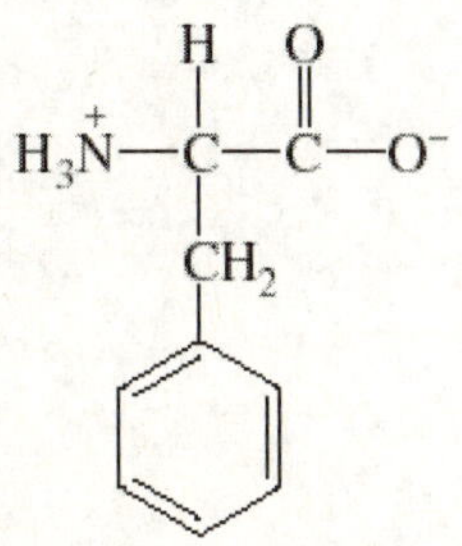

    D.

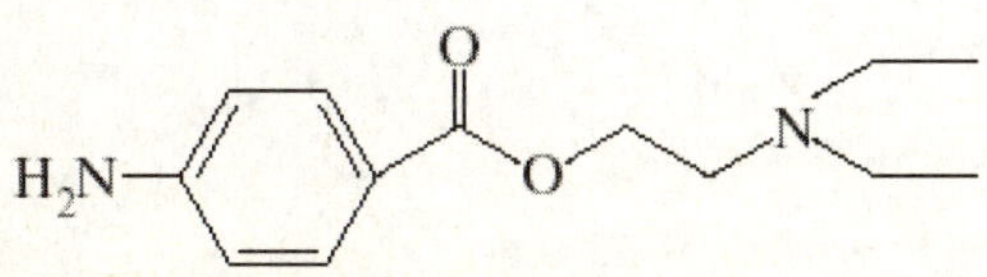

30. Which of the following does **not** describe a pair of enantiomers?
    A. biologically equivalent
    B. mirror images
    C. contain chiral carbon atoms
    D. "handed"

Answers

1. C	**2.** D	**3.** D	**4.** D	**5.** B	**6.** B	**7.** E	**8.** A	**9.** A	**10.** C
11. A	**12.** B	**13.** C	**14.** A	**15.** B	**16.** D	**17.** C	**18.** B	**19.** D	**20.** B
21. C	**22.** A	**23.** C	**24.** B	**25.** E	**26.** C	**27.** D	**28.** A	**29.** A	**30.** A

Chapter 4 – Solutions to Odd-Numbered Problems

Practice Problems

4.1 Describe the difference between a Lewis structure and a condensed structure in terms of atoms and bonds shown in the structures.

The Lewis structure shows all the bonds and lone pairs in a molecule. The condensed structure shows as few bonds as possible. Condensed structures may or may not show lone pairs and are less useful in drawing cycloalkanes.

4.3 Draw condensed structural formulas for the ten straight-chain alkanes in Table 4.1.

In the condensed structural formula, the atoms are listed from left to right in the order in which they are connected in the molecule, and it is understood that the octet of each atom is completed according to its preferred covalent bonding pattern.

For example; CH_4 is the condensed formula derived from:

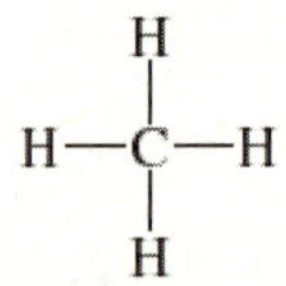

Using similar rationale, the rest of the series is:

CH_3CH_3
$CH_3CH_2CH_3$
$CH_3CH_2CH_2CH_3$
$CH_3CH_2CH_2CH_2CH_3$
$CH_3CH_2CH_2CH_2CH_2CH_3$
$CH_3CH_2CH_2CH_2CH_2CH_2CH_3$
$CH_3CH_2CH_2CH_2CH_2CH_2CH_2CH_3$
$CH_3CH_2CH_2CH_2CH_2CH_2CH_2CH_2CH_3$
$CH_3CH_2CH_2CH_2CH_2CH_2CH_2CH_2CH_2CH_3$

4.5 Explain why it is not possible to draw a skeletal structure for methane.

Skeletal structures show mainly bonds between carbon atoms. Recall that the structure of methane is:

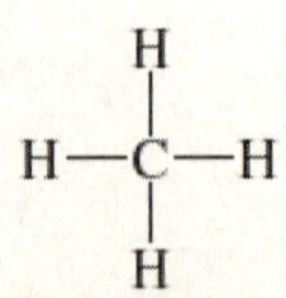

Since methane has only one carbon, it is not possible to draw its skeletal structure.

4.7 Cyclohexane, C_6H_{12}, is used to clean greasy machines and parts in many industrial settings. Draw the Lewis structure and the skeletal structure of cyclohexane without referring back to Table 4.2.

From the name, two important facts are given. First, the prefix "cyclo" indicates that the carbon atoms form a ring. From the Greek prefix "hex," it is known that there are six carbon atoms. Place the six C atoms in a ring and attach as many H atoms as necessary to give each carbon an octet. This works out to 2 H per 1 C atom. The resulting structure is:

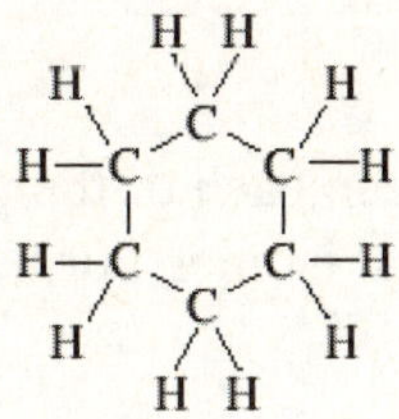

which can be represented with a skeletal structure of:

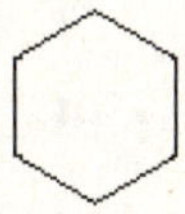

4.9 Stearidonic acid [18:4] is an essential omega-3 fatty acid found in black currant and the cyanobacterial spirulina. Draw stearidonic acid in skeletal structure.

[18:4] indicates that the fatty acid has 18 C and 4 double bonds. The omega-3 indicates that the first double bond is between carbon 3 and 4 when the chain is numbered with the C1 of the chain furthest from the C of carboxylic acid group.

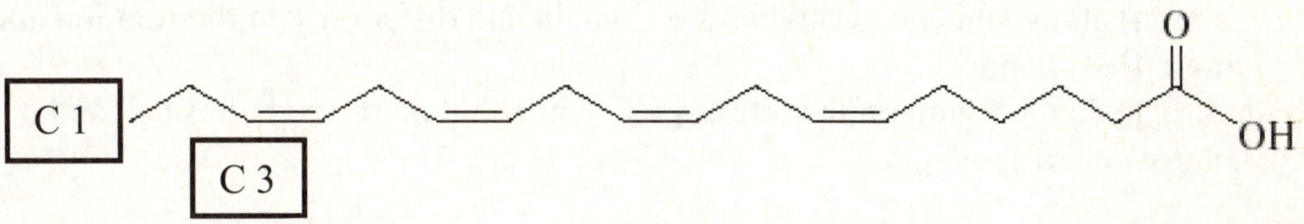

4.11 Identify the family of hydrocarbon present by the functional group found in the following:

a. $H_3CC{\equiv}CH$

b. 

c. 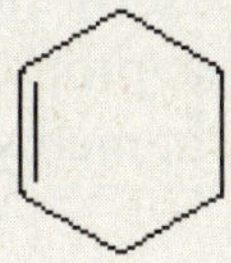

a. Contains a C to C triple bond: alkyne

b. Contains C to C double bonds (3): alkene
c. Contains a C to C double bond: alkene which is part of a ring , more specifically a cycloalkene

4.13 **a.** Describe what conformational isomers have in common.
b. Describe how conformational isomers are different.

a. Conformational isomers share the same molecular formula (same types and numbers of atoms) and are identical in how the atoms are connected to each other (connectivity).
b. Conformational isomers differ in the rotation of atoms about one or more of the single bonds.

4.15 Determine the relationship between each of the pairs of compounds shown below. Are they structural isomers, conformational isomers, or neither?

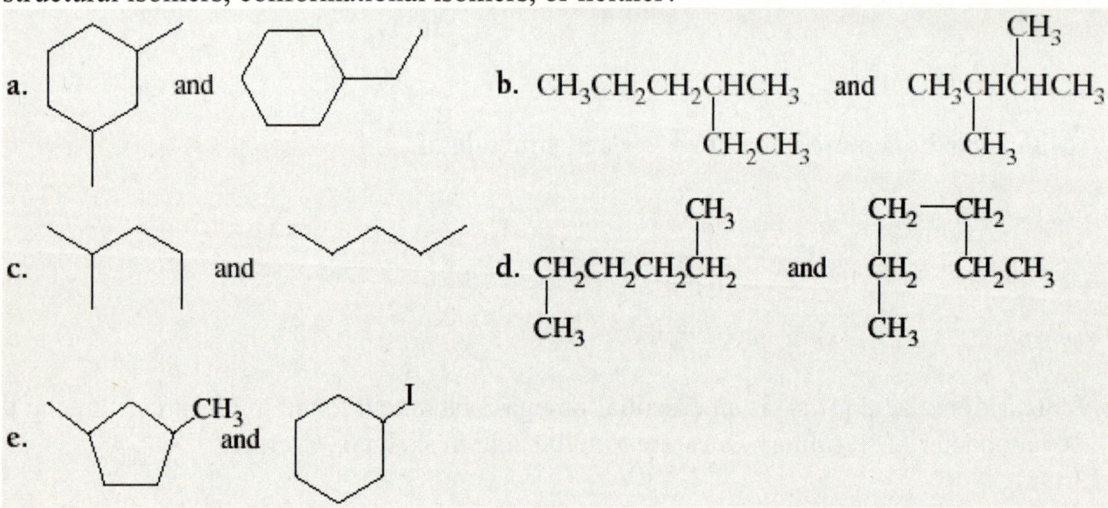

a. structural isomers: same number and types of atoms but differ in connectivity
b. not related: the figure on the left has 7 C atoms while the one on the right has only 6
c. not related: the figure on the left has 6 C atoms while the one on the right has only 5
d. conformation isomers: each has a 6 C chain and differ only in the rotation about the single C—C bonds
e. not related: the figure on the left has 7 C while that on the right has 6 C and a heteroatom of I

4.17 Draw the condensed structural formula for each of the following alkyl groups:
a. propyl
b. methyl

See Table 3.1 for the meaning of the prefix.
a. propyl: indicates 3 C

$$CH_3CH_2CH_2—$$

b. methyl: indicates 1 C

CH_3—

4.19 Draw the skeletal structure for each of the following compounds:

a. 2,3-dimethylpentane
b. 2-ethyl-1,4-dimethylcyclohexane
c. 1,2-dichlorohexane
d. 3-ethylhexane
e. 1-bromo-3-iodocyclopentane
f. propylcyclopentane

Step 1: Find the longest continuous chain of carbon atoms. This is the parent chain. Name the parent chain according to the alkane names given in Table 4.1. (Hint: Count from each end to every other end to make sure you find the longest chain.)
Step 2: Identify the groups bonded to the main chain but not included in the main chain. These groups are known as **substituents** and, in the case of branched-chain alkanes, they are called **alkyl** groups. The name of each alkyl group is derived from the alkane with the same number of carbon atoms by changing the "-ane" ending of the name of the corresponding alkane to "-yl."

a. 2,3-dimethylpentane: longest continuous carbon chain contains 5 carbon atoms with methyl groups bonded to carbon 2 and carbon 3

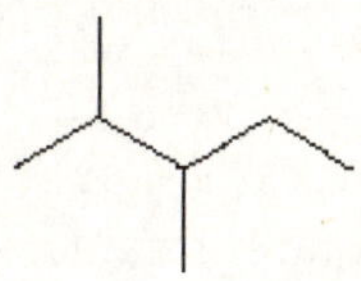

b. 2-ethyl-1,4-dimethylcyclohexane: longest continuous carbon chain contains 6 carbon atoms bonded to form a ring with methyl groups bonded to carbon 1 and carbon 4 and an ethyl group to carbon 2

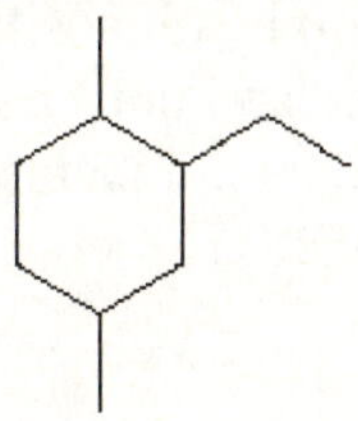

c. 1,2-dichlorohexane: longest continuous carbon chain contains 6 carbon atoms with chloro groups bonded to carbon 1 and carbon 2

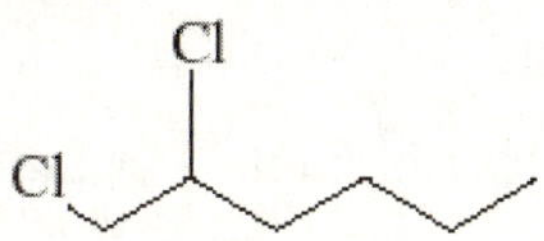

d. 3-ethylhexane; longest continuous carbon chain contains 6 carbon atoms with an ethyl group bonded to carbon 3

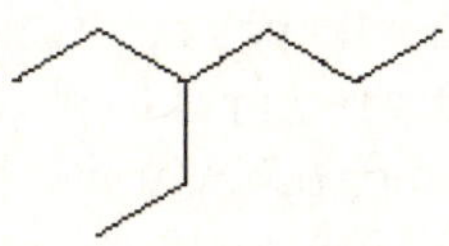

e. 1-bromo-3-iodocyclopentane: longest continuous carbon chain contains 5 carbon atoms bonded to form a ring with a bromo group bonded to carbon 1 and an iodo group bonded to carbon 3

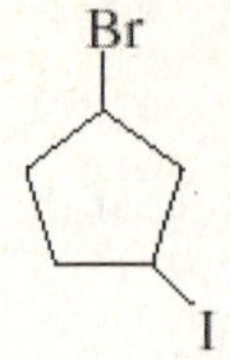

f. propylcyclopentane: longest continuous carbon chain contains 5 carbon atoms bonded to form a ring with a propyl group bonded to a carbon of the ring

4.21 Draw the skeletal structure and give the correct IUPAC name for three cycloalkane structural isomers with the molecular formula C_5H_{10}.

The basic structures will be a 5 C ring, then a 4 C ring with a branch, then a 3 C ring. In the last structure the branches can be from the same carbon atom or from two different carbon atoms, giving 4 structural isomers total.

To name cylcoalkanes:

Step 1: The ring serves as the parent name.
Step 2: Identify the substituents.
Step 3: Number the carbons in the ring. Carbon 1 will always have a substituent.
Step 4: Assign numbers to the substituents. On a ring bearing a single substituent, that substituent is assumed to be on position #1 of the ring. If only one substituent is present, a one is implied and need not be listed. When more than one substituent is present, the ring should be numbered to give the substituents the lowest possible combination of numbers.

cyclopentane methylcyclobutane 1,2-dimethylcyclopropane 1,1-dimethylcyclopropane

4.23 Determine if each of the cycloalkanes below can exist as cis-trans stereoisomers. For those that can, use wedge and dash structures to draw the two isomers. Label each of the isomers you drew as the cis stereoisomer or the trans stereoisomer.

a 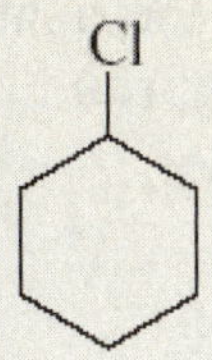

b. 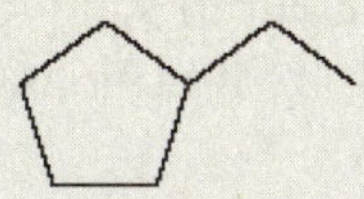

c 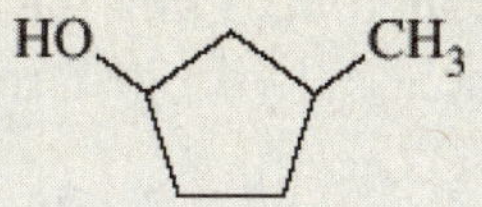

d 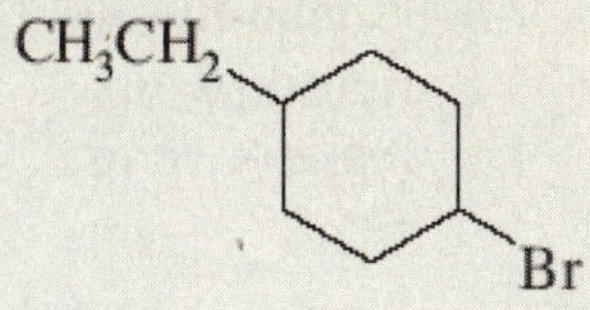

In order to have cis-trans stereoisomers in cycloalkanes, there must be two groups attached to two different carbons of the ring. This eliminates the possibility of cis-trans stereoisomers for a and b. The structures in c and d do exhibit cis-trans stereoisomerism as shown below. In the trans isomer the substituents are on opposite faces of the ring while in the cis isomer they are on the same face of the ring. These relative positions are indicated with the wedge-shaped bonds and the dashed bonds.

c. 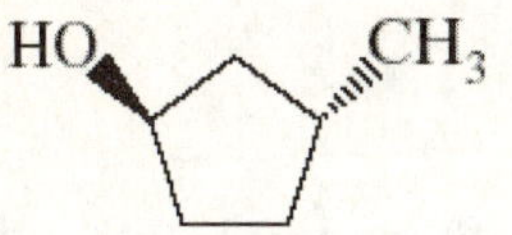

trans

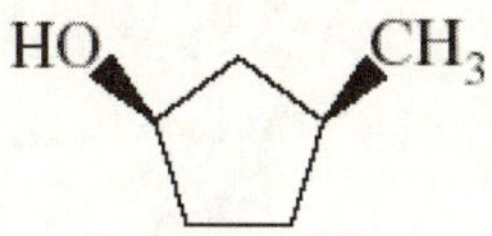

cis

d. 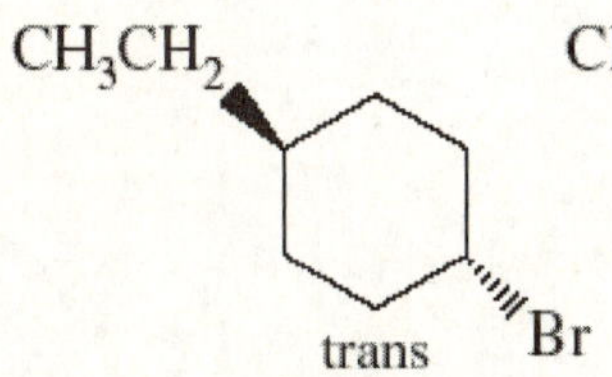

trans

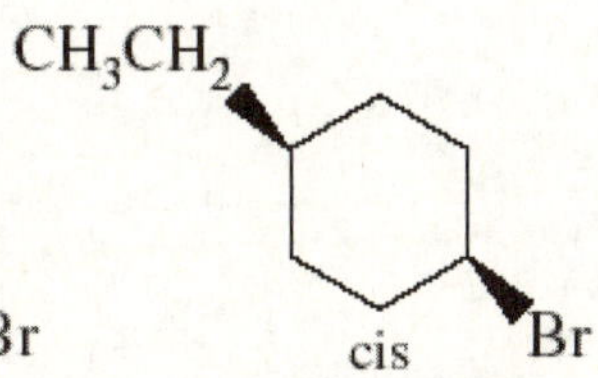

cis

4.25 Mark the chiral carbons in the following molecules, if any, with an asterisk (*).

a. 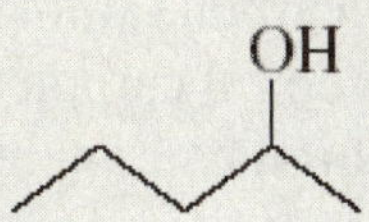

b. 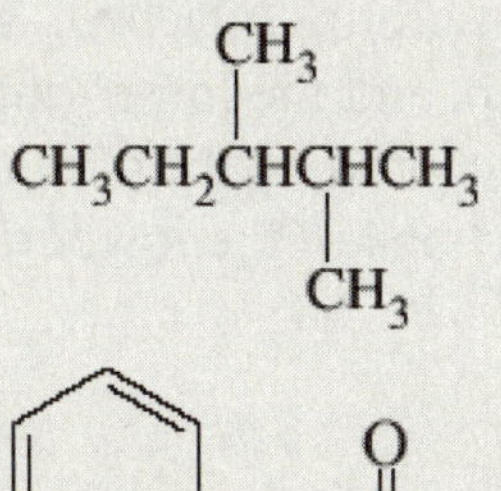

c. 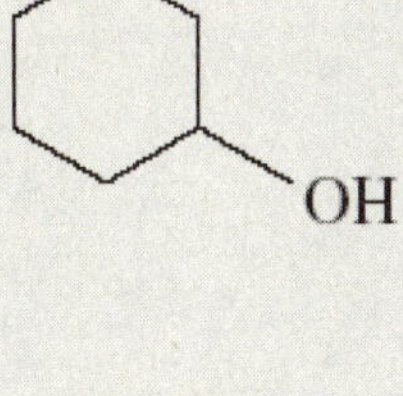

d. 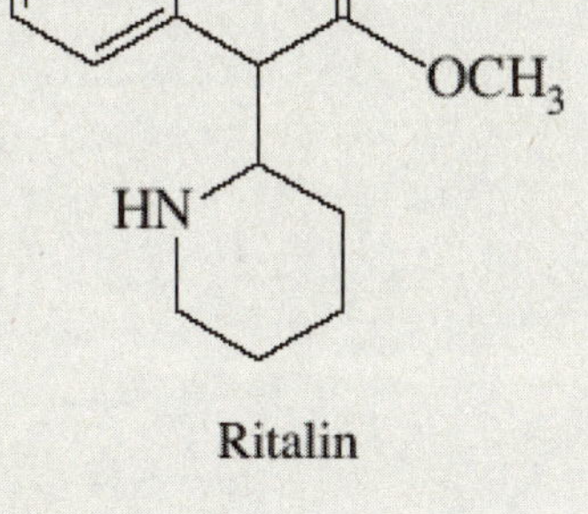

Ritalin

Locate the tetrahedral carbons in the molecule. (Any carbons with double or triple bonds cannot be chiral centers since they do not have four groups attached.) Determine if four different groups are attached to the tetrahedral carbons. If they are, the carbon atom is a chiral center.

a. All carbon atoms are tetrahedral. Only the carbon atom second from the right has four different groups attached (hydrogen, methyl, hydroxyl, and propyl) and is therefore chiral.

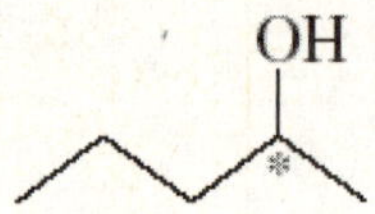

b. All carbon atoms are tetrahedral. Only the middle carbon atom (third from the right or left) has four different groups attached (hydrogen, methyl, ethyl, and isopropyl) and is therefore chiral.

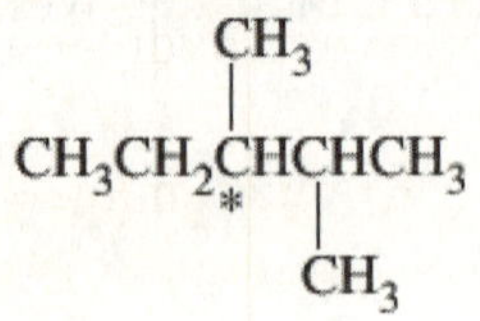

c. All carbon atoms are tetrahedral except the one bonded to the oxygen with a double bond. Only the carbon atom bonded to the OH has four different groups attached (hydrogen, hydroxyl, and two different portions of the ring) and is therefore chiral.

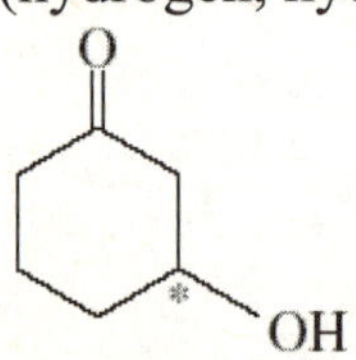

d. None of the ring carbons of the aromatic ring are tetrahedral. The C=O is not tetrahedral. Of the remaining carbon atoms only the two centrally located carbon atoms have four different groups attached. The upper central carbon (aromatic ring, hydrogen, the C=O group, and the lower ring) and the lower central carbon (the entire upper complex group anchored by the previously described chiral carbon, hydrogen and the two different portions of the lower ring).

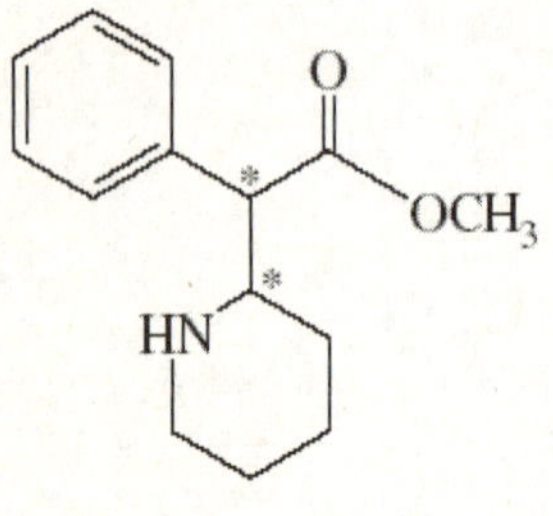

Ritalin

Additional Problems

4.27 Identify the following formulas as organic or inorganic compounds

a. KCl	**b.** C_4H_{10}	**c.** CH_3CH_2OH
d. H_2SO_4	**e.** $CaCl_2$	**f.** CH_3CH_2Cl

Organic compounds are composed primarily of carbon and hydrogen, but may also include oxygen, nitrogen, sulfur, phosphorus, and a few other elements. Compounds that do not contain carbon and hydrogen are classified as inorganic compounds.

a. KCl: inorganic
b. C_4H_{10}: organic
c. CH_3CH_2OH: organic
d. H_2SO_4: inorganic
e. $CaCl_2$: inorganic
f. CH_3CH_2Cl: organic

4.29 Alkanes are also referred to as *saturated hydrocarbons*. Explain the meaning of the term "hydrocarbon." Why are alkanes called "saturated" hydrocarbons?

Hydrocarbons are organic compounds composed only of carbon and hydrogen.

Saturated refers to the fact that each carbon is bonded to the maximum number of hydrogen atoms.

4.31 Give the molecular formula and draw the Lewis structure of cyclopentane.

See Table 4.2.

C_5H_{10}

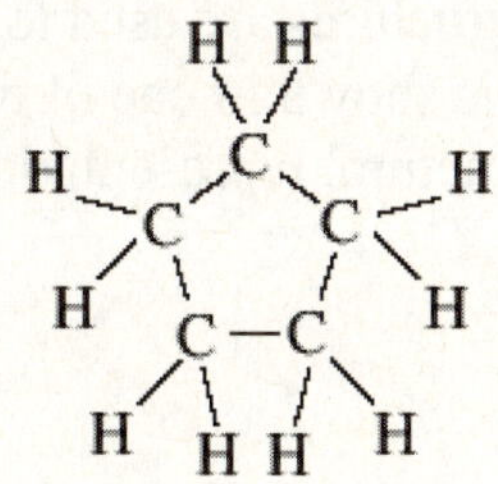

4.33 Give the name of the alkane with 10 carbons.

See Table 4.1.

The prefix *dec* indicates 10, the ending *-ane* indicates an alkane, therefore, decane is the 10 carbon alkane.

4.35 Convert each of the Lewis structures shown below into a condensed structural formula.

a.

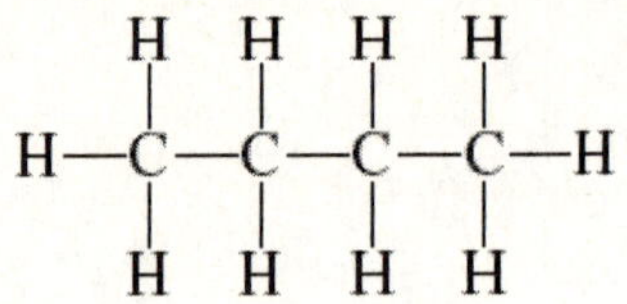

b.

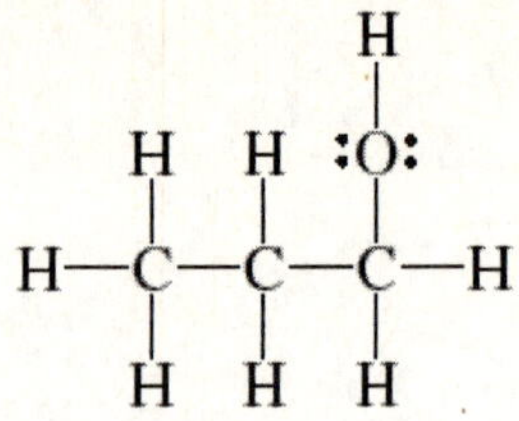

c.

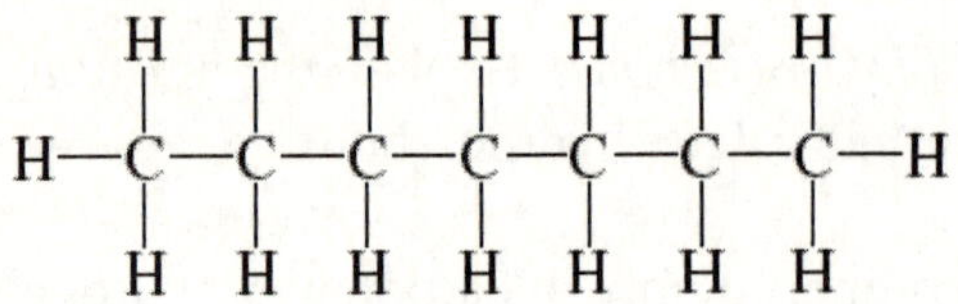

In the condensed structural formula, the atoms are listed from left to right in the order in which they are connected in the molecule, and it is understood that the octet of each atom is completed according to its preferred covalent bonding pattern.

a. $CH_3CH_2CH_2CH_3$

b. $CH_3CH_2CH_2OH$

c. $CH_3CH_2CH_2CH_2CH_2CH_2CH_3$

4.37 Lewis structures, condensed structural formulas, and skeletal structures are used to represent the structure of an organic compound. Each compound below is shown in one of these representations. Convert each compound into the other two structural representations not shown.

a.

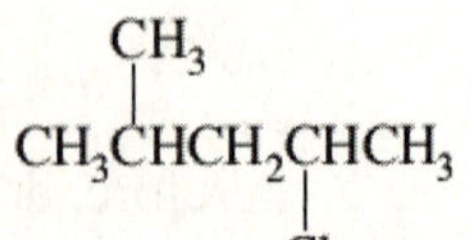

b.

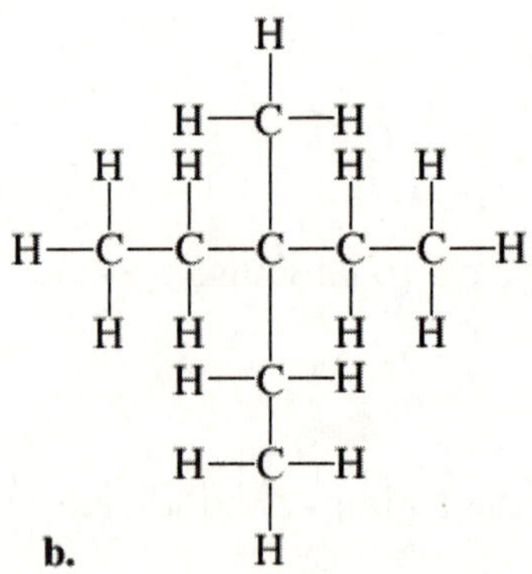

c.

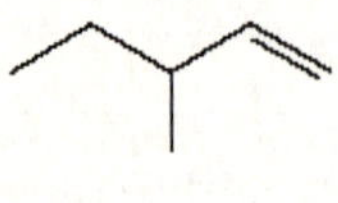

a. Shown as a condensed structure

skeletal structure

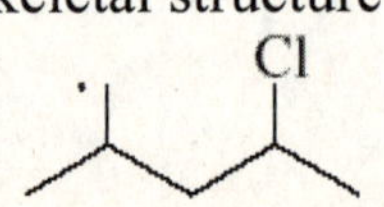

Lewis structure

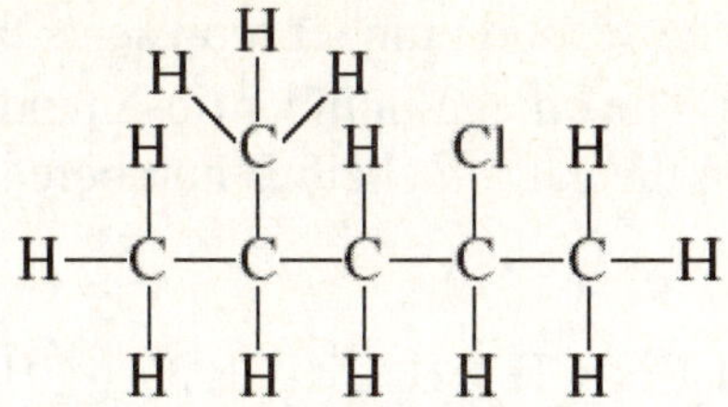

b. Shown as a Lewis structure

skeletal structure

condensed structure

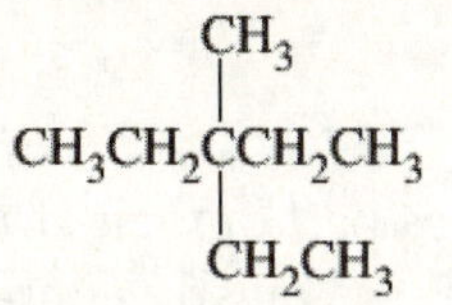

c. Shown as a skeletal structure

Lewis structure

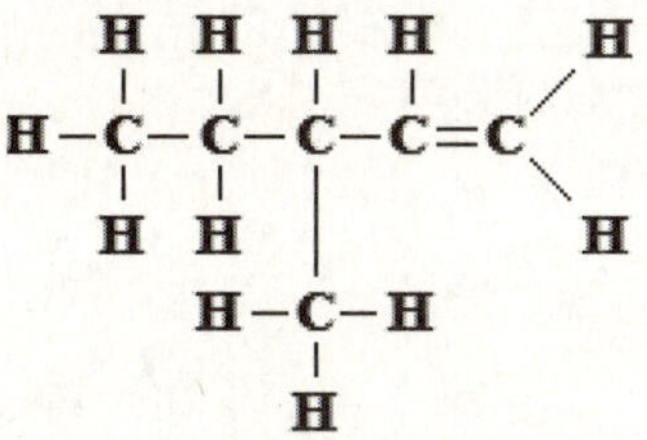

condensed structure

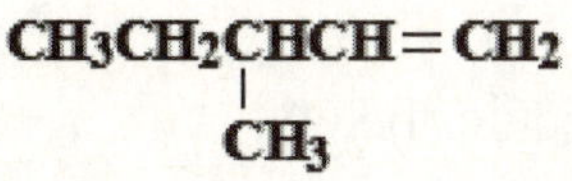

4.39 Explain the structural difference between a saturated and an unsaturated fatty acid.

Saturation depends on the number of hydrogen atoms present in a molecule relative to the maximum number it could contain. Double (triple) bonds do not contain the maximum number of hydrogen atoms and thus determine unsaturation. An unsaturated fatty acid contains one or more carbon–carbon double bonds, but a saturated fatty acid contains no double bonds.

4.41 Give the name and omega number of the fatty acid shown here.

$CH_3CH_2CH_2CH_2CH_2HC{=}CHCH_2HC{=}CH_2CH_2CH_2CH_2CH_2CH_2CH_2CH_2COOH$

[XX:X] the first number in the brackets in the name of a fatty acid represents indicates that the total number of C atoms in the fatty acid. The second number represents the number of carbon to carbon double bonds in the fatty acid. The omega-number (ω-X) indicates that the position of the first double bond is located when the carbon chain is numbered from the C **furthest** from carboxylic acid group.

$CH_3CH_2CH_2CH_2CH_2HC{=}CHCH_2HC{=}CH_2CH_2CH_2CH_2CH_2CH_2CH_2CH_2COOH$
The longest carbon chain contains 18 C atoms. [18:X]
The longest carbon chain contains 2 carbon to carbon double bonds. [18:2]
The first double bond occurs between C 6 and C 7. [18:2], ω-6
The name of the polyunsaturated fatty acid that meets the above criteria is linoleic. (See Table 4.4.)

linoleic acid, [18:2], ω-6

4.43 Despite the fact that fatty acids have a polar –COOH group as part of their structure, they are mainly nonpolar compounds. Explain why these compounds are mostly nonpolar.

Most common fatty acids contain a carbon chain that is 12–20 carbon atoms long. Only the –COOH group is polar, the rest of the hydrocarbon chain which contains 11–19 carbon atoms that are nonpolar (tetrahedral C centers with nonpolar C—C and C—H bonds). The nonpolar portion predominates due to its sheer size.

For example:

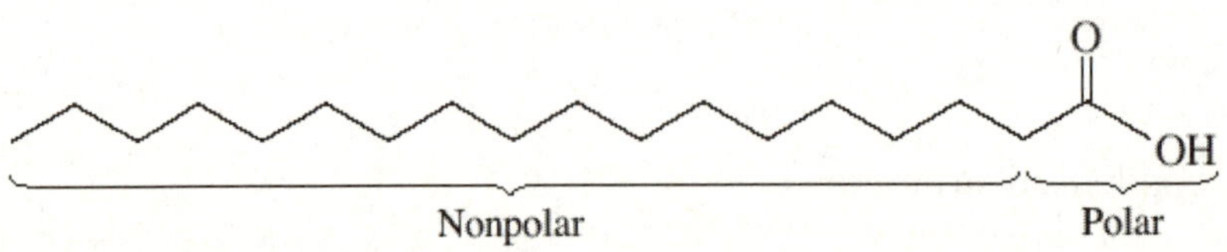

4.45 Identify all of the functional groups in each of the molecules below.

a.

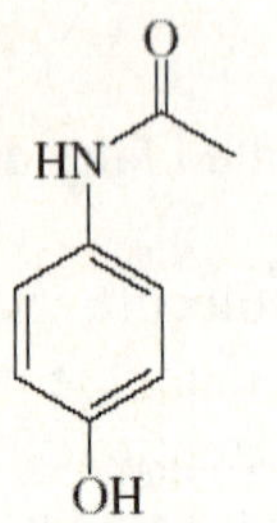

Acetaminophen

b.

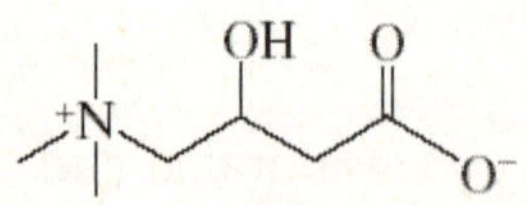

Carnitine

Use Table 4.6 to identify the arrangement of atoms that constitute the functional groups. Each of the functional groups is enclosed in a box or circle below.

a.

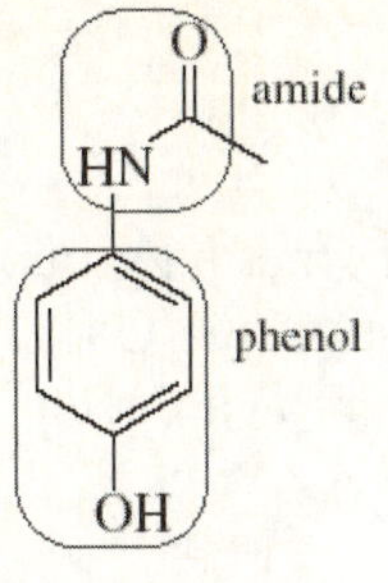

Acetaminophen

b.

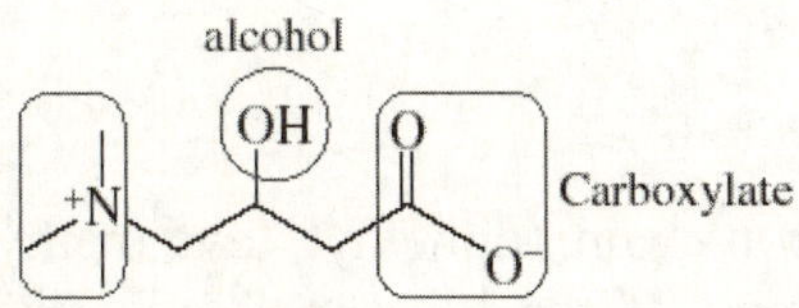

Amine Carnitine

4.47 Zantac™ is used to treat the overproduction of stomach acid for patients with chronic heartburn or gastric ulcers. Identify the four functional groups circled in the structure of Zantac below.

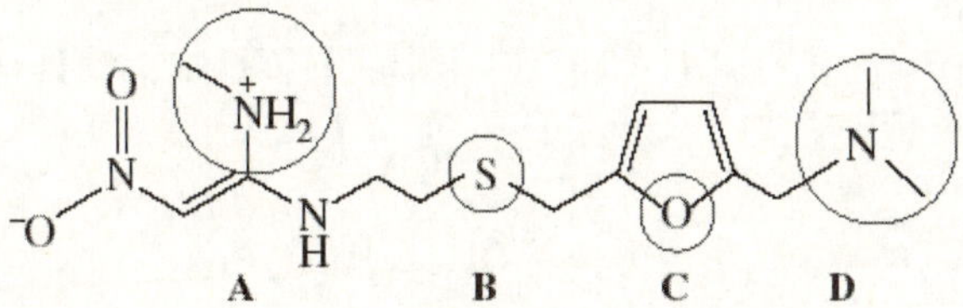

Zantac

Use Table 4.6 to identify the arrangement of atoms that constitute the functional groups. A = protonated amine, B = sulfide (thioether), C = ether, and D = amine

4.49 Write the condensed formula for each of the following molecules

a. 3-ethylhexane
b. 1,3-dichloro-3-methylheptane
c. 4-isopropyl-2,3-dimethylnonane

Use the IUPAC rules to convert the name to a condensed structure.

Step 1: Use the prefix of "-ane" in the IUPAC name to find the longest continuous chain of carbon atoms. This is the parent chain and can be determined using Table 4.1.
Step 2: Identify the groups bonded to the parent chain but not included in the parent chain. These groups are known as **substituents** and, in the case of branched-chain alkanes, they are called **alkyl** groups. The name of each alkyl group is derived from the alkane with the same number of carbon atoms by changing the "-ane" ending of the name of the corresponding alkane to "-yl."

a. 3-ethylhexane: hex, 6 carbon chain; ethyl, ethyl group substituent (CH_3CH_2); 3- the ethyl group is on C 3. Fill in hydrogens to give each carbon an octet.

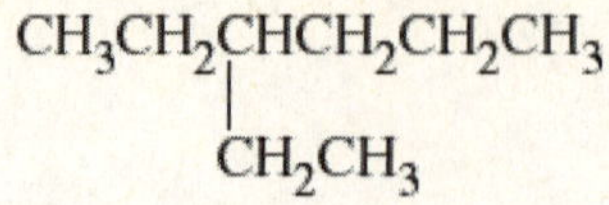

b. 1,3-dichloro-3-methylheptane: hept, 7 carbon chain; methyl, methyl group substituent (CH_3); 3- the ethyl group is on C 3; dichloro, two chloro substiutents (C l); 1,3-, one Cl is on C 3 and the other on C 1. Fill in hydrogens to give each carbon an octet.

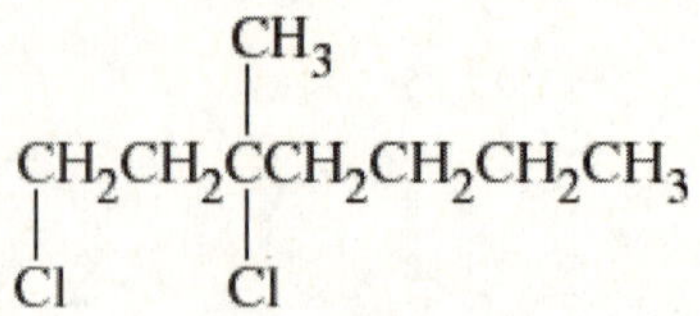

c. 4-isopropyl-2,3-dimethylnonane: non, 9 carbon chain; dimethyl, two methyl group substituents (CH_3); 2,3- one methyl group is on C 2, the other on C 3; isopropyl, isopropyl substituent (CH_3CHCH_3), 4-, the isopropyl group is on C 4. Fill in hydrogens to give each carbon an octet.

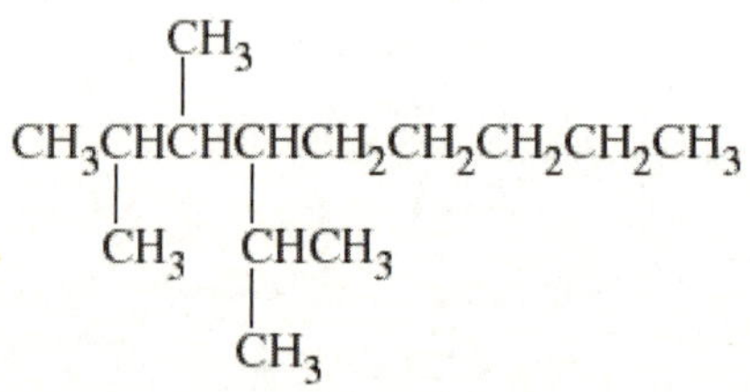

4.51 Using condensed structural formulas, draw three conformational isomers of hexane.

Recall:

1. Conformational isomers share the same molecular formula (same types and numbers of atoms) and are identical in how the atoms are connected to each other (connectivity).
2. Conformational isomers differ in the rotation of atoms about one or more of the single bonds.

See Table 4.1 for one conformer of hexane

$$CH_3CH_2CH_2CH_2CH_2CH_3$$

To generate other conformers, rotate sections of the chain out of the line of carbon atoms given above. Here are two of many possibilities.

Ethyl group rotated upward.

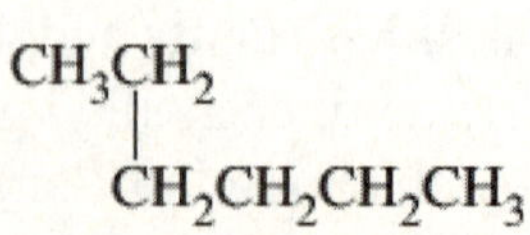

Ethyl group rotated upward and a methyl group rotated downward.

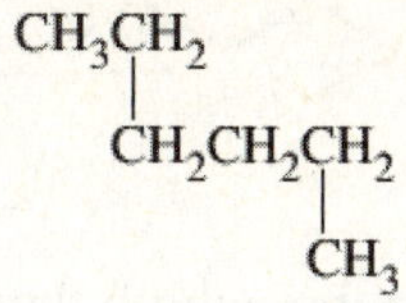

4.53 Draw the two alkane structural isomers with molecular formula C_4H_{10}. Give the IUPAC name of each compound.

Draw the isomer with all four carbons in the longest chain.

butane

Then branching the chain, e.g., switch a methyl group from the end of the chain with one of the hydrogen atoms on the carbon in the middle of the chain.

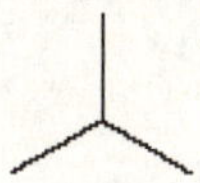

2-methylpropane

These are the only two structural isomers possible.

4.55 How many structural isomers are possible for the molecular formula $C_5H_{11}F$? Draw the skeletal structure and give the IUPAC name of each compound.

Naming:
Step 1: Find the longest continuous chain of carbon atoms. This is the parent chain. Name the parent chain according to the alkane names given in Table 4.1. (Hint: Count from each end to every other end to make sure you find the longest chain.)
Step 2: Identify the groups bonded to the main chain but not included in the main chain. These groups are known as **substituents** and, in the case of branched-chain alkanes, they are called **alkyl** groups. The name of each alkyl group is derived from the alkane with the same number of carbon atoms by changing the "-ane" ending of the name of the corresponding alkane to "-yl." Halogen substituents are named as fluoro, chloro, bromo, and iodo.

Draw the isomer with all five carbons in the longest chain. Put the F on an end carbon atom.

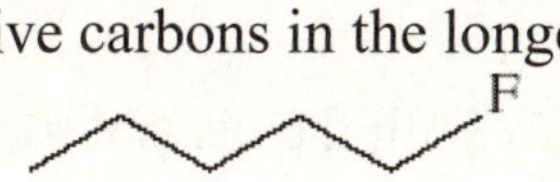

1-fluoropentane

Move the F one carbon into the chain.

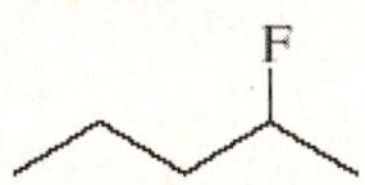

2-fluoropentane

Move the F one more carbon into the chain. Any further and the structure will be the same as one of the two given above.

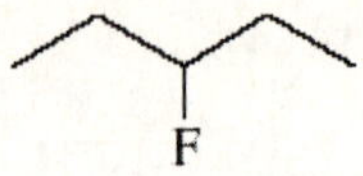

3-fluoropentane

With the F on the end carbon, branching the chain using methyl groups. With one methyl branch:

1-fluoro-2-methylbutane

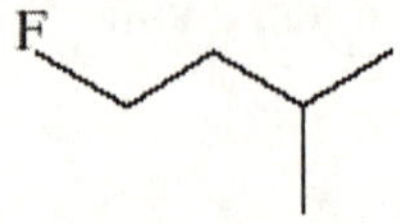

1-fluoro-3-methylbutane

With two methyl branches:

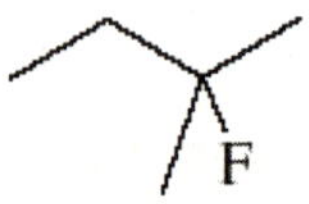

2-fluoro-2-methylbutane

1-fluoro-2,2-dimethylpropane

With the F on the second carbon, and branching the chain using methyl groups. With one methyl branch the last isomer is generated:

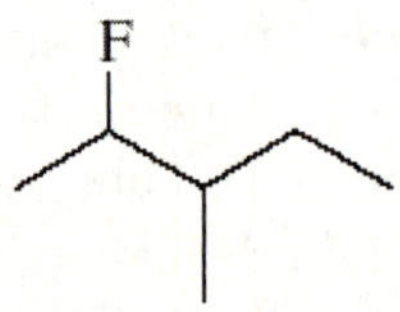

2-fluoro-3-methylbutane

Any further movement of the F or branching will produce a structure that is the same as one already drawn. Naming each isomer provides a check on whether structures are really isomers or are the same compound. All of the structures above have unique names and thus are really isomers.

4.57 Using wedge-and-dash bonds, draw both the cis and trans stereoisomers for each of the following compounds.

a. 1, 3-dimethylcyclohexane

b. 1-bromo-2-ethylcyclopentane

Determine the generic structure of the compound using the given IUPAC name. Organic chemists use types of bonds, known as wedge-and-dash bonds, to show the three-dimensional nature of a molecule on the flat page. The wedge bond ( ) is used to represent atoms that project out of the textbook page toward the viewer. The dash bond ( ) is used to represent atoms that project behind (or into) the textbook page away from the viewer.

a. 1, 3-dimethylcyclohexane: cyclo, a ring, and hex, a ring with six carbon atoms; dimethyl, two methyl (CH_3) groups; 1,3- , one methyl group attached at C 1 and the other at C 3. The cis structure will show both methyl groups on the same side of the ring while the trans structure will show one above the ring and the other below the ring.

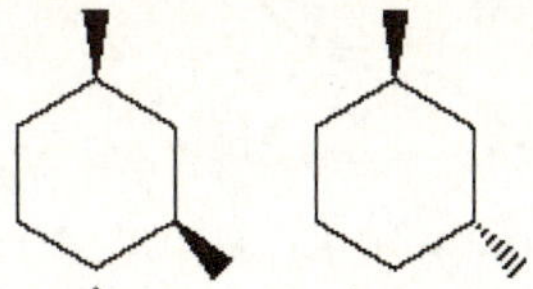

cis trans

b. 1-bromo-2-ethylcyclopentane: cyclo, a ring, and pent, a ring with five carbon atoms; one ethyl (CH_3CH_2) group; 2-, ethyl group attached at C 2; bromo, one bromo (Br) group, 1-, bromo group attached at C 1. The cis structure will show both the methyl group and bromo groups on the same side of the ring while the trans structure will show one above the ring and the other below the ring.

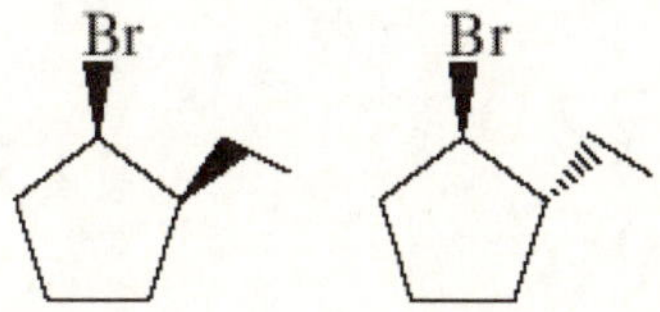

cis trans

4.59 For each of the following compounds, indicate whether or not it can exist as cis-trans stereoisomers. If it can exist as the two isomers, draw an expanded structural formula of both.

a. $H_2C{=}CHCH_2CH_3$ b. 

c. $CH_3CH_2CH{=}CHCH_2CH_3$ d. 

If three groups attached to the carbon atoms of the double are the same, cis-trans isomers are not possible. Examine each structure for the types of groups attached to the carbon atoms of the double bond. In the cis isomer both groups are on the same side of the double bond, while in the trans isomer the groups are on the opposite side of the double bond.

a. Two H atoms are attached to the carbon on the left of the double bond and one H on the carbon atom on the right side of the double bond. Cis-trans isomers are not possible.

b. There is one ethyl group and one H atom attached to each carbon of the double bond. Cis-trans isomers are possible, cis on the left and trans on the right.

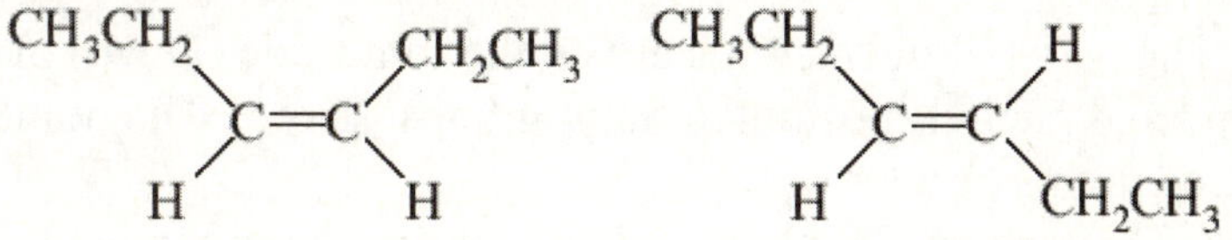

c. There is one ethyl group and one H atom attached to each carbon of the double bonds. Cis-trans isomers are possible, cis on the left and trans on the right.

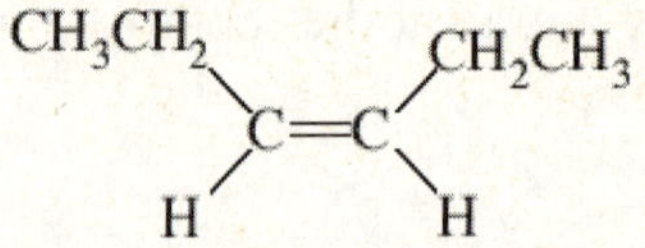

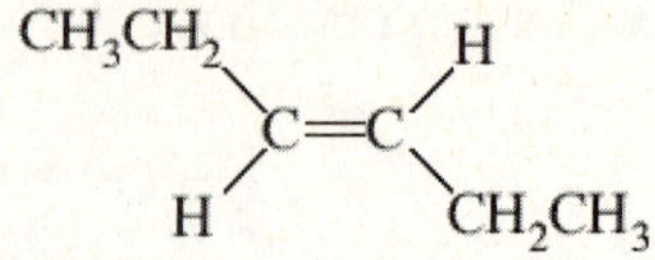

d. There is one ethyl group and one H atom attached to one carbon of the double bond and one methyl group and one H atom attached to the other carbon of the double bond. Cis-trans isomers are possible, cis on the left and trans on the right.

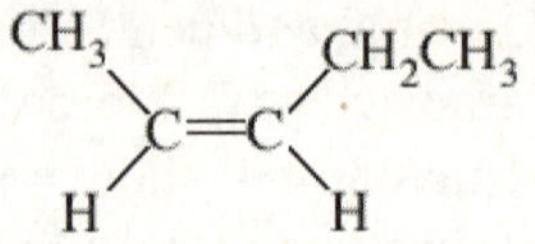

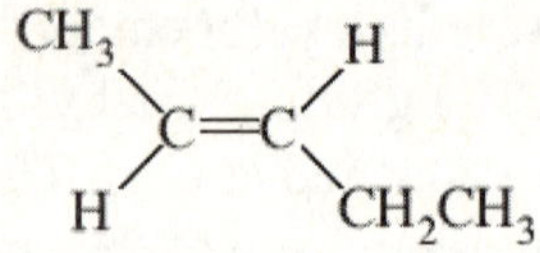

4.61 Determine whether each of the following is the cis or the trans stereoisomer of the compound shown.

a. 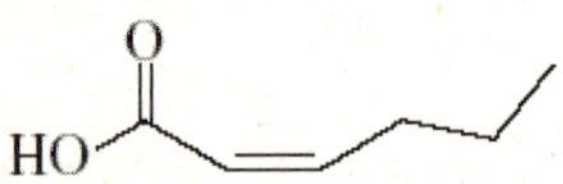

b. 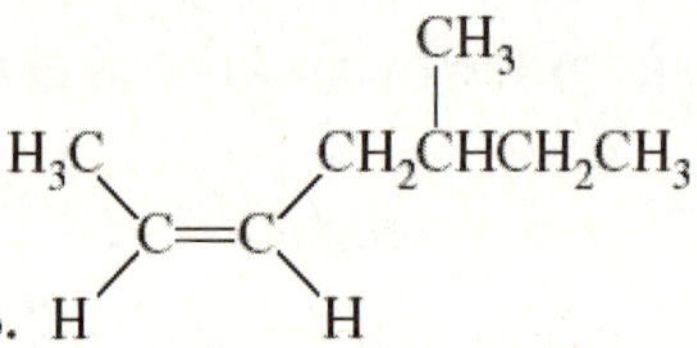

c. 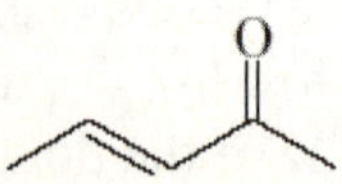

d. 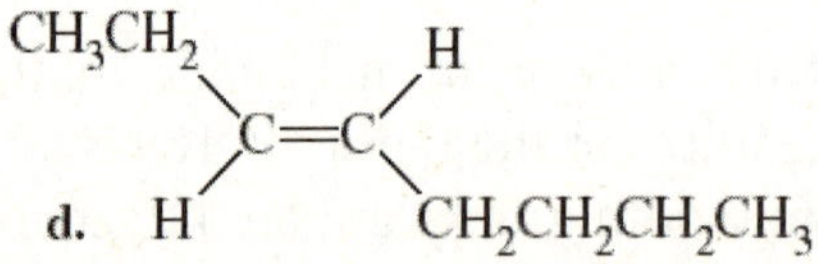

In the cis isomer both groups are on the same side of the double bond, while in the trans isomer the groups are on the opposite sides of the double bond. Examine each structure for the relative positions of the bonded groups.

a. cis: The carboxylic acid group and the butyl group are on the same side (up) of the double bond.
b. cis: The two hydrogen atoms are on the same side (down) of the double bond as are the methyl and methyl substituted butyl groups. (Note: The nomenclature of this group is not covered in this chapter.)
c. trans: The carboxylic acid group and the methyl group are on opposite sides (one up and one down) of the double bond.
d. trans: The two hydrogen atoms are on opposite sides of the double bond as are the ethyl and butyl groups (one up and one down).

4.63 Mark the chiral carbons in the following molecules, if any, with an asterisk (*).

a.

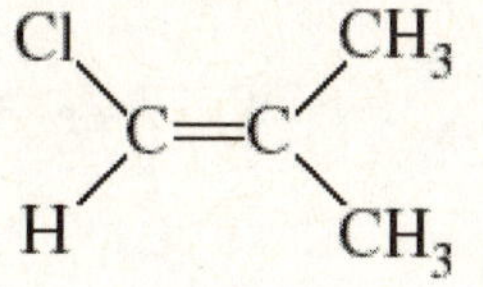

b.

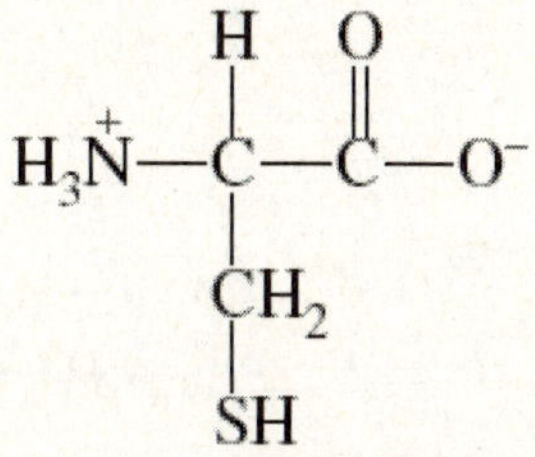

The amino acid cysteine

c.

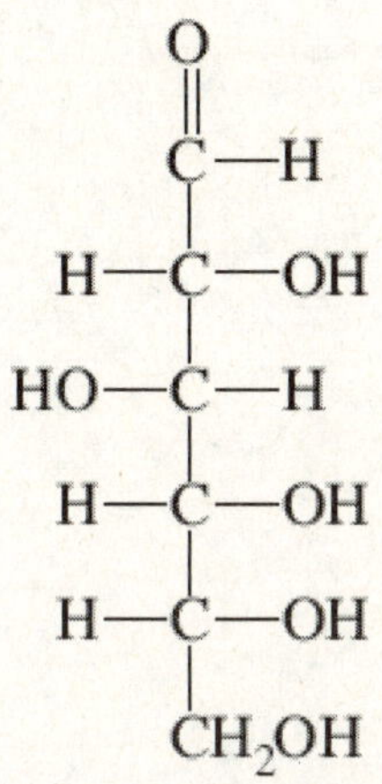

The carbohydrate D-glucose

d.

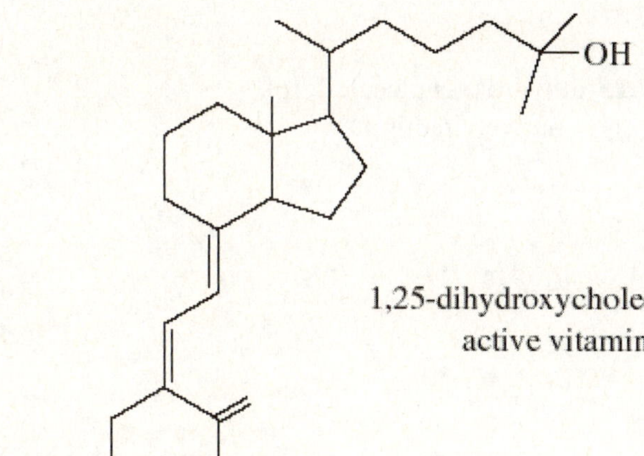

1,25-dihydroxycholecalciferol, active vitamin D

Examine each tetrahedral carbon atom (carbon atom with 4 charge clouds) for the presences of 4 different bonded groups. If this criterion is met, the C will be chiral.

a. none

b.

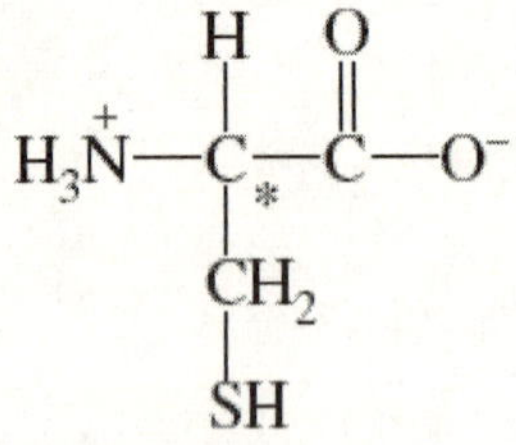

The amino acid, cysteine

c.

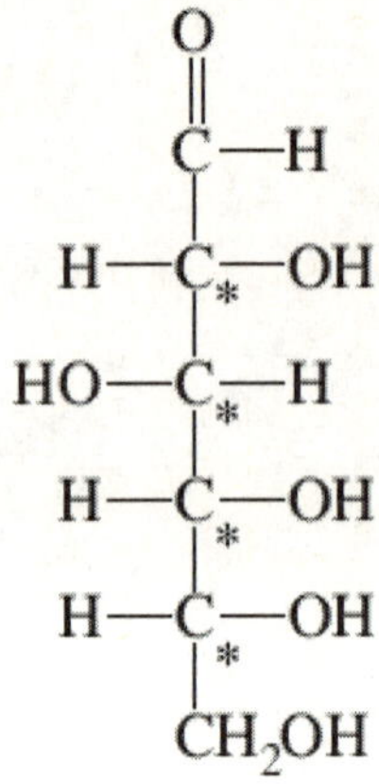

The carbohydrate, D-glucose

d.

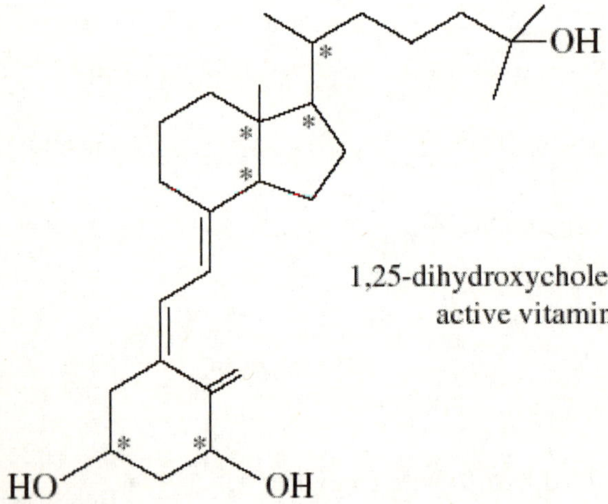

1,25-dihydroxycholecalciferol, active vitamin D

4.65 Are the following compounds structural isomers, cis-trans isomers, or enantiomers?

a. 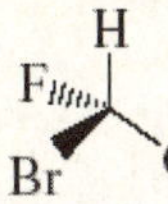 and 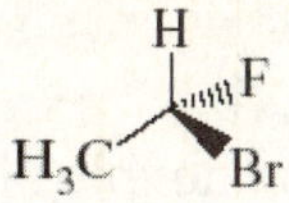

b. 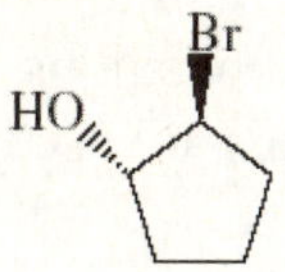 and 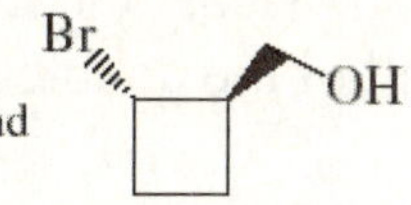

c.  and 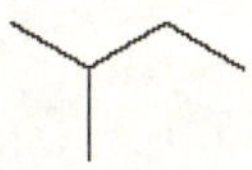

Recall the definition of each type of isomer relative to the alkanes and cycloalkanes given:

Structural isomers have the same number and types of atoms but differ in connectivity.

In order to have cis-trans stereoisomers in cycloalkanes there must be two groups attached to two different carbons of the ring.

Enantiomers are nonsuperimposable mirror images. Look for tetrahedral carbon atoms (carbon atom with 4 charge clouds) connected to 4 different bonded groups.

a. Enantiomers: Tetrahedral carbon atom bonded to four different groups. From orientation given, the mirror image effect is evident.
b. Structural isomers: Have the same number and type of atoms (same molecular formula, C_5H_9OBr) but differ in connectivity.
c. Structural isomers: Have the same number and type of atoms (same molecular formula, C_5H_{12}) but differ in connectivity.

Challenge Problems

4.67 Certain ω-3 fatty acids can only be found in animal sources, like fatty fish. Two of these are eicosapentaenoic acid (EPA) [20:5] and docosaheanaenoic acid (DHA) [22:6] both of which are omega-3 fatty acids. DHA has been shown to be important in healthy brain development, so it has recently been added to infant formulas. Breastmilk is rich in DHA as long as the mother maintains a healthy diet that includes fish. Draw skeletal structures of the fatty acids EPA and DHA.

[XX:X] the first number in the brackets in the name of a fatty acid indicates the total number of C atoms in the fatty acid. The second number represents the number of carbon-to-carbon double bonds in the fatty acid. The omega-number (ω-X) indicates the position of the first double bond (C-X) when the carbon chain is numbered from the C **furthest** from carboxylic acid group.

eicosapentaenoic acid (EPA) [20:5]

The longest carbon chain contains 20 C atoms. [20:X]
The longest carbon chain contains 5 carbon-to-carbon double bond. [20:5]
Omega-3: The first double bond occurs between C 3 and C 4. [20:5], ω-3.
Begin placing the double bonds at this C atom and placing the rest at positions three carbon atoms removed from this atom and repeat until all of the double bonds are used.

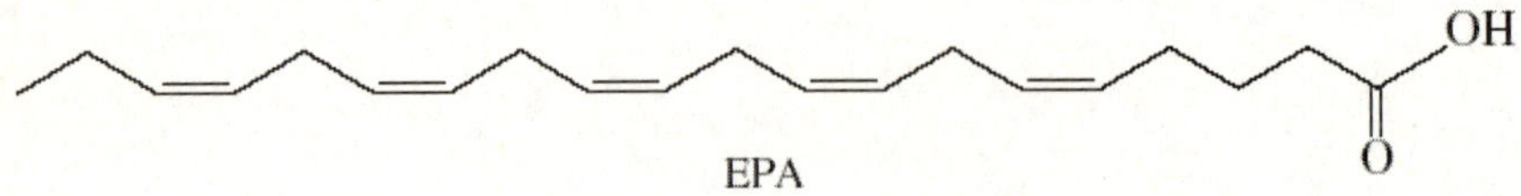

EPA

docosaheanaenoic acid (EPA) [22:6]

The longest carbon chain contains 22 C atoms. [22:X]
The longest carbon chain contains 6 carbon-to-carbon double bonds. [22:6]
Omega-3: The first double bond occurs between C 3 and C 4. [22:6], ω-3
Begin placing the double bonds at this C atom and placing the rest at positions three carbon atoms removed from this atom and repeat until all of the double bonds are used.

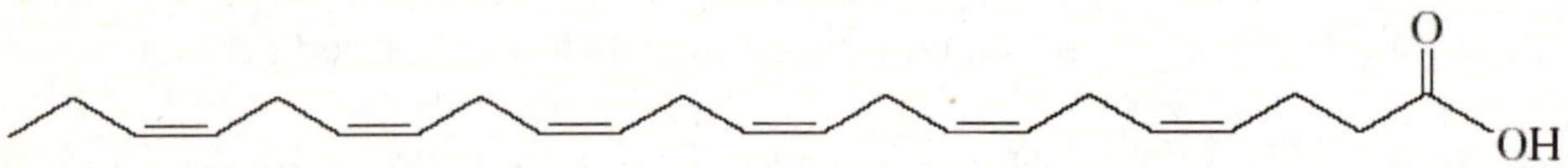

DHA

Chapter

5

Carbohydrates: Life's Sweet Molecules

5.1 Classes of Carbohydrates

Learning Objectives

Upon completion of this material, a student should be able to do the following:

A. define the following key terms:

carbohydrates	monosaccharides	disaccharides
oligosaccharides	polysaccharides	

B. describe the importance of carbohydrates as an energy source

C. differentiate between simple and complex carbohydrates

D. identify the various classes of carbohydrates based on numbers of monosaccharides in their structure

5.2 Monosaccharides

Learning Objectives

Upon completion of this material, a student should be able to do the following:

A. define the following key terms:

carbonyl	alcohols	primary (1°) alcohol
secondary (2°) alcohol	tertiary (3°) alcohol	aldehydes
ketones	aldose	ketose
glucose	fructose	Fischer projection
D-sugar	L-sugar	diastereomers
glycolysis	galactose	epimer
mannose		

B. identify a carbohydrate by its molecular formula and functional groups

C. recognize a carbonyl group in a formula

D. identify the functional groups aldehyde, ketone, and alcohol

E. distinguish primary, secondary, and tertiary alcohols

F. classify a monosaccharide as an aldose or ketose

G. identify chiral centers in a formula for a monosaccharide

H. draw and interpret a Ficsher projection

I. classify a monosaccharide as D- or L- based on the Fischer projection

J. draw enantiomers and diasteromers of a given Fischer projection

K. recognize the Fischer projections and common name of some of the relevant monosaccharides

5.3 Oxidation and Reduction Reactions

Learning Objectives

Upon completion of this material, a student should be able to do the following:

A. define the following key terms:

oxidation	reduction	reducing agent
oxidizing agent	Benedict's test	reducing sugar

B. identify the substances in an inorganic oxidation-reduction reaction undergoing oxidation and reduction
C. identify the substances in an organic oxidation-reduction reaction undergoing oxidation and reduction
D. identify the substances in an oxidation reduction reaction that are acting as the oxidizing and reducing agents
E. interpret the results of Benedict's test when applied to reducing sugars
F. predict the product of the oxidation of a reducing sugar
G. predict the product of the reduction of an aldose or a ketose

5.4 Ring Formation – The Truth About Monosaccharide Structure

Learning Objectives

Upon completion of this material, a student should be able to do the following:

A. define the following key terms:

hemiacetal	anomeric carbon	anomers
pyranose	furanose	

B. recognize pyranose and furanose rings
C. draw the ring structure (pyranose or furanose) of a linear monosaccharide.
C. identify anomeric carbon atoms
D. identify α and β anomers

5.5 Disaccharides

Learning Objectives

Upon completion of this material, a student should be able to do the following:

A. define the following key terms:

glycosidic bond	glycosides	condensation
dehydration	hydrolysis	maltose
lactose	sucrose	

B. locate a glycosidic bond in the structure of a disaccharide
C. relate the formation of a glycosidic bond in a disaccharide as an example of a condensation (a.k.a. hydrolysis) reaction
D. relate the production of monosaccharides from a disaccharide as an example of a hydrolysis reaction
E. recognize that a hydrolysis reaction is the reverse of a dehydration reaction
F. identify reactions as dehydration or hydrolysis reactions
G. name the type of glycosidic bond present in a disaccharide
H. describe the composition of maltose, lactose, and sucrose in terms of:
   1. monosaccharide composition
   2. type of glycosidic bond

3. common names
4. sources and uses

5.6 Polysaccharides

Learning Objectives

Upon completion of this material, a student should be able to do the following:

A. define the following key terms:
   storage polysaccharides   structural polysaccharides
B. characterize the polysaccharides amylose, amylopectin, glycogen, cellulose, and chitin in terms of
   1. function (storage or structure)
   2. type of glycosidic bond(s) and monosaccharide composition
   3. source (plant or animal) and physiological location (if possible)

5.7 Carbohydrates and Blood

Learning Objectives

Upon completion of this material, a student should be able to do the following:

A. define the following key terms:
   heparin   glycosaminoglycans
B. distinguish the ABO blood types based on carbohydrate composition
C. identify compatibility within the ABO blood types
D. cite the importance and medical usage of heparin
E. recognize the unique structural features of the disaccharide in heparin and the relationship to its function

Practice Test for Chapter 5

1. The following could be a representation of a:

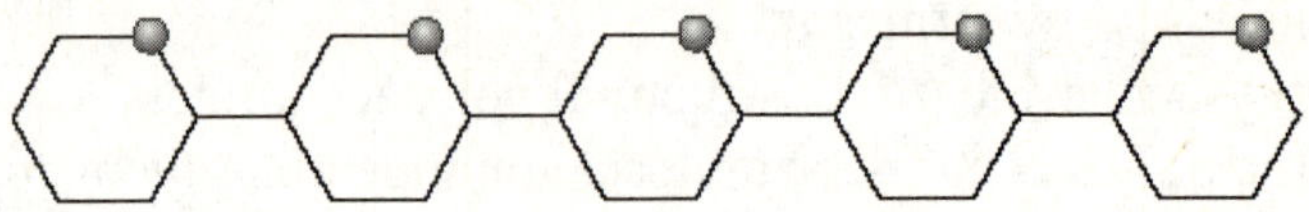

   A. monosaccharide.
   B. disaccharide.
   C. oligosaccharide.
   D. polysaccharide.

2. The molecule represents epinephrine (adrenaline). Epinephrine contains:

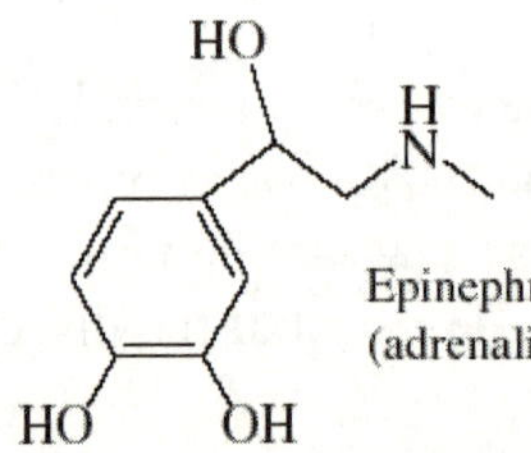

Epinephrine
(adrenaline)

   A. one primary alcohol substituent.
   B. one secondary alcohol substituent.
   C. three secondary alcohol substituents.
   D. one tertiary alcohol substituent.

3. Which of the following contains a ketone functional group?

A.

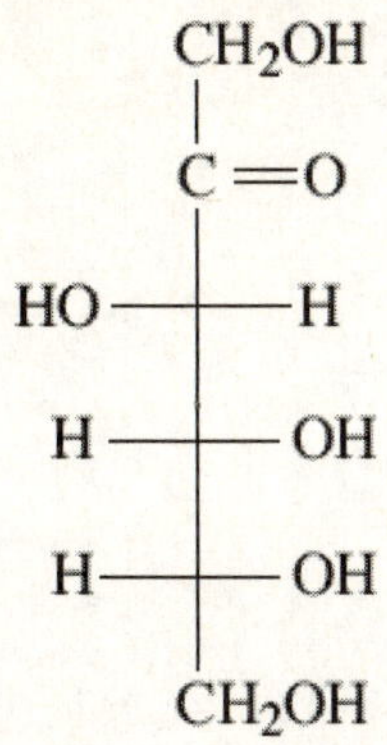

B.

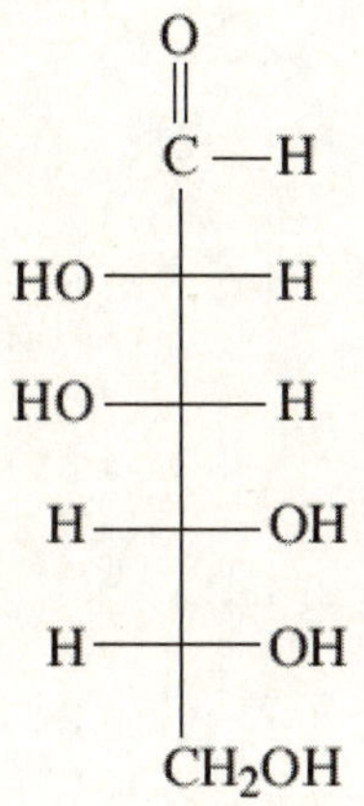

C.

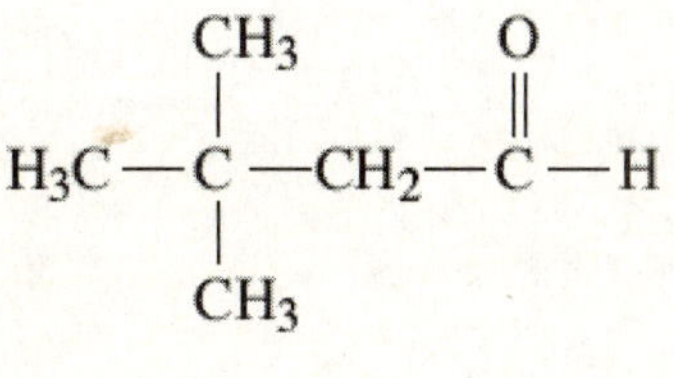

D.

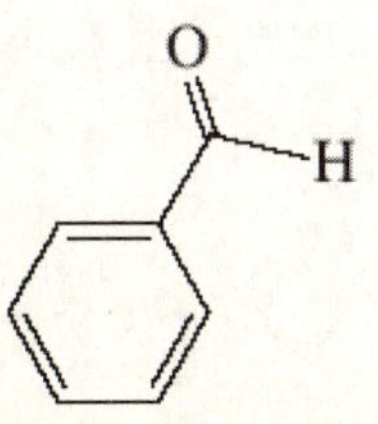

4. How many chiral centers are in the following monosaccharide?

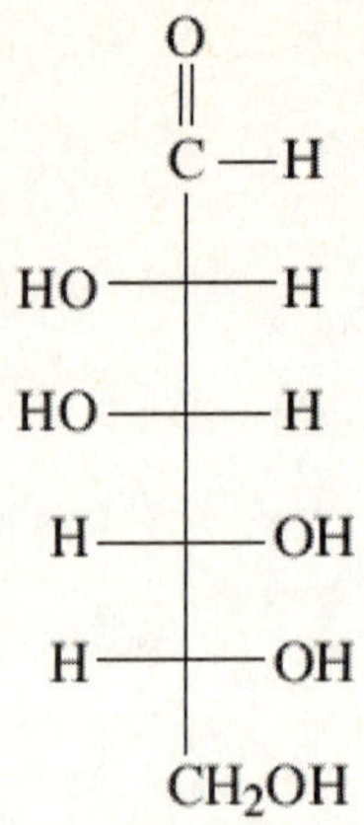

A. 2
B. 3
C. 4
D. 5
E. 6

5. In a Fischer projection, the difference between the D-sugar and the L-sugar is:
   A. the presence of an aldehyde versus a ketone functional group.
   B. the number of chiral carbon atoms.
   C. the position of the —OH group closest to the carbonyl atom.
   D. the position of the —OH group furthest from the carbonyl carbon atom.
   E. the position of the —OH group on the chiral carbon atom furthest from the carbonyl carbon atom.

6. Which of the following would be classified as an aldohexose?
A.

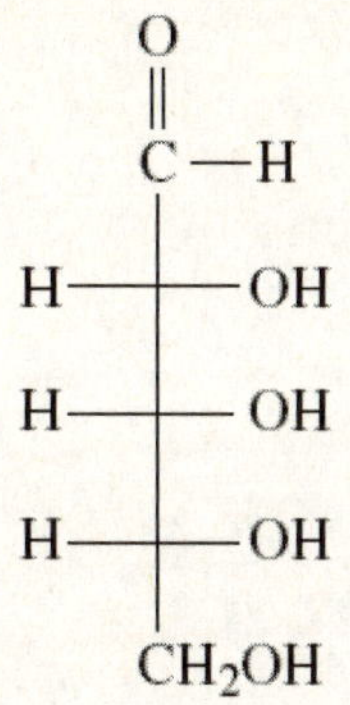

B.

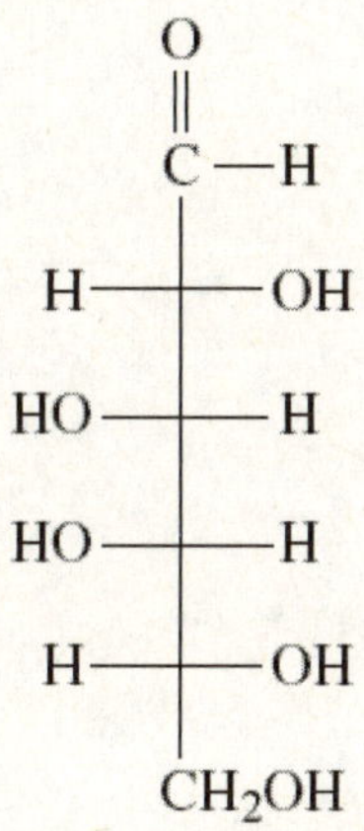

C.

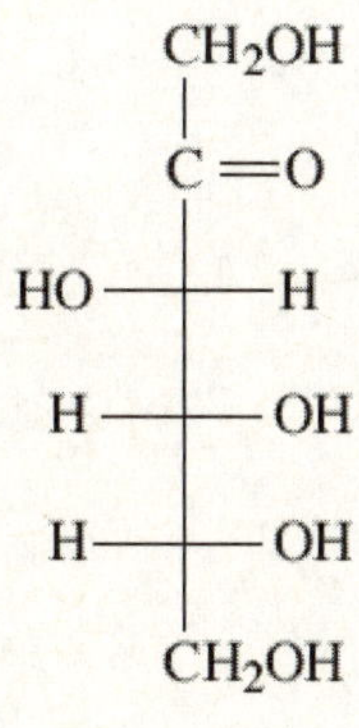

D. None of these is an aldohexose.

7. Which of the following represents the L-sugar of an aldose?

A.

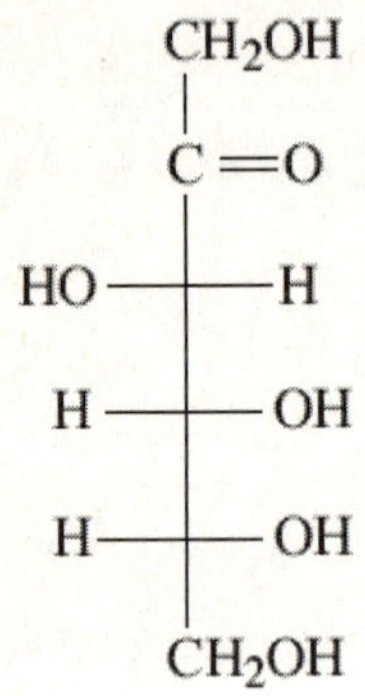

B.

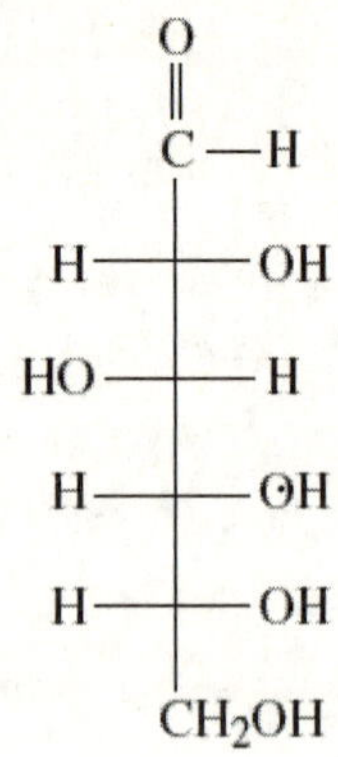

C.

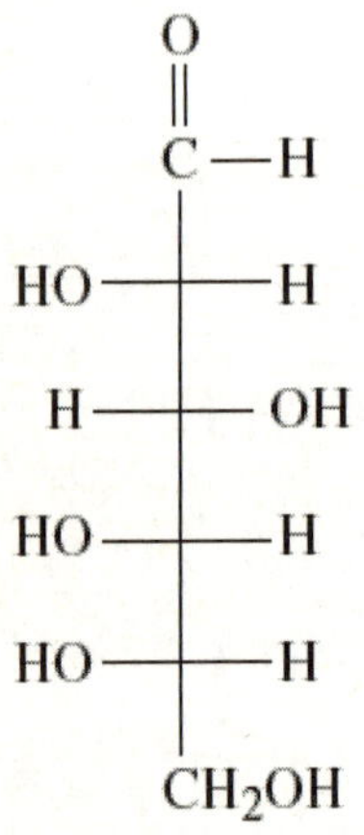

D.

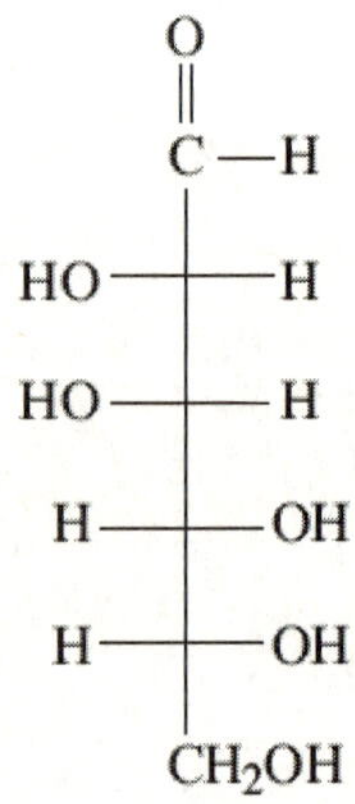

E. None of these is the L-sugar of an aldose.

8. Which of the following represents an enantiomer of:

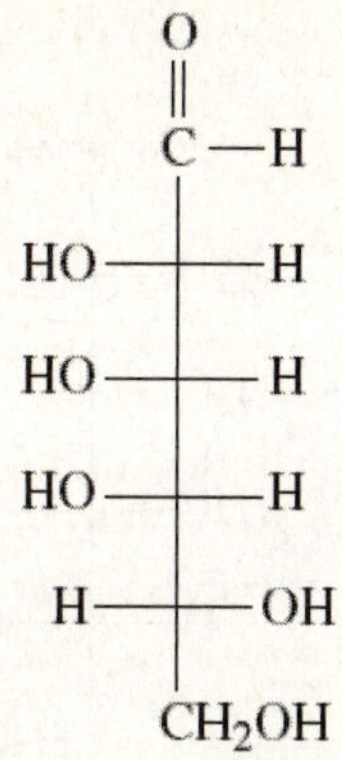

A.

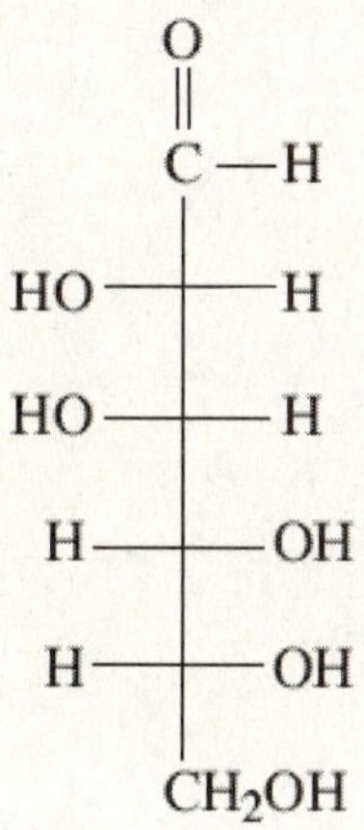

B.

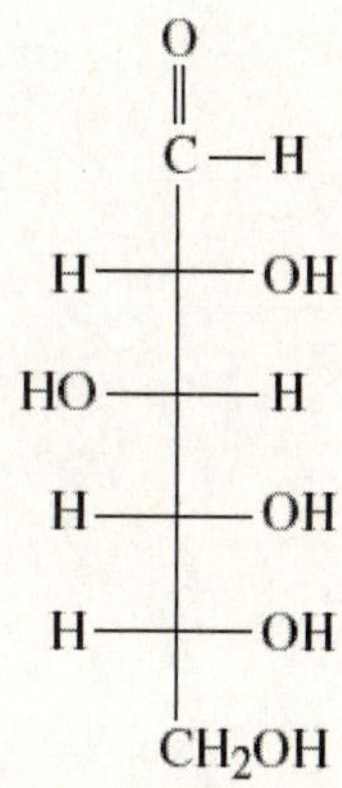

C.

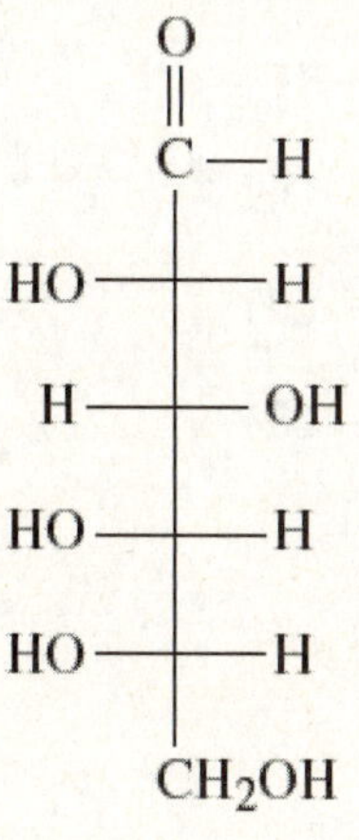

D.

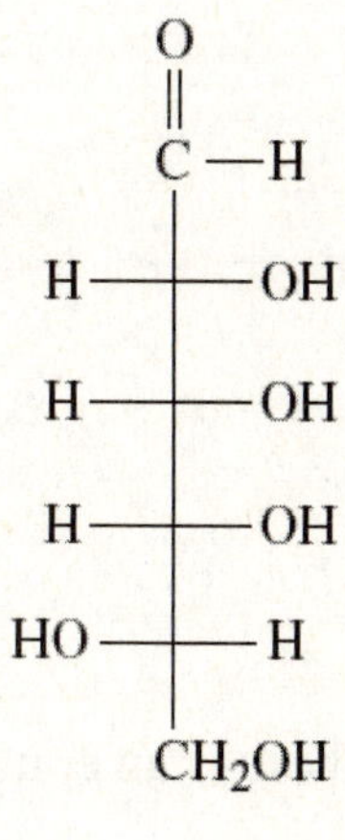

E. None of these is an enantiomer.

9. Which of the following represents an epimer of:

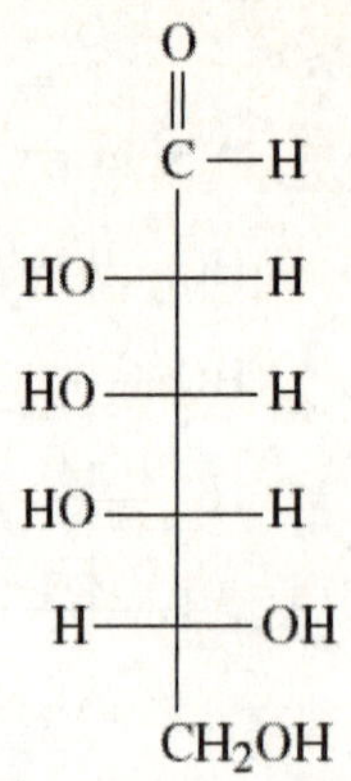

A.

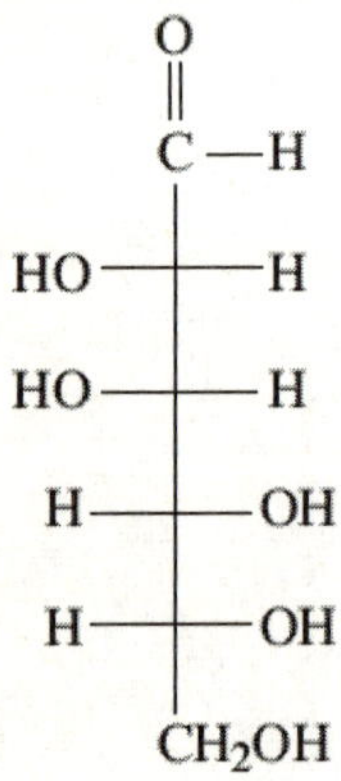

B.

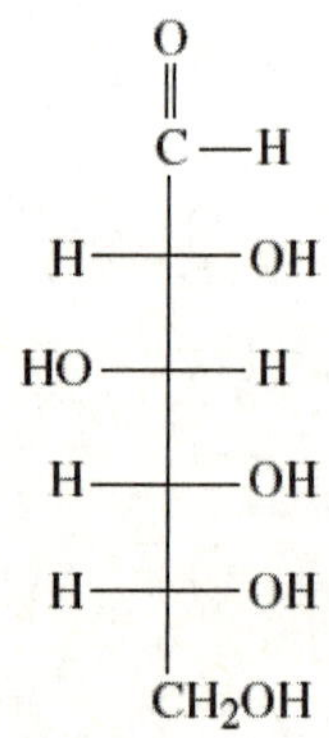

C.

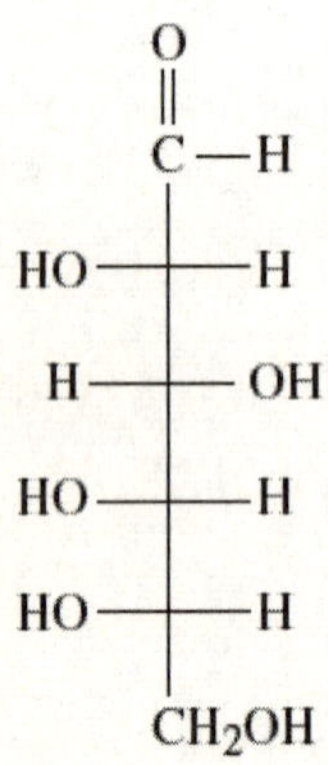

D.

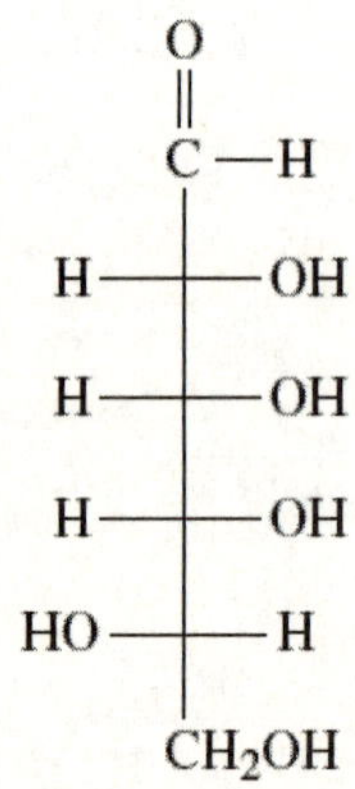

E. None of these is an epimer.

10. Which of the following represents a diastereomer of:

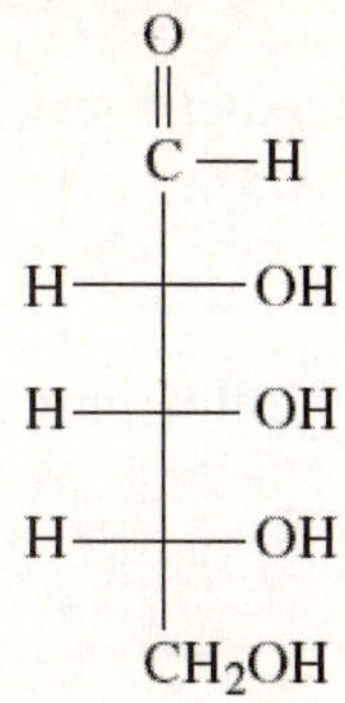

A.

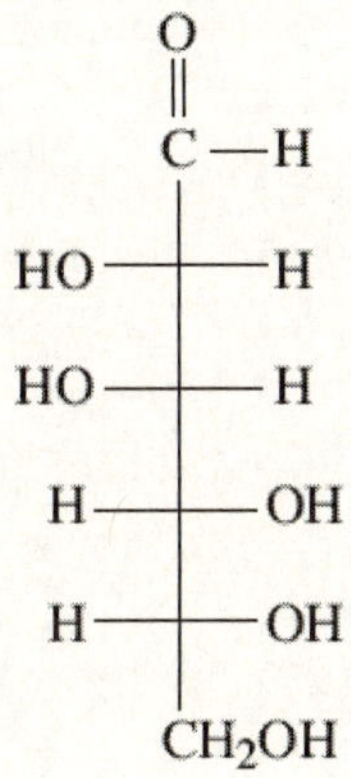

B.

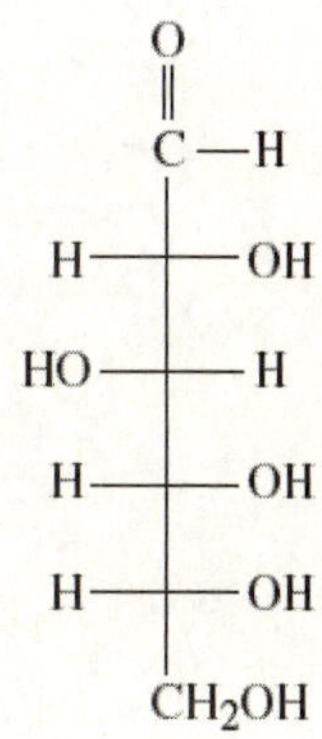

C.

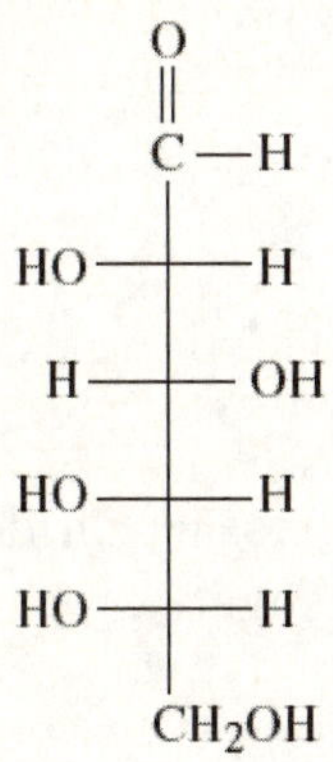

D.

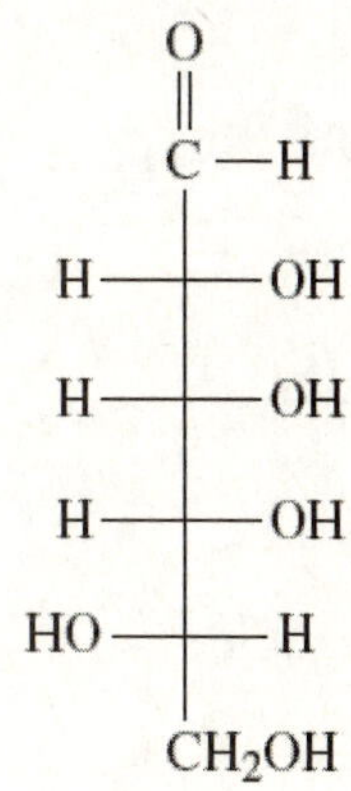

E. None of these is a diastereomer.

11. Which of the following changes would represent a reduction?
    A. Mg forms MgO in air.
    B. Cu^{2+} ions form Cu_2O during a Benedict's test.
    C. The biomolecule NADH forms NAD^+.
    D. CH_3OH forms HCOOH.

12. What is the product formed when the following monosaccharide is treated with Benedict's reagent?

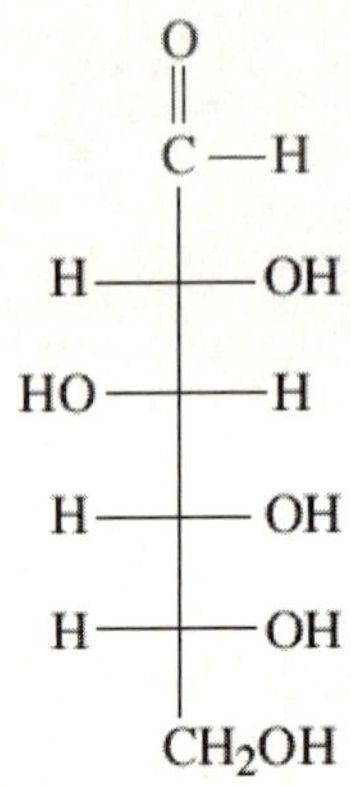

A.

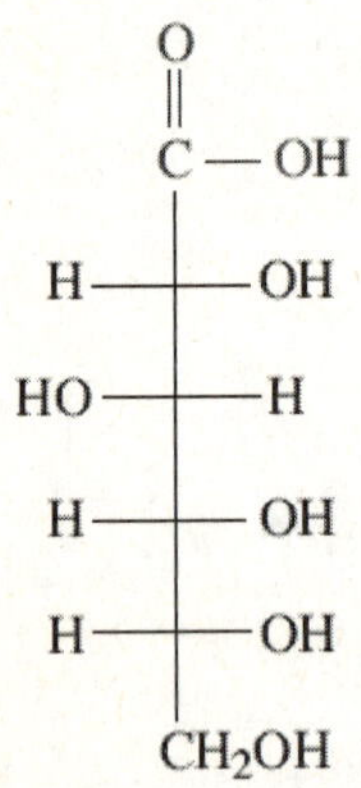

B.

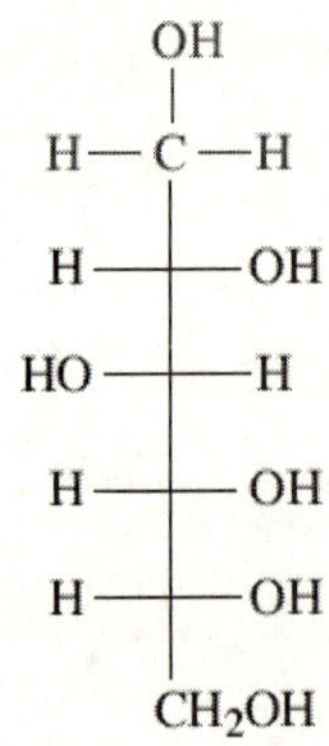

C.

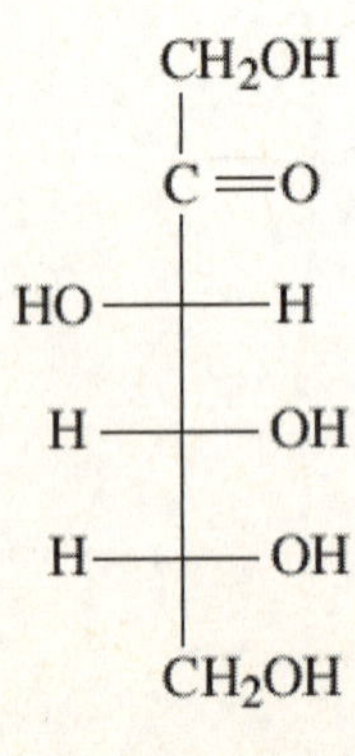

D. The given monosaccharide is a nonreducing sugar.

13. The following compound would be designated as:

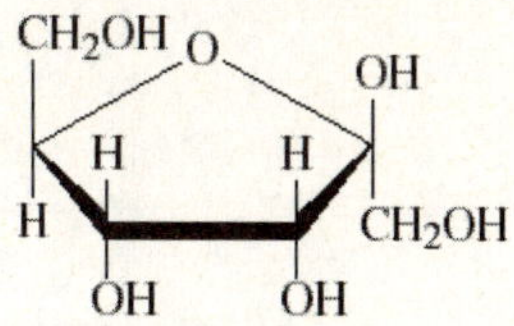

A. pyranose.
B. furanose.
C. nonreducing sugar.
D. both A and C
E. both B and C

14. Which of the following would be designated as an α anomer?
A.

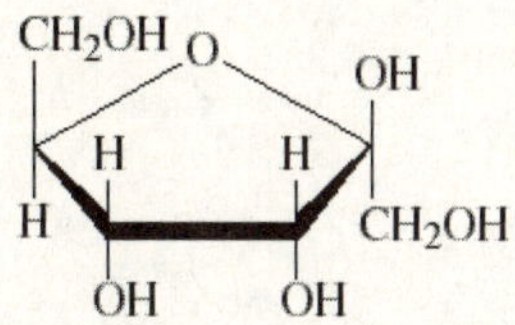

B.

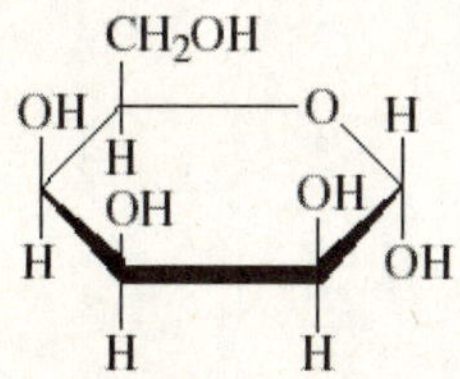

C.

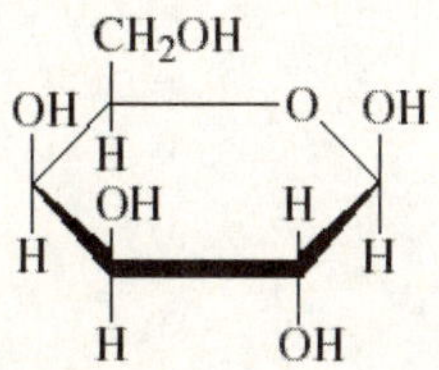

D. All are β anomers.

15. When an open chain monosaccharide forms a ring, a(n) ________________ functional group is formed. Which of the following should be used in the blank?
A. aldehyde
B. ketone
C. tertiary alcohol
D. hemiacetal

16. What is the correct ring structure for the α anomer in pyranose ring form for the monosaccharide shown below?

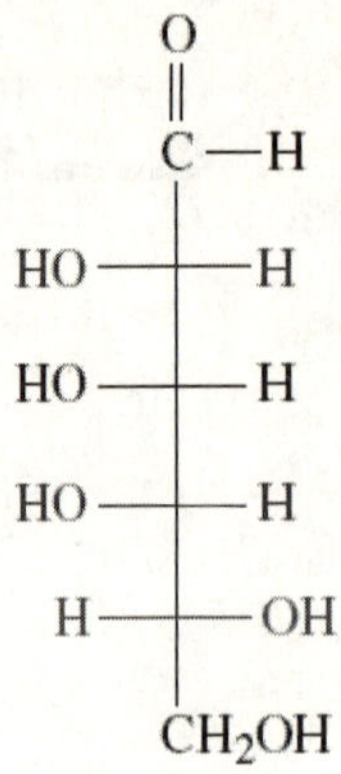

A.

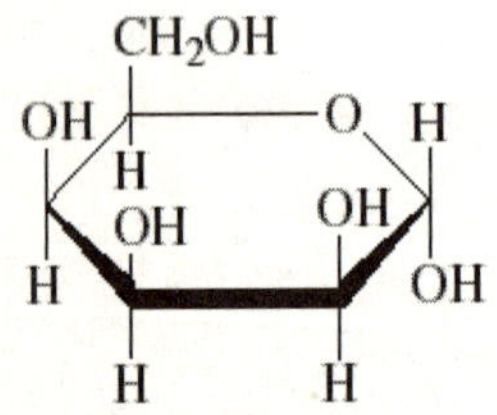

B.

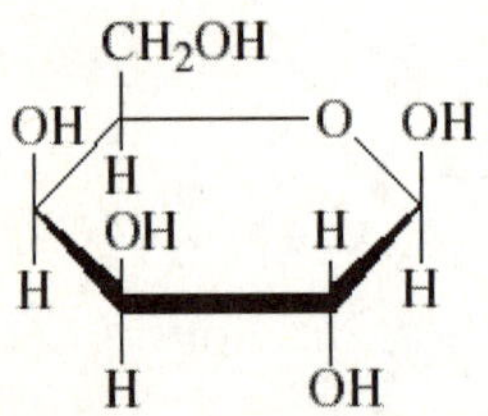

C.

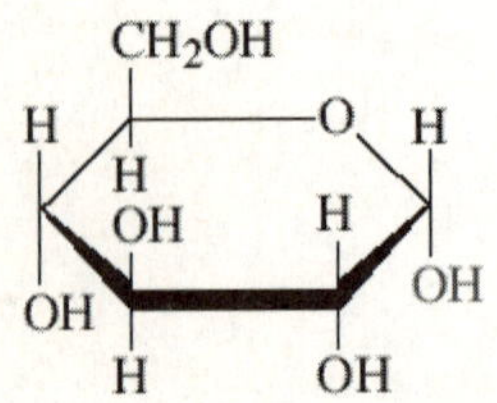

D.

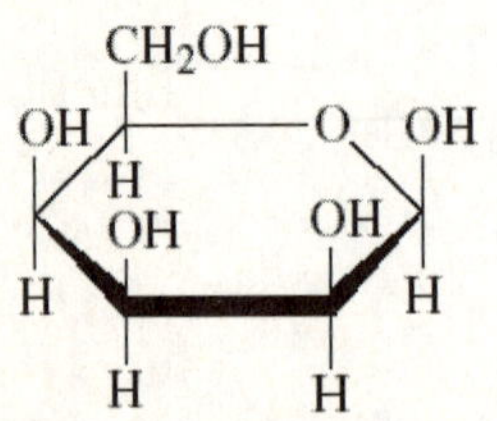

17. What will be the difference in the ring forms of the following two monosaccharides?

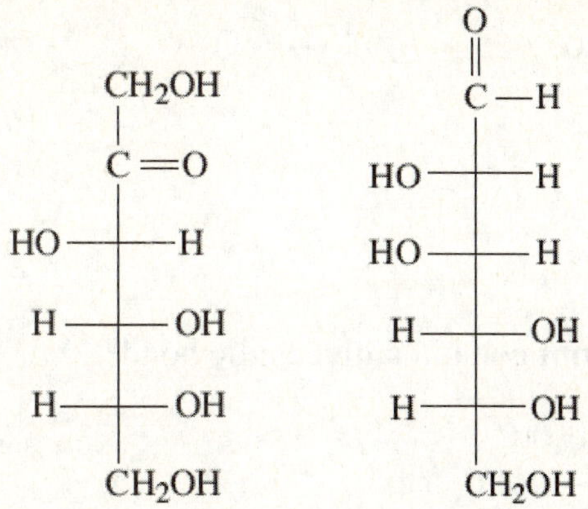

A. the number of atoms in the ring
B. the number of anomers possible
C. the placement of the —OH on C1 relative to the ring (above versus below)
D. the product being a reducing or nonreducing sugar

18. Which of the following in **not** a characteristic of fructose?
A. also called fruit sugar
B. a component of sucrose
C. epimer of glucose
D. sweetest monosaccharide

19. A monosaccharide is a component of both lactose and sucrose and is commonly known as blood sugar. This monosaccharide is:
A. mannose.
B. galactose.
C. glucose.
D. ribose.

20. Consider the following structure.

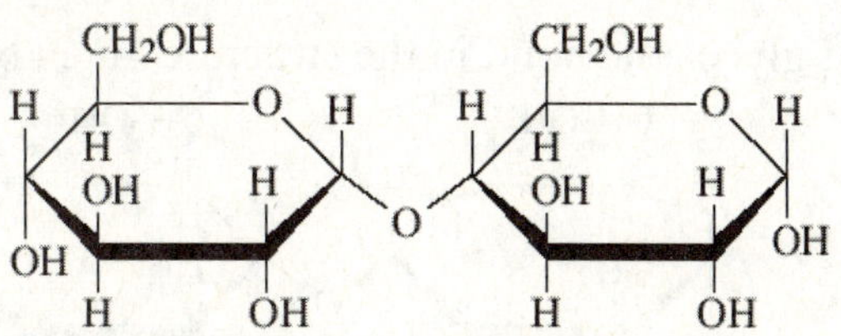

This structure represents:
A. a nonreducing oligosaccharide.
B. a reducing disaccharide.
C. the product of a hydrolysis reaction.
D. both B and C

21. The following reaction symbolically represents a:

A. hemiacetal reaction.
B. dehydration reaction.
C. condensation reaction.
D. hydrolysis reaction.

22. Which of the following does **not** contain a glycosidic bond?

A.

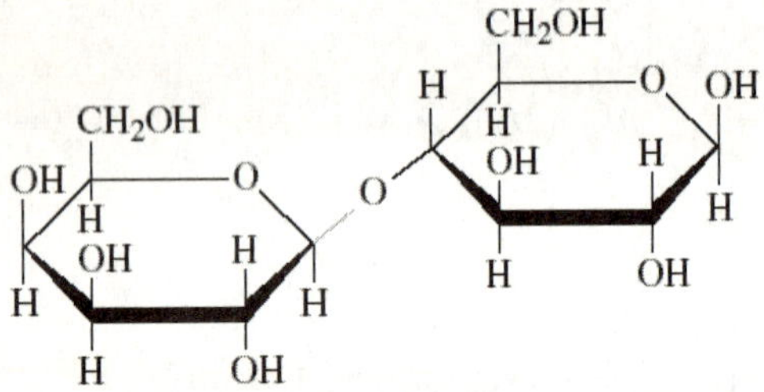

B.

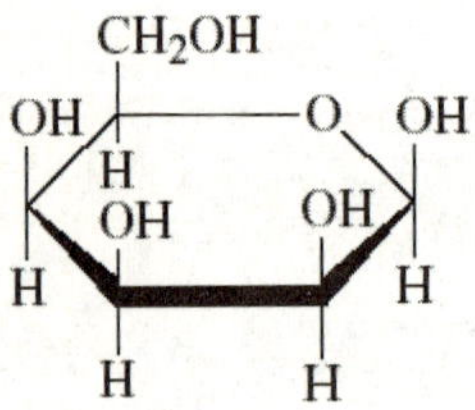

C.

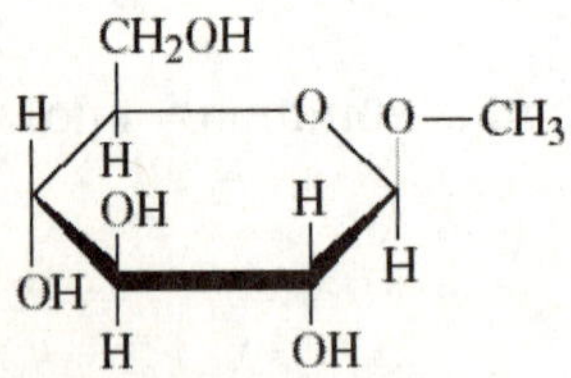

D. B and C
E. A, B, and C contain glycosidic bonds.

23. Determine the type of glycosidic bond in the structure given below.

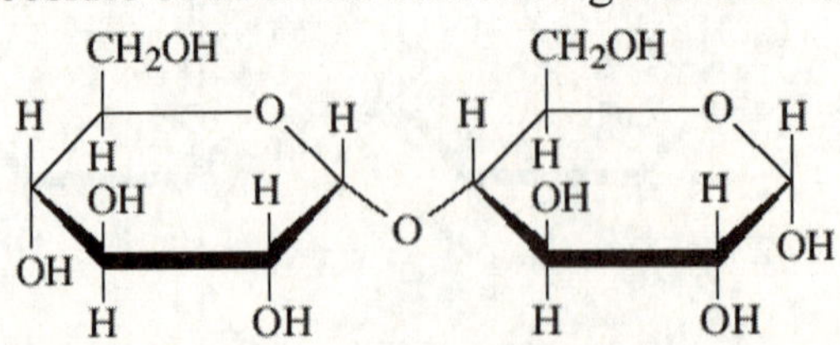

A. α (1→4)
B. β (1→4)
C. α (1→6)
D. β (1→6)

24. The product of the following reaction would be:

glucose + galactose $\longrightarrow$

A. maltose.
B. lactose.
C. sucrose.
D. amylose.

25. Which of the following describes sucrose?
A. a reducing sugar
B. undergoes hydrolysis to produce one type of monosaccharide
C. most abundant disaccharide in nature
D. found in mammalian milk
E. All of the above describe sucrose.

26. The product(s) of the hydrolysis reaction of the following sugar would be:

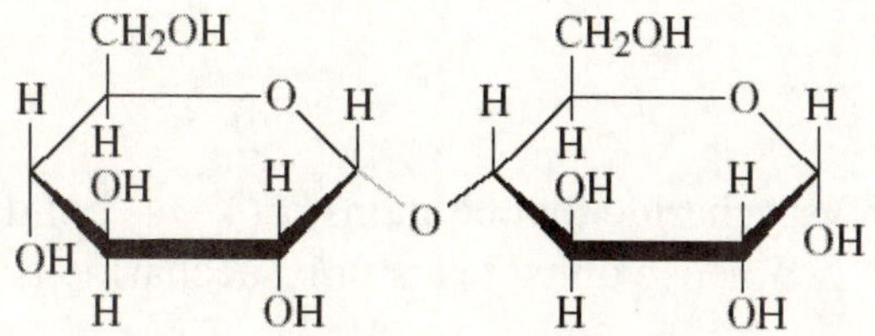

A.

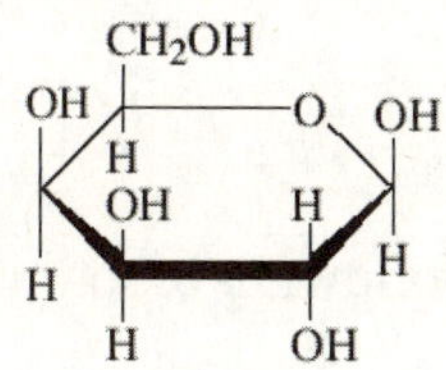

and

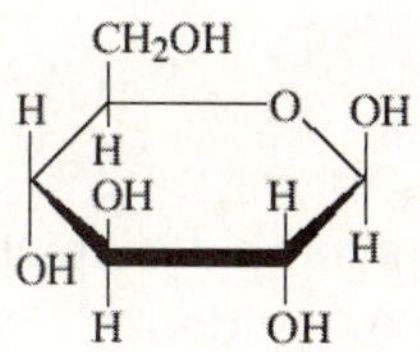

B.

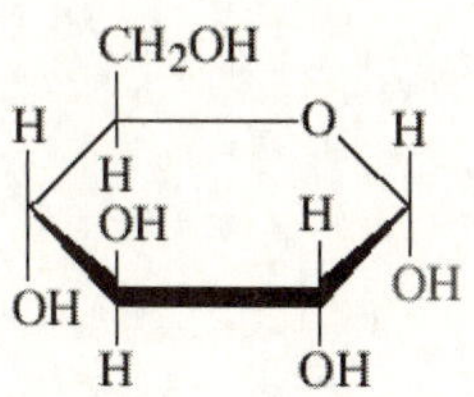

and

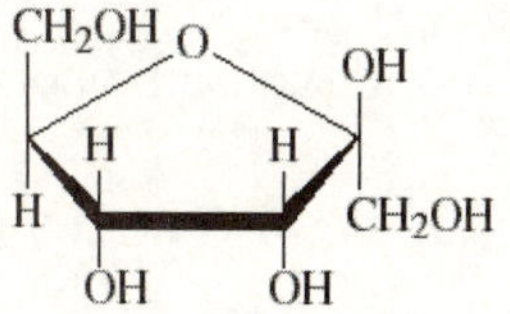

C.

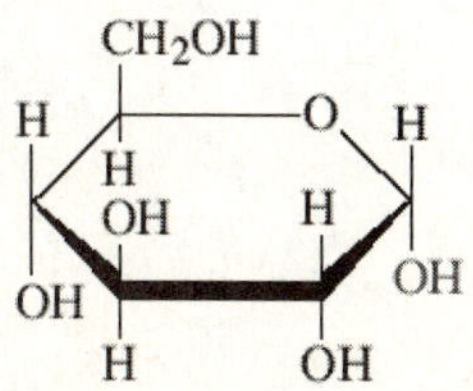

and

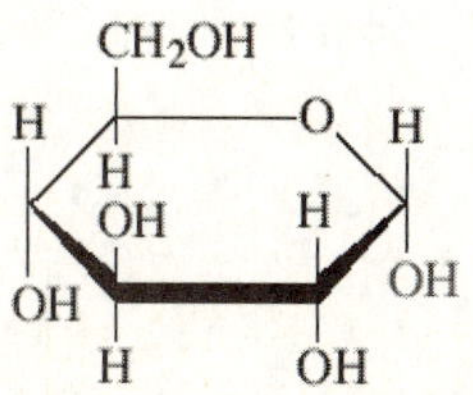

D. Anomers cannot be determined from given information.

27. Consider the segment of a polysaccharide shown below:

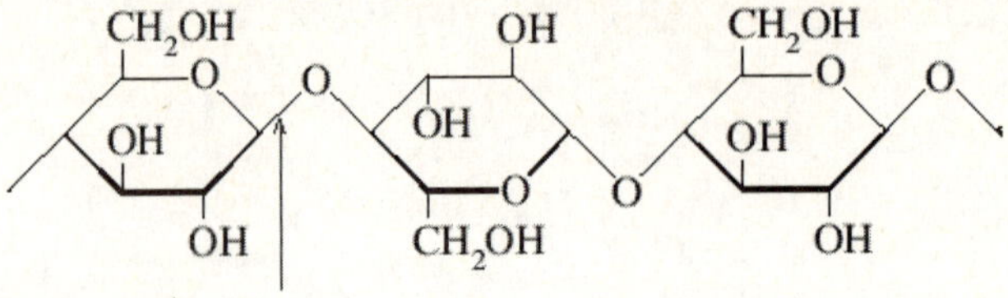

The arrow in the figure represents a(n):
A. α (1→4) bond.
B. β (1→4) bond.
C. α (1→6) bond.
D. β (1→6) bond.

28. Which of the following is an example of a structural polysaccharide?
A. amylose.
B. amylopectin.
C. glycogen.
D. chitin.

29. Consider a storage polysaccharide that contains α (1→4) bonds, is found in starch, and has branches about every 25 glucose units. This polysaccharide is:
A. amylose.
B. amylopectin.
C. glycogen.
D. cellulose.

30. Type O blood is available for transfusion. Recipients with the following blood types need a transfusion. Which can receive type O blood?
A. A
B. B
C. AB
D. A, B, and AB
E. Only recipients with type O blood would be compatible.

Answers

1. C	**2.** B	**3.** A	**4.** C	**5.** E	**6.** B	**7.** C	**8.** D	**9.** A	**10.** E
11. B	**12.** A	**13.** B	**14.** B	**15.** D	**16.** A	**17.** A	**18.** D	**19.** C	**20.** B
21. D	**22.** B	**23.** A	**24.** B	**25.** C	**26.** C	**27.** B	**28.** D	**29.** B	**30.** D

Chapter 5 – Solutions to Odd-Numbered Problems

Practice Problems

5.1 Classify each of the following alcohols as a primary (1°), secondary (2°), or tertiary (3°) alcohol.

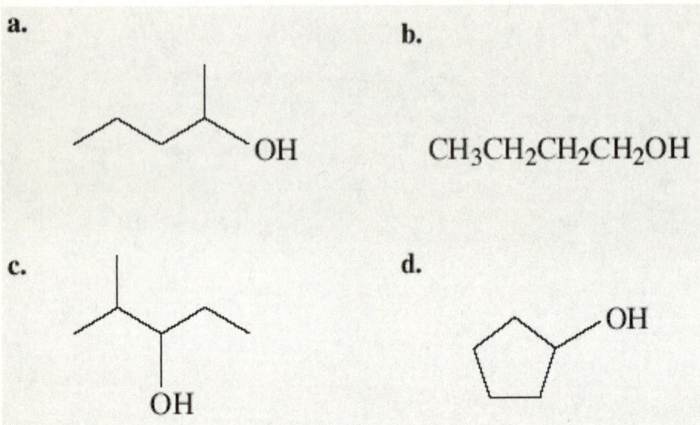

A primary (1°) alcohol has *one* alkyl group attached to the alcoholic carbon, a secondary (2°) alcohol has *two* such alkyl groups, and a tertiary (3°) alcohol has *three.*

a. secondary (2°) alcohol: C to which the —OH is attached is bonded to the second carbon (C2) of the pentane chain. C2 is bonded to two other carbon atoms.
b. primary (1°) alcohol: C to which the —OH is attached is bonded to the first carbon (C1) of the butane chain. C1 is bonded to one other carbon atom.
c. secondary (2°) alcohol: C to which the —OH is attached is bonded to the third carbon (C3) of the pentane chain. C3 is bonded to two other carbon atoms.
d. secondary (2°) alcohol: C to which the —OH is attached is bonded to a carbon of the cyclopentane ring. This carbon is bonded to two other carbon atoms.

5.3 Identify each of the following compounds as an aldehyde or a ketone

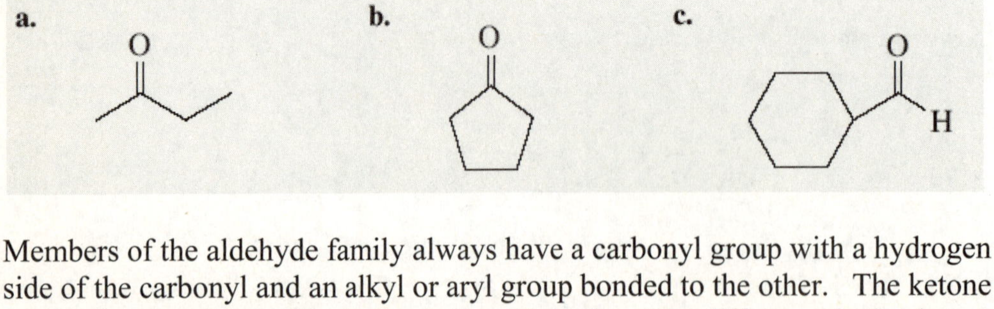

Members of the aldehyde family always have a carbonyl group with a hydrogen atom bonded to one side of the carbonyl and an alkyl or aryl group bonded to the other. The ketone family of organic compounds is structurally similar to the aldehydes. The difference is that ketones have an alkyl or aryl group on *both* sides of the carbonyl.

a. ketone: The carbonyl carbon (C=O) has a methyl group on the left and an ethyl group on the right.

b. ketone: The carbonyl carbon (C=O) has one carbon of the cyclohexane ring bonded on either side.

c. aldehyde: The carbonyl carbon (C=O) has a cyclohexane ring on the left and a hydrogen atom on the right.

5.5 How are disaccharides related to monosaccharides?

Two monosaccharides link via glycosidic bond to form a **di**saccharide.

5.7 What functional groups are present in aldoses?

As indicated by the prefix (**ald**) in the name, one functional group is an **ald**ehyde. Like all monosaccharides, an aldose also contains multiple alcohol (—OH) groups.

5.9 Draw the Fischer projections of the enantiomers of the following aldohexoses:

a.

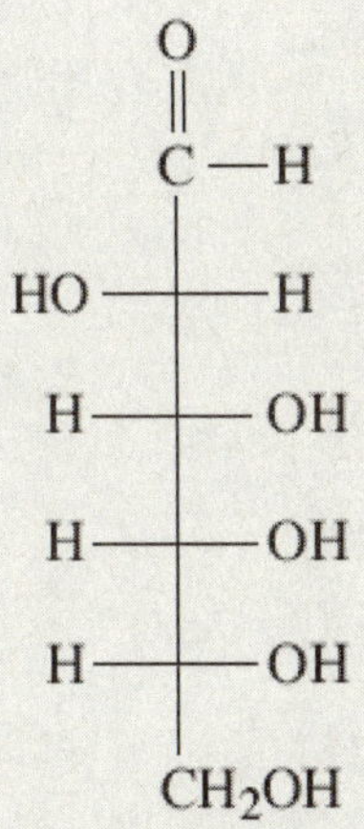

D-Altrose

b.

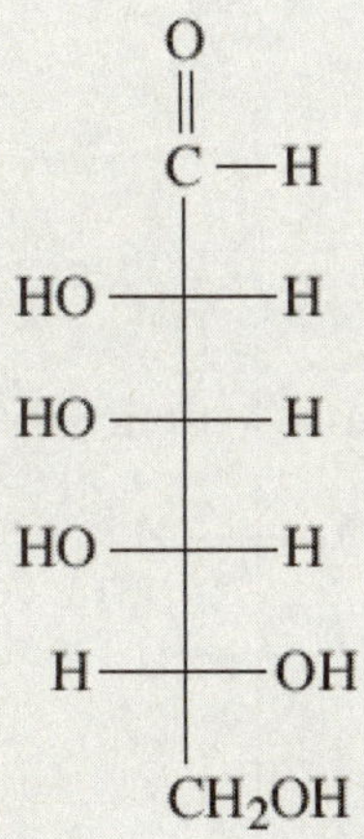

D-Talose

Recall that an enantiomer is a mirror image. Draw the mirror image of the given aldohexoses. Note: In the mirror image the —OH on the chiral carbon furthest from the C=O is pointing to the left (rather than the right) so the mirror images are the L-sugar.

a.

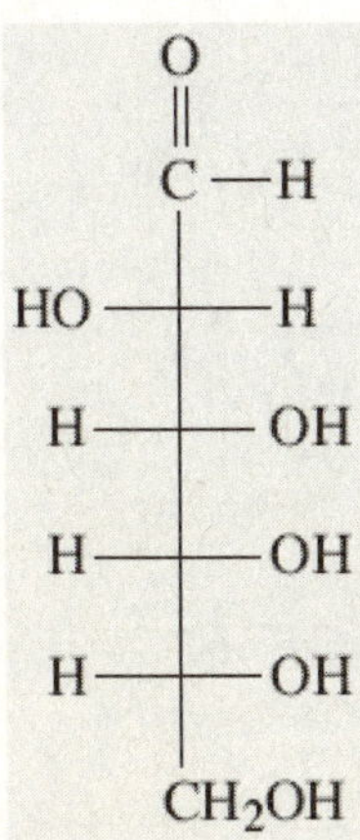

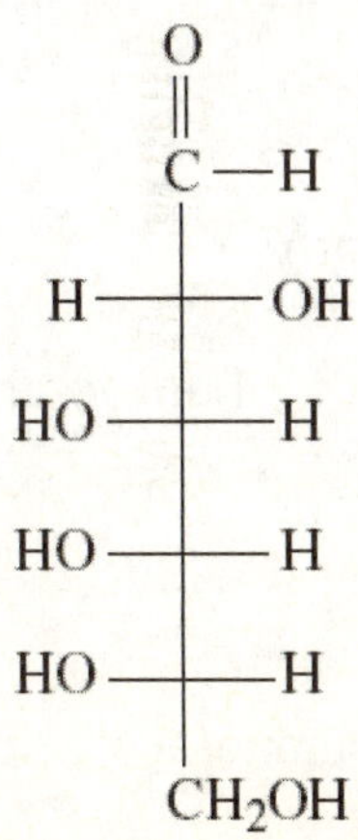

b.

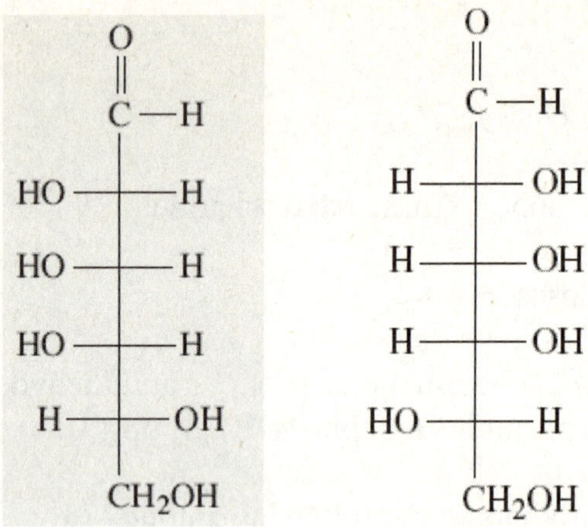

5.11 Classify structures A, B, and C below as being either an enantiomer or a diastereomer of D-galactose.

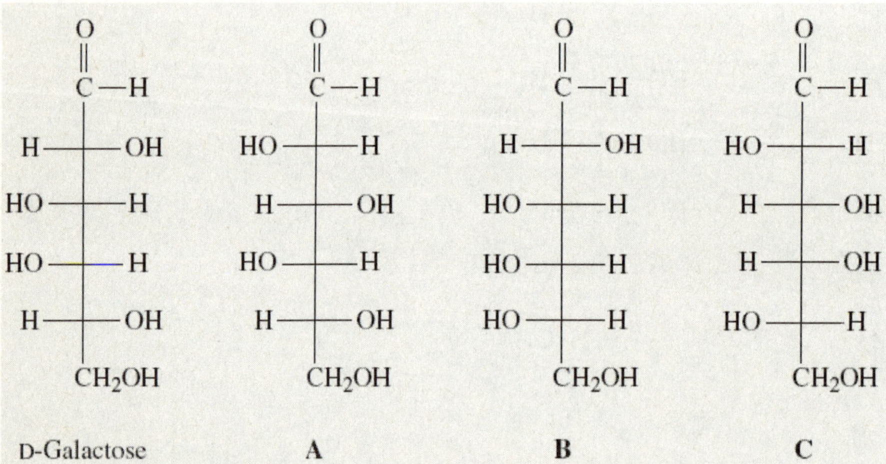

Recall that an enantiomer is a mirror image.

Stereoisomers that are not enantiomers are called diastereomers. Diastereomers are stereoisomers that are not exact mirror images.

A. diastereomer (not a mirror image)
B. diastereomer (not a mirror image)
C. enantiomer (is an exact mirror image)

5.13 Identify the monosaccharide that fits each of the descriptions below:

a. also referred to as dextrose
b. also called fruit sugar
c. used to treat urinary tract infections

Review the properties of the important monosaccharides.

a. D-glucose
a. D-fructose
a. D-mannose

5.15 Are the substances shown in bold undergoing oxidation or reduction?

a. **D-glucose** forms a brick red solution when subjected to the Benedict's test.
b. **Aluminum** metal forms aluminum oxide (Al_2O_3).
c. A carboxylic acid (**RCOOH**) is reacted to form an alcohol (RCH_2OH).

During the Benedict's test an aldehyde group reduces copper ions from Cu^{2+} to Cu^{1+} (copper gained one electron). The reacting Cu^{2+} ions are soluble, coloring the initial reaction solution blue. The aldehyde group is oxidized by Cu^{2+}, forming a sugar acid, while Cu^{2+} becomes reduced by the aldehyde, forming Cu^{1+} ions, which are precipitated as red copper (I) oxide.

a. D-glucose is an aldose and is therefore oxidized in a Benedict's test.

A reactant is oxidized by losing one or more electrons; another reactant is simultaneously reduced, gaining one or more electrons.

b. Aluminum metal loses 3 electrons to form the Al^{3+} ion in Al_2O_3. Al is oxidized.

An organic molecule is oxidized if it gains oxygen *or* loses hydrogen and is reduced if it either gains hydrogen *or* loses oxygen.

c. The formula for the alcohol contains more hydrogen atoms (and less oxygen atoms) than the carboxylic acid. The carboxylic acid is reduced.

5.17 Write the products if (a) carbon 1 is oxidized and (b) if carbon 1 is reduced for D-ribose.

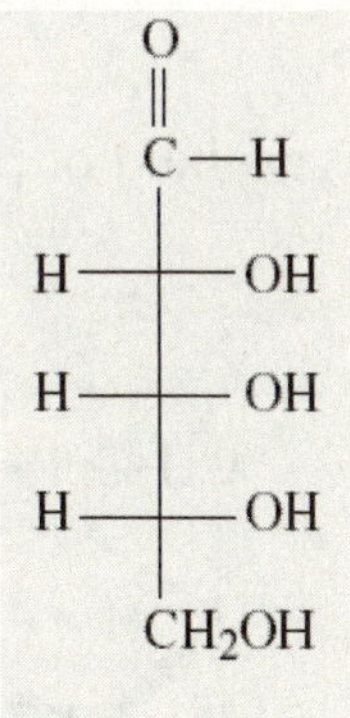

D-Ribose

Carbon 1 is the carbonyl carbon atom. Recall:

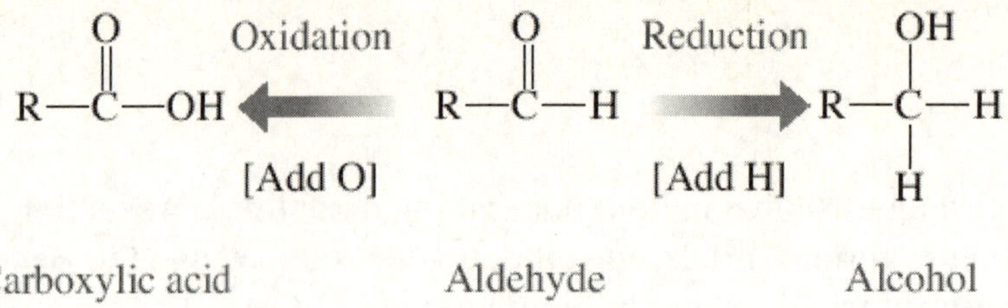

a. The carbonyl carbon forms the corresponding carboxylic acid when oxidized.

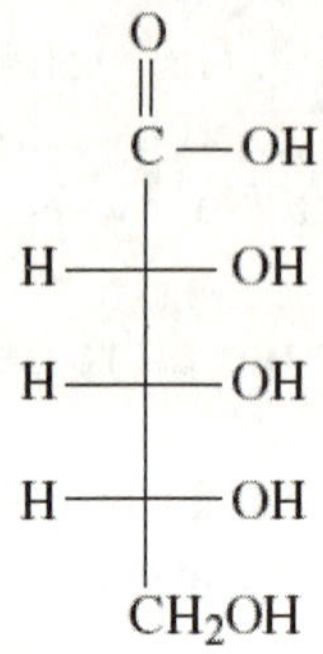

b. The carbonyl carbon forms the corresponding alcohol when reduced.

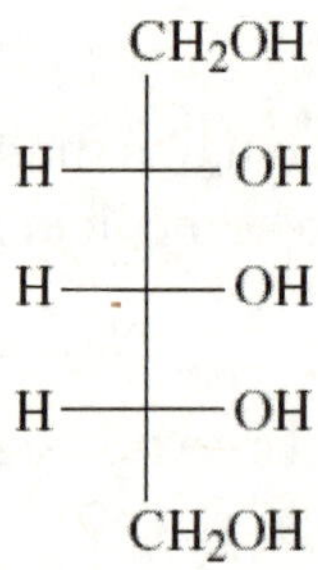

5.19 Identify the following carbohydrates as the α or β anomer.

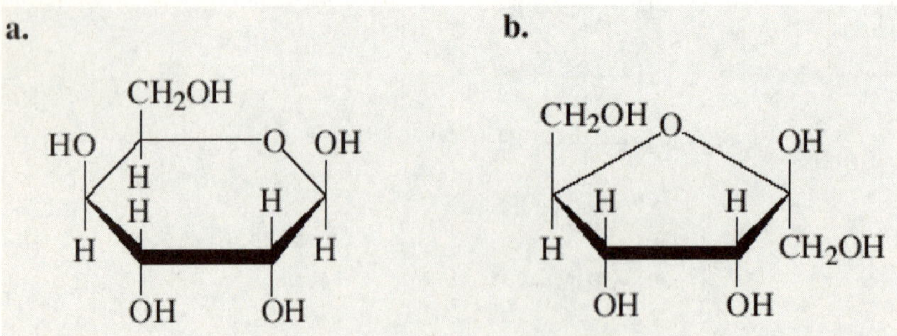

In any monosaccharide, the carbonyl carbon that reacts to form the hemiacetal in the reaction is referred to as the **anomeric carbon**. (Note that in the ring form, the anomeric carbon is the only carbon bonded directly to two oxygen atoms.)

Anomers are distinguished by the positioning of the –OH group on the anomeric carbon relative to the position of the carbon outside the ring.

- In the α anomer, the –OH on the anomeric carbon is trans to the carbon outside the ring (they are on opposite sides of the ring).
- In the β anomer, the –OH on the anomeric carbon is cis to the carbon outside the ring (they are on the same side of the ring).

a. beta (β): the —OH on the anomeric carbon is cis to the —CH_2OH group on the upper left of the ring

b. beta (β): the —OH on the anomeric carbon is cis to the —CH_2OH group on the upper left of the ring

5.21 Draw the α or β anomer of D-talose in pyranose ring form.

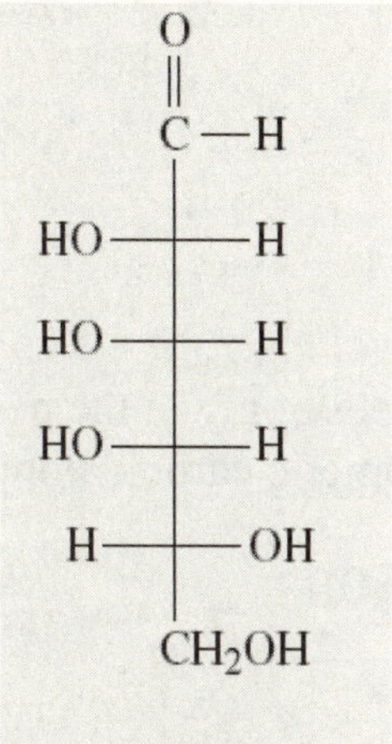

D-Talose

The hydroxyl on carbon 5 (C5) can curl around to react with the carbonyl at carbon 1 (C1) to produce the hemiacetal functional group. Two ring structures are possible. When this occurs, five carbons and an oxygen form a ring, and one of the carbons (carbon 6) remains outside the ring. See below for D-glucose.

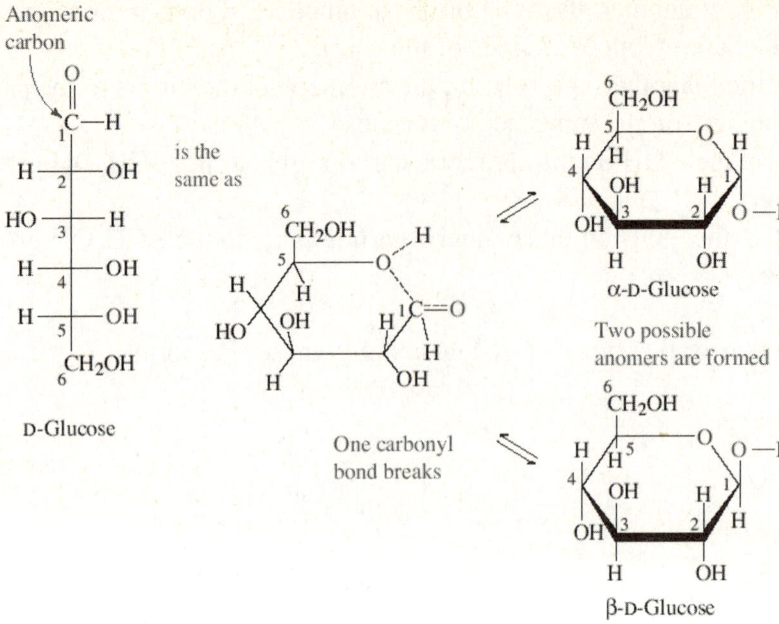

D-talose will differ from glucose only in the positioning of the —OH groups (excluding the newly formed one on the anomeric carbon atom). In D-talose all of the —OH groups will be above the plane of the ring.

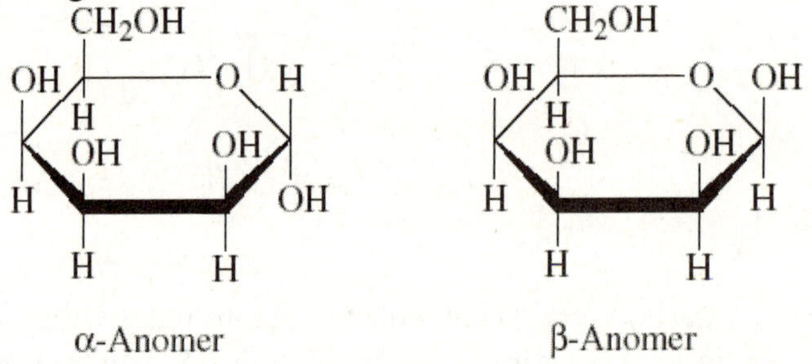

5.23 Identify the following reactions as a condensation or hydrolysis:

a. two monosaccharides reacting to form a disaccharide

b. the formation of a glycosidic bond

c. a reaction in which one molecule breaks into two and an –H and –OH are added

The formation of glycosidic bonds is an example of another common chemical reaction during which a molecule of water is eliminated as two molecules (monosaccharides) are joined. When a small molecule like water is produced, this reaction is referred to as a condensation reaction.

A condensation reaction operating in reverse is a hydrolysis reaction where water is consumed as a reactant.

a. condensation
b. condensation
c. hydrolysis

5.25 Name the glycosidic bond present in mannobiose shown below.

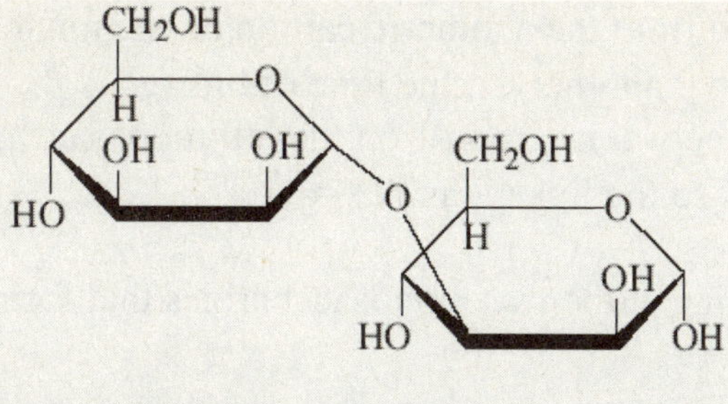

Mannobiose

Recall that when one hydroxyl on one monosaccharide reacts with a hydroxyl on another monosaccharide, a glycosidic bond forms. These glycosidic bonds connect monosaccharides to each other.

The convention for naming glycosidic bonds is to first specify the anomer bonded (α or β) and then the carbons bonded (with numbers that indicate their ring position separated by a right arrow) using the monosaccharides in the formula from left to right.

The bond formed from the anomeric carbon on the monosaccharide on the left is pointing downward (from the α anomer) so the type of linkage is α.
The bond forms between carbon 1 on the left monosaccharide and carbon 3 on the right monosaccharide, so the linkage is $\alpha(1\rightarrow3)$.

5.27 For each of the following disaccharides, name the glycosidic bond and draw the monosaccharide units produced by hydrolysis:

a.

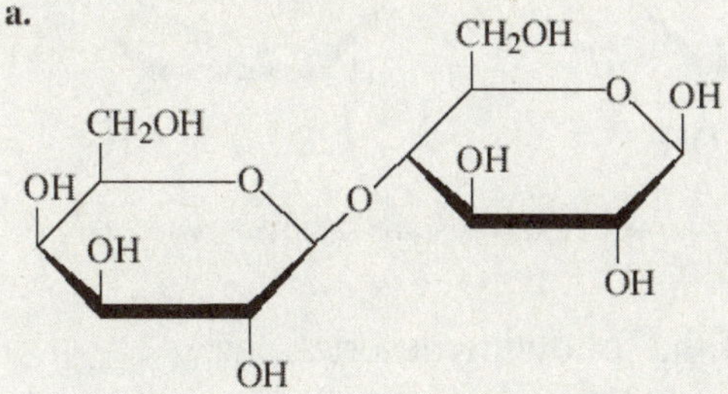

b.

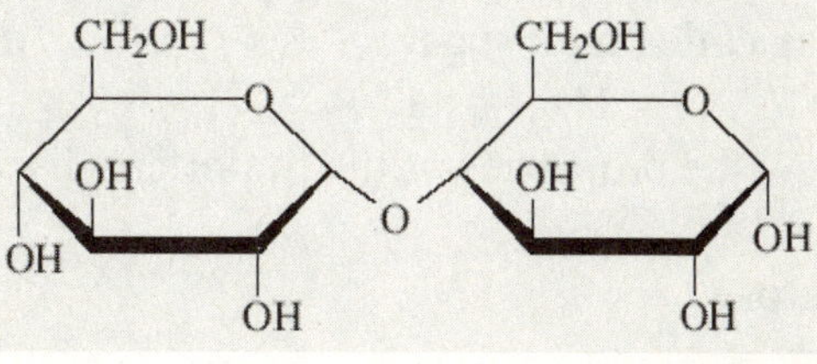

Recall that when one hydroxyl on one monosaccharide reacts with a hydroxyl on another monosaccharide, a glycosidic bond forms. These glycosidic bonds connect monosaccharides to each other.

The convention for naming glycosidic bonds is to first specify the configuration of the anomeric carbon used in the bond (α or β) and then the carbons bonded (with numbers that indicate their ring position separated by a right arrow) using the monosaccharides in the formula from left to right.

a. The bond formed from the anomeric carbon on the monosaccharide on the left is pointing upward (from the β anomer) so the type of linkage is β.
The bond forms between carbon 1 on the left monosaccharide and carbon 4 on the right monosaccharide, so the linkage is β(1→4).

Verified by looking at the two monosaccharides that formed the bond:

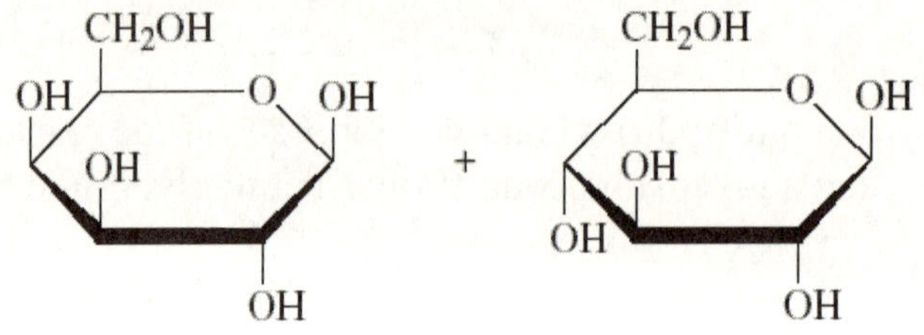

D-Galactose D-Glucose

b. The bond formed from the anomeric carbon on the monosaccharide on the left is pointing downward (from the α anomer) so the type of linkage is α.
The bond forms between carbon 1 on the left monosaccharide and carbon 4 on the right monosaccharide, so the linkage is α(1→4).

Verified by looking at the two monosaccharides that formed the bond:

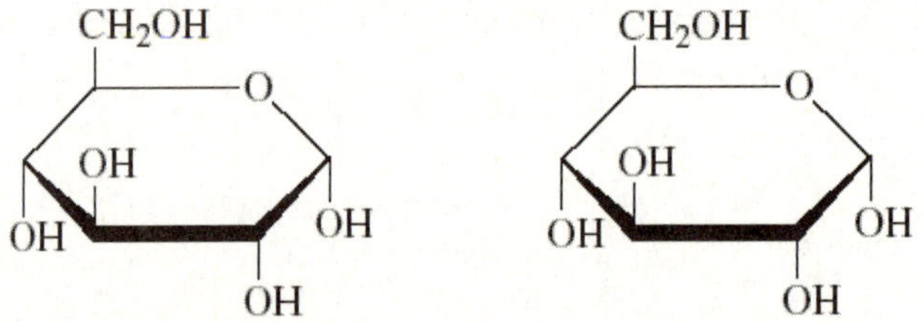

Both monosaccharides are D-Glucose

5.29 Identify disaccharides that fit each of the following descriptions:
a. ordinary table sugar
b. found in milk and milk products
c. also called malt sugar
d. hydrolysis gives galactose and glucose

Review the properties of the important disaccharides.

a. sucrose
b. lactose
c. maltose
d. lactose

5.31 Describe the similarities and differences in the following polysaccharides:
a. amylose and amylopectin
b. amylopectin and glycogen

Review the properties of the important polysaccharides.

a. Both are storage polysaccharides. They both contain α(1→4) glycosidic bonds and only D-glucose. Amylopectin also contains branching α(1→6).
b. Both are storage polysaccharides. They both contain α(1→4) glycosidic bonds, branching α(1→6) and only D-glucose. Branching occurs more often (about every 12 glucose units) in glycogen than amylopectin (about every 25 glucose units).

5.33 Give the name of one or more polysaccharides that matches each of the following descriptions:

a. not digestible by humans
b. the storage form of carbohydrates in plants
c. contains only α(1→4) glycosidic bonds
d. glucose polysaccharide with the most branching

Review the properties of the important polysaccharides.

a. cellulose and chitin (structural polysaccharides)
b. amylose and amylopectin (components of starch)
c. amylose
d. glycogen (storage polysaccharide in animals)

5.35 Explain whether the following blood types could donate to a person with type B blood:
a. A b. AB

Refer to Table 5.3.

TABLE 5.3 COMPATIBILITY OF BLOOD GROUPS

Blood Group	Can Receive Blood Types	Cannot Receive Blood Types
A	A,O	B, AB
B	B, O	A, AB
AB[a]	A, B, AB, O	Can receive all blood types
O[b]	O	A, B, AB

a. Type A blood is not compatible with type B blood.
b. Type AB blood is not compatible with type B blood.

Additional Problems

5.37 Write the molecular formula for a carbohydrate containing four carbons.

The general molecular formula for carbohydrates is: $C_n(H_2O)_n$ where n is 3–6 for monosaccharides.

If $n = 4$: $C_n(H_2O)_n$ becomes $C_4(H_2O)_4$ or $C_4H_8O_4$.

5.39 Explain the difference between an oligosaccharide and a polysaccharide.

The difference lies in the number of monosaccharides that are linked together. Oligosaccharides conventionally are limited to the linking of 3–9 monosaccharides by glycosidic bonds. Polysaccharides contain 10 or more linked monosaccharides.

5.41 Classify each of the following as primary, secondary, or tertiary alcohols.

a.

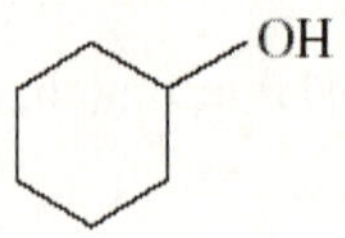

b.

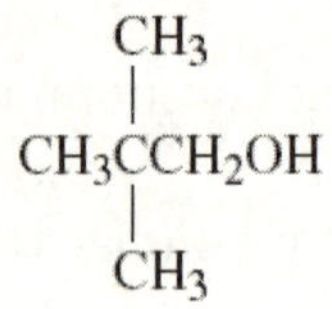

c.

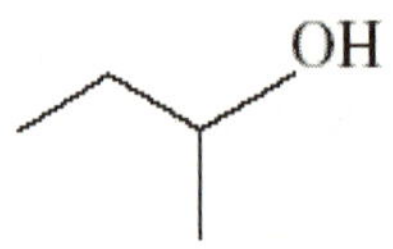

d.

A primary (1°) alcohol has *one* alkyl group attached to the alcoholic carbon, a secondary (2°) alcohol has *two* such alkyl groups, and a tertiary (3°) alcohol has *three.*

a. secondary (2°) alcohol: C to which the —OH is bonded to two carbon atoms of the cyclohexane ring.
b. primary (1°) alcohol: C to which the —OH is bonded to only one carbon atom of the propane chain
c. secondary (2°) alcohol: C to which the —OH is bonded to two carbon atoms of the butane chain.
d. primary (1°) alcohol: C to which the —OH is bonded to only one carbon atom of the butane chain.

5.43 How are the following pairs of carbohydrates, shown in Fischer projection, related to each other? Are they structural isomers, enantiomers, diastereomers, or epimers?

a.

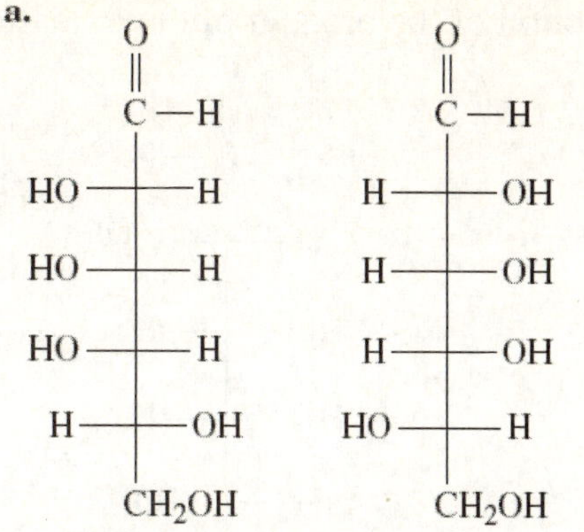

b.

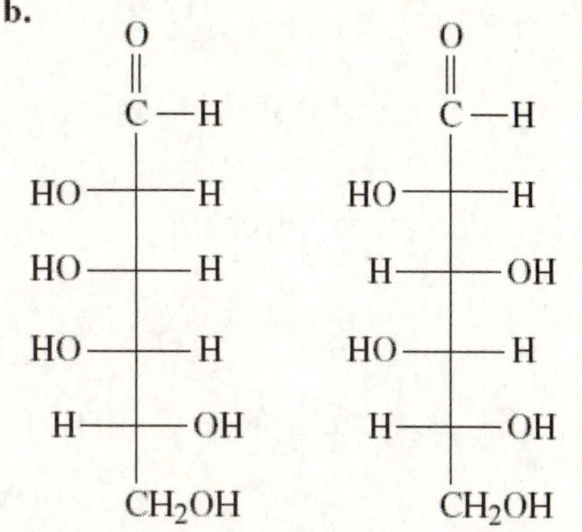

Recall that an enantiomer is a mirror image.

Stereoisomers that are not enantiomers are called diastereomers. Diastereomers are stereoisomers that are not exact mirror images.

Diastereomers that differ in just one chiral center (as compared with more than one chiral center) are called epimers.

a. The two structures are exact mirror images: enantiomers
b. The two structures are not exact mirror images and differ at only one chiral center (carbon 3): diastereomers but more specifically epimers.

5.45 Draw the Fischer projection of the product of the oxidation of D-galactose at C1.

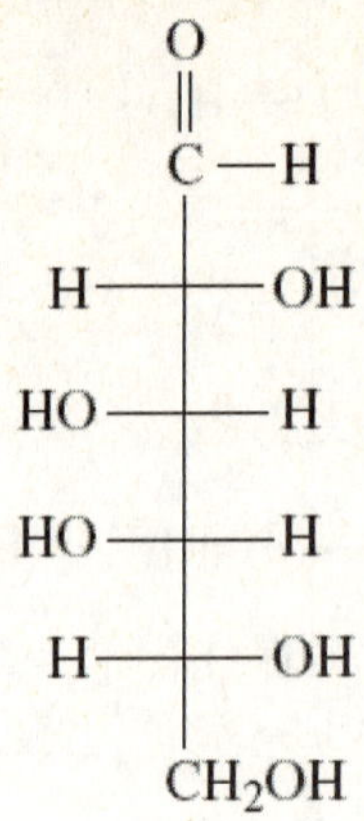

D-Galactose

Carbon 1 is the carbonyl carbon atom. Recall:

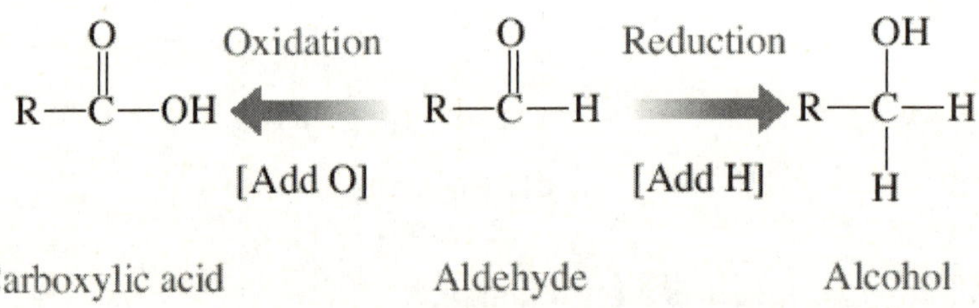

The aldehyde group is oxidized to the corresponding carboxylic acid.

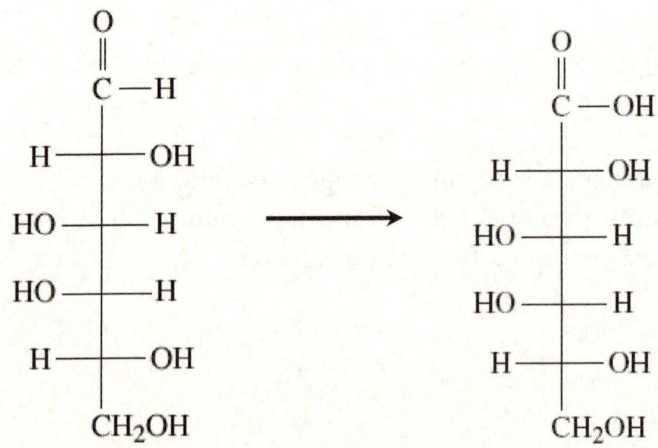

5.47 Draw the Fischer projection of the product of reduction reaction of D-galactose at C1.

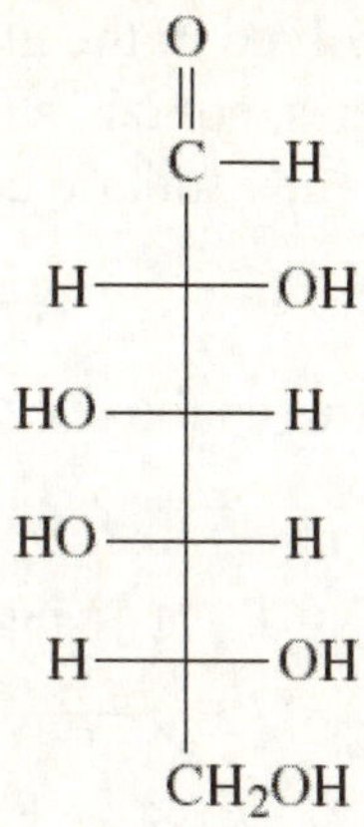

D-Galactose

Carbon 1 is the carbonyl carbon atom. Recall:

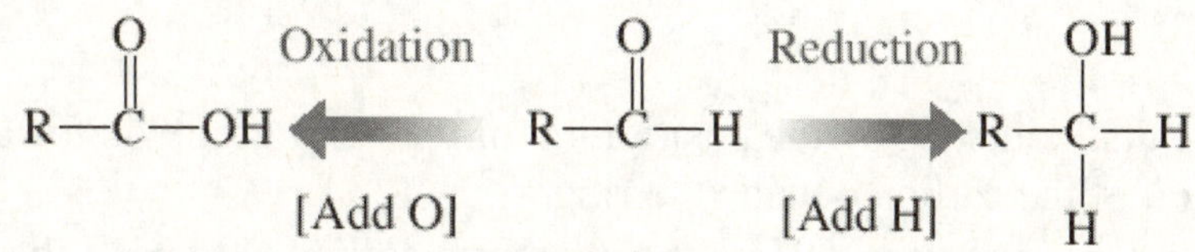

Carboxylic acid Aldehyde Alcohol

The aldehyde group is reduced to the corresponding alcohol.

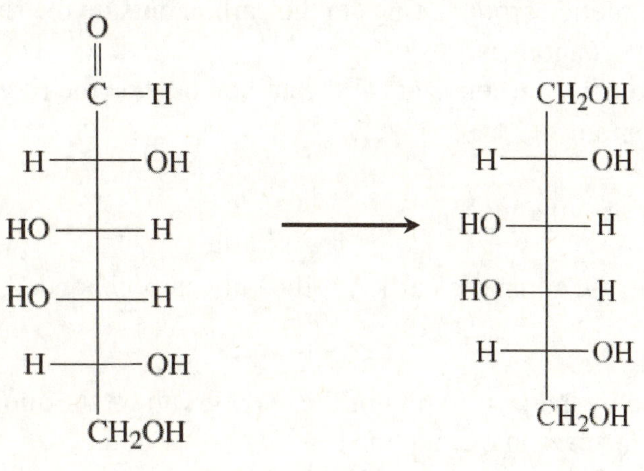

5.49 Will the following carbohydrates produce a positive Benedict's test?
a. D-glucose **b.** lactose **c.** sucrose **d.** starch

A reducing sugar will produce a positive test with Benedict's solution. A reducing sugar must have a free anomeric carbon (one not involved in a glycosidic bond). Examine the structures of the given carbohydrates for a free anomeric carbon.

a. Contains a free anomeric carbon: positive Benedict's test
b. Contains a free anomeric carbon on the D-galactose unit: positive Benedict's test
c. Both anomeric carbons are involved in the glycosidic bond: negative Benedict's test
d. Starch, like most polysaccharides, contain a single reducing end and one (unbranched) or many (branched) nonreducing ends and are considered nonreducing sugars: negative Benedict's test

5.51 Identify the following carbohydrates as the α or β anomer.

a.

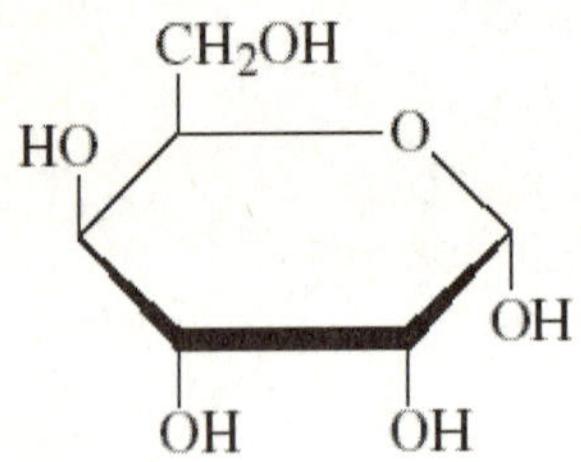

b.

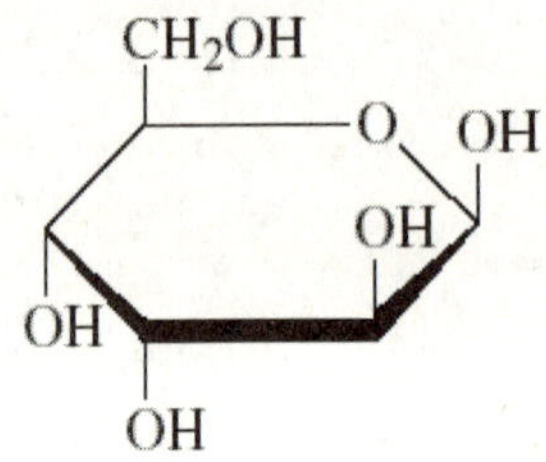

(Note that in the ring form, the anomeric carbon is the only carbon bonded directly to two oxygen atoms.)

Anomers are distinguished by the positioning of the –OH group on the anomeric carbon relative to the position of the carbon outside the ring.

- In the α anomer, the –OH on the anomeric carbon is trans to the carbon outside the ring (they are on opposite sides of the ring).
- In the β anomer, the –OH on the anomeric carbon is cis to the carbon outside the ring (they are on the same side of the ring).

a. The –OH group on the anomeric carbon is trans to the carbon outside the ring (they are on the opposite sides of the ring): α anomer
b. The –OH group on the anomeric carbon is cis to the carbon outside the ring (they are on the same side of the ring): β anomer

5.53 Draw the α and β anomer of D-mannose.

(Note that in the ring form, the anomeric carbon is the only carbon bonded directly to two oxygen atoms.)

Anomers are distinguished by the positioning of the –OH group on the anomeric carbon relative to the position of the carbon outside the ring.

- In the α anomer, the –OH on the anomeric carbon is trans to the carbon outside the ring (they are on opposite sides of the ring).
- In the β anomer, the –OH on the anomeric carbon is cis to the carbon outside the ring (they are on the same side of the ring).

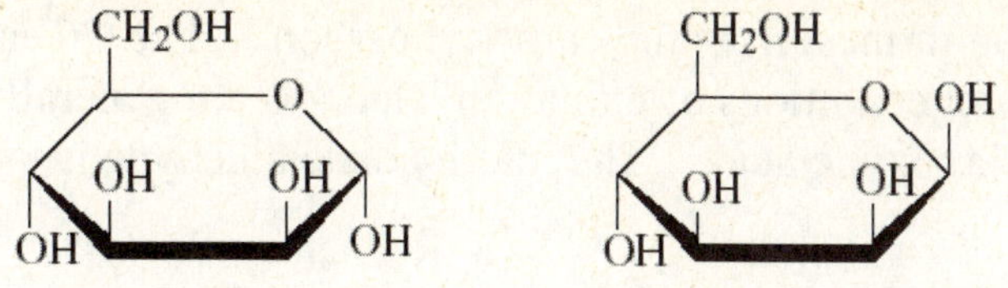

α-D-Mannose β-D-Mannose

5.55 Draw the product of the 1→4 condensation and name the glycosidic bond:

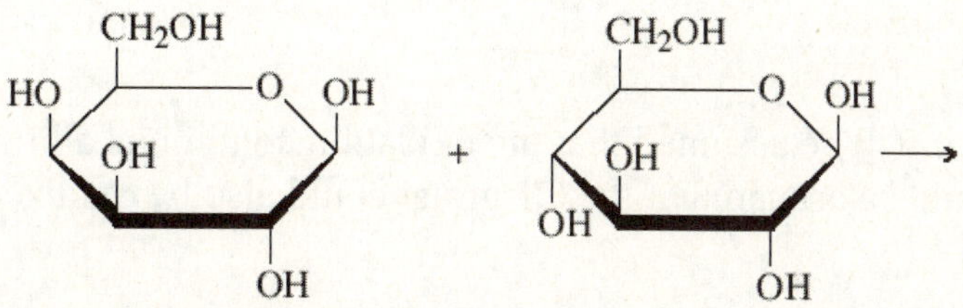

(1→4): The bond will form between the anomeric carbon atom (C1) on the left monosaccharide and C4 on the right monosaccharide. Since the –OH on the left anomeric carbon atom is cis to the carbon outside the ring (they are on the same side of the ring), it is the β anomer and the bond will be β(1→4).

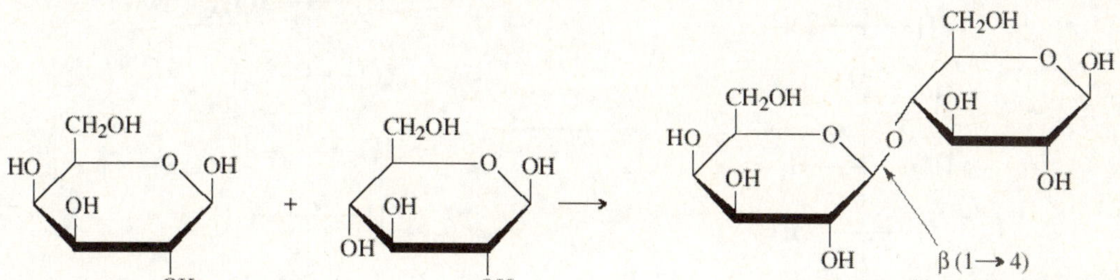

5.57 Give the name of one or more carbohydrates that match the following description:

a. ordinary table sugar
b. dextrose
c. an L-sugar found in the ABO blood types
d. found in starch
e. found in the cell walls of fungi

Review the properties of the monosaccharides, disaccharides, and polysaccharides.

a. sucrose
b. glucose
c. L-fructose
d. amylose and amylopectin
e. chitin

Challenge Problems

5.59 Carbohydrates are also easily oxidized at C6. Provide the product of the oxidation of (a) D-galactose at C6 and (b) D-talose at C6.

C6 is the position furthest from the carbonyl carbon. The C6 carbon atom carries an alcohol substituent. Under conditions of oxidation, alcohols are generally oxidized to the corresponding carboxylic acid . The intermediate aldehyde is usually not isolated.

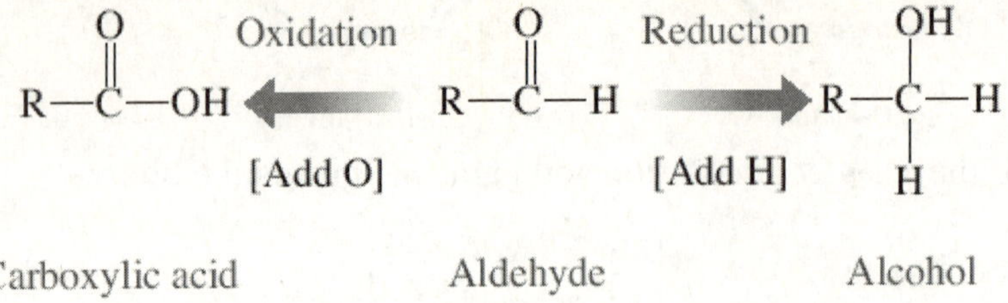

In both cases it will be assumed that the oxidation conditions allow only for reaction at the C6 carbon. It could be argued that the C1 group could also be oxidized.

a.

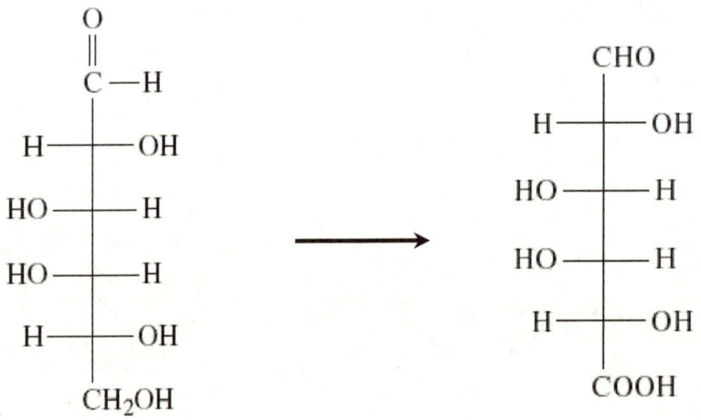

D-Galactose

b.

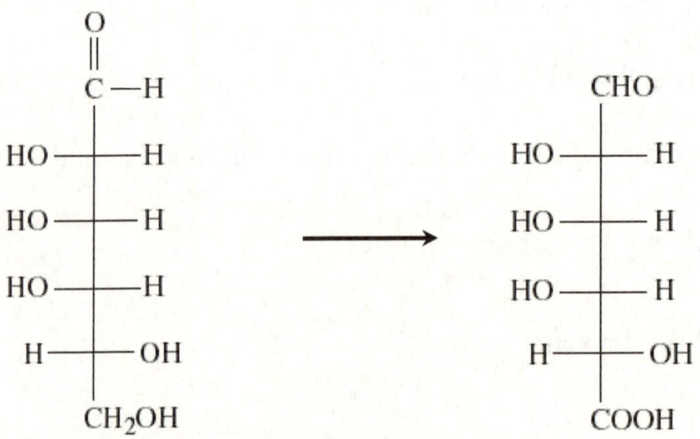

D-Talose

5.61 The structure of sucralose, found in the artificial sweetener Splenda® is shown in the figure. Its advertising slogan is "made from sugar, so it tastes like sugar." [http://www.splenda.com/page.jhtml?id=splenda/faqs/nocalorie.inc#q1] It consists of a chlorinated disaccharide made up of galactose and fructose. In the molecule shown

a. identify the galactose unit and the fructose unit.
b. identify the type of glycosidic bond present.
c. determine if Splenda is a reducing sugar.

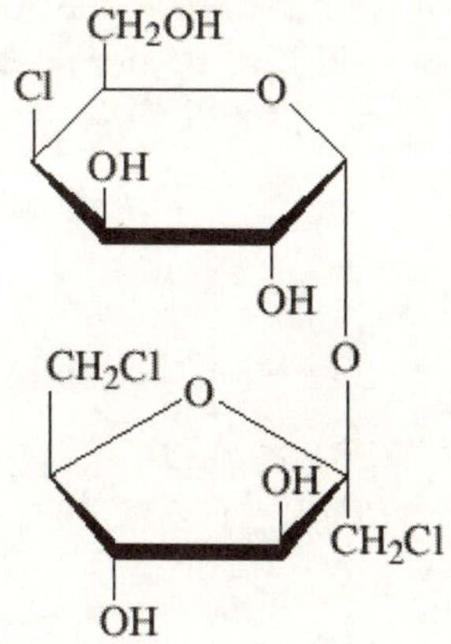

Review the structure of the monosaccharides glucose, a pyranose or 6-membered ring, and fructose, a furanose, or a 5-membered ring.

a. Top monosaccharide unit is galactose, and the lower unit is fructose.

The convention for naming glycosidic bonds is to first specify the configuration of the anomeric carbon used in the bonded (α or β) and then the carbons bonded (with numbers that indicate their ring position separated by a right arrow) using the monosaccharides in the formula proceeding here from top to bottom. In this case like that of sucrose, the second monosaccharide uses the –OH of the anomeric carbon. This must also be specified in the name with the anomer of the second monosaccharides also being specified.

b. The galactose unit had the –OH bond in trans position to the carbon outside the ring (α) and the fructose unit had the –OH bond in the cis position to the carbon outside the ring (β). The glycosidic bond formed between C1 of galactose and C2 of fructose, the bond is therefore α,β (1 →2).

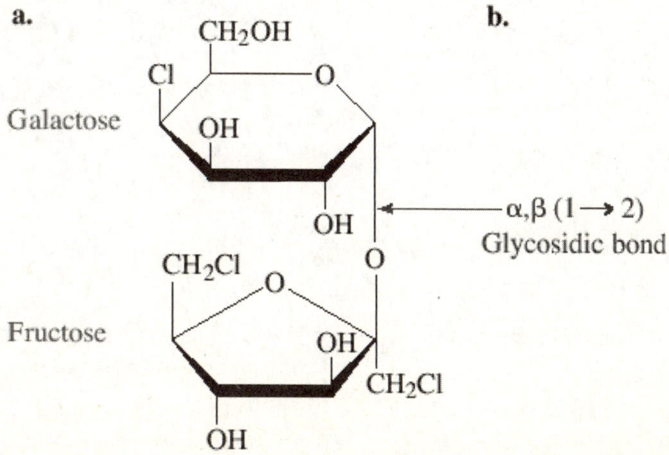

A reducing sugar must have a free anomeric carbon (one not involved in a glycosidic bond).

c. The anomeric carbon of each monosaccharide unit was used in forming the glycosidic bond leaving no anomeric carbon. Sucralose is, therefore, a nonreducing sugar.

Chapter

6

Intermolecular Forces: State Changes, Solubility, and Cell Membranes

6.1 Types of Intermolecular Forces

Learning Objectives

Upon completion of this material, a student should be able to do the following:

A. define the following key terms:

London forces	dipole-dipole attraction	hydrogen bonding
ion-dipole attraction		

B. explain the difference between a chemical bond and an intermolecular force

C. determine the type of intermolecular force based on a chemical formula

D. rank the types of intermolecular forces by strength of attraction

E. assign intermolecular forces to small organic molecules

F. demonstrate how ions interact with polar molecules

6.2 Intermolecular Forces and Solubility

Learning Objectives

Upon completion of this material, a student should be able to do the following:

A. define the following key terms:

golden rule of solubility	triglyceride	esterification
hydration	salt	amphipathic
hydrophobic	hydrophilic	micelle

B. use the golden rule of solubility to predict the solubility of nonpolar compounds in water

C. use the golden rule of solubility to predict the solubility of polar compounds in water

D. use the golden rule of solubility to predict the solubility of ionic compounds in water

E. identify triglycerides

F. write a chemical equation for the formation of a triglyceride

G. identify the ester groups in triglycerides

H. describe soaps in terms of their amphipathic, hydrophilic, and hydrophobic nature

I. describe the solubility of a fatty acid in water (micelle formation).

6.3 Intermolecular Forces and Changes of State

Learning Objectives

Upon completion of this material, a student should be able to do the following:

A. define the following key terms:

melt	evaporate	change of state
phase transition	boiling point	

B. describe the effect of intermolecular forces on phase transitions

C. rank compounds according to boiling point based on molecular structure in terms of molar mass and molecular structure
D. based on boiling points determine relative strength of intermolecular forces

6.4 Fats, Oils, and Margarine—Solid to Liquid and Back Again: Melting

Learning Objectives

Upon completion of this material, a student should be able to do the following:

A. define the following key terms:

fat	oil	hydrogenation
catalyst		

B. distinguish fats and oils in terms of properties and chemical structure
B. draw the structure of a triglyceride
C. describe the reaction that can convert an oil to a fat
D. provide the products of the complete hydrogenation of a fatty acid.
E. describe the effect of a catalyst on a chemical reaction

6.5 Intermolecular Forces and the Cell Membrane

Learning Objectives

Upon completion of this material, a student should be able to do the following:

A. define the following key terms:

phospholipids	fluid mosaic model

B. recognize and be able to label the portions of a phospholipid
C. distinguish structural differences between a triglyceride and a phospholipid.
D. draw a phospholipid bilayer and identify the polar and nonpolar parts.
E. describe the structural features of the fluid mosaic model of a cell membrane
F. identify cholesterol as a lipid
F. describe the role of cholesterol (its positioning and function) in a phospholipid bilayer.

Practice Test for Chapter 6

1. The type of intermolecular force depicted by the arrow in the image is:

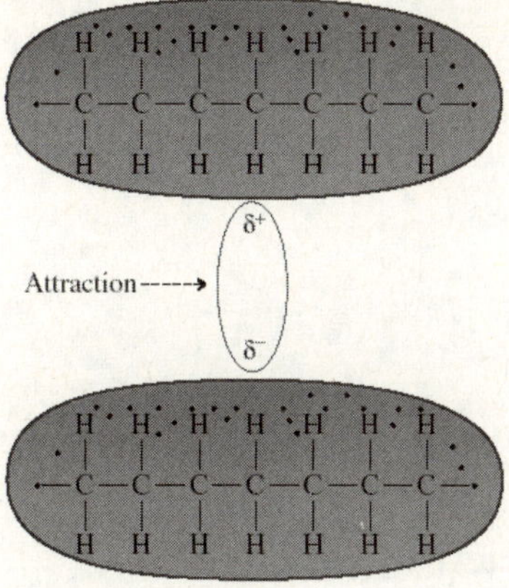

A. London force.
B. dipole-dipole attraction.
C. hydrogen bonding.
D. ion-dipole attraction.

2. What type(s) of intermolecular attraction would exist between molecules of the following type?

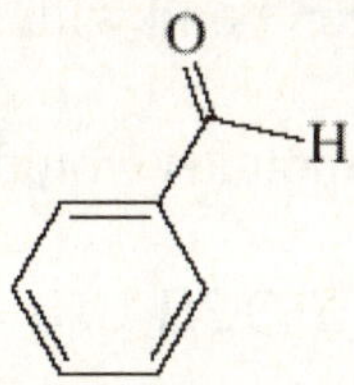

A. London forces
B. dipole-dipole attraction
C. hydrogen bonding
D. A and B
E. A, B, and C

3. How many hydrogen bonds could the following molecule form with water?

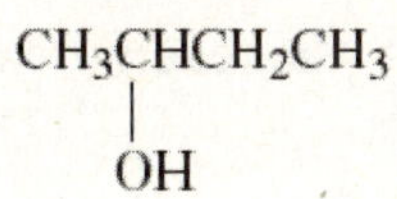

A. none
B. 1
C. 2
D. 3
E. 4

4. Identify the strongest attractive force that could exist between the following pair of molecules.

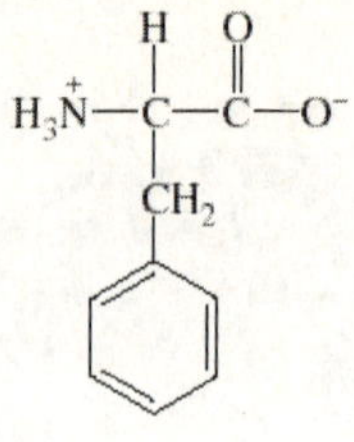

and

$CH_3CHCH_2CH_3$ (with OH on the second carbon)

OH

A. London forces
B. dipole-dipole attraction
C. hydrogen bonding
D. ion-dipole attraction.

5. Which of the following is a correct description of the manner in which the molecule shown below would interact with water?

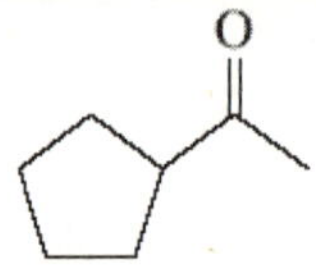

A. This molecule would be a hydrogen bond donor and water a hydrogen bond acceptor.
B. This molecule would be a hydrogen bond acceptor and water a hydrogen bond donor.
C. Both this molecule and water function as hydrogen bond acceptors and donor.
D. This molecule would not hydrogen bond with water.

6. Which of the following pairs of compounds would most likely **not** be soluble in each other?
A. NH_3 and CH_3OH
B. $CH_3CH_2CH_2CH_3$ and $CH_3CH_2CH_2OCH_2CH_3$
C. NH_4NO_3 and $CH_3CH_2OCH_2CH_3$
D. $CH_3CH_2CH_2OH$ and $CH_3CH_2NH_2$

7. Compare the solubility of the following substances in water. In which of the pairs of substances would the compound listed on the **left** probably be **more soluble** in water?
A. $CH_3CH_2CH_2OH$ and CH_3OH
B. $CH_3CH_2CH_2OCH_2CH_3$ and $CH_3CH_2NH_2$
C. KCl and CO_2
D. $CH_3CH_2CH_3$ and $CH_3CH_2CH_2CH_2NH_2$

8. When glycerol and a fatty acid react, the reaction is called a(n):
A. esterification reaction.
B. condensation reaction.
C. hydrogenation reaction.
D. esterification or condensation reaction.
E. triglyceride reaction.

9. The "like" in the expression "like dissolves like" refers to molecular:
A. polarity.
B. size.
C. shape.
D. all of the above

10. Consider the structure of a soap. Which of the following statements is(are) true?
    A. Soaps are amphipathic compounds.
    B. A soap contains a hydrophobic head.
    C. A soap contains a hydrophilic tail.
    D. All of the above statements are true.

11. A soap is able to dissolve nonpolar grease and oil in water by:
    A. forming a micelle around the oil with the soap tails in the interior.
    B. forming a micelle around the oil with the soap heads in the interior.
    C. forming an ion-dipole attraction to the oil.
    D. surrounding the oil with carboxylate group of the soap.

12. In a micelle formed from water and a soap,
    A. hydrophobic heads are oriented toward the surface of the micelle.
    B. hydrophobic tails are oriented toward the surface of the micelle.
    C. hydrophilic heads are oriented toward the surface of the micelle.
    D. hydrophilic tails are oriented toward the surface of the micelle.

13. In "dry" cleaning, an oil stain can be removed by dissolving the oil directly in the liquid being used. Which of the following might make a dry cleaning liquid?
    A. CH_3CH_2OH
    B. $CH_3CH_2CH_2CH_2CH_2CH_3$
    C. $HOCH_2CH_2CH_2CH_2OH$
    D. All would be equally good candidates.

14. Consider the bond formed when a triglyceride is produced. Which of the following general bonding schemes has the same type of bond enclosed in the circle?
    A.

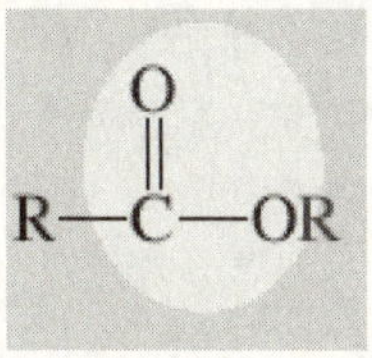

ester

B.

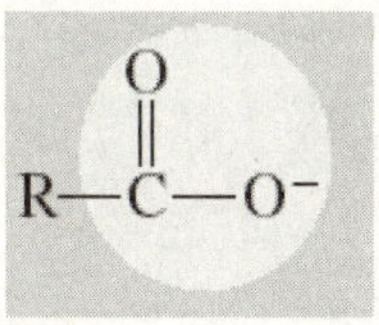

carboxylate

C.

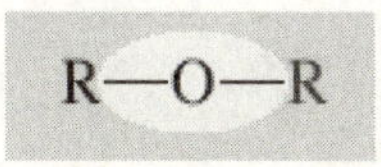

ether

D.

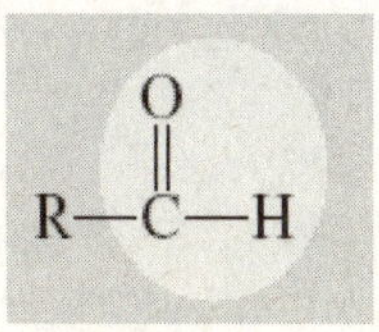

aldehyde

15. In the reaction producing a triglyceride, the reactants would be:
    A. 1 alcohol molecule and 1 fatty acid molecule .
    B. 1 alcohol molecule and 3 fatty acid molecules.
    C. 1 glycerol molecule and 3 fatty acid molecules.
    D. 3 glycerol molecules and 1 fatty acid molecule.

16. Which of the following substances would have the highest boiling point?
    A. $CH_3CH_2CH_3$
    B. $CH_3CH_2CH_2CH_3$
    C. $CH_3CH_2CH_2CH_2CH_3$
    D. $CH_3CH_2CH_2CH_2CH_2CH_3$

17. Which of the following substances would have the lowest boiling point?
    A.

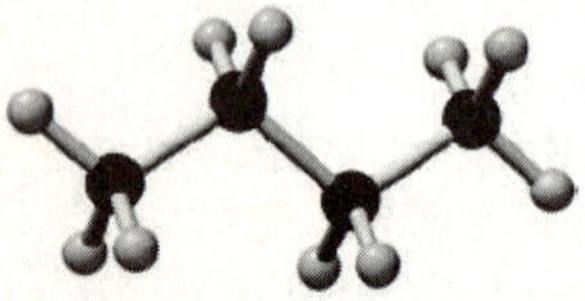

B.

C.

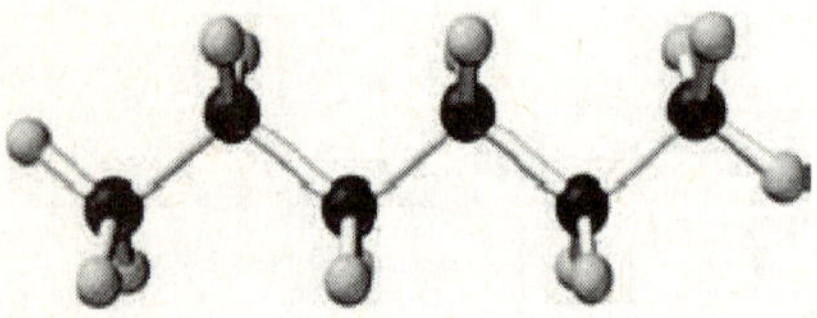

D.

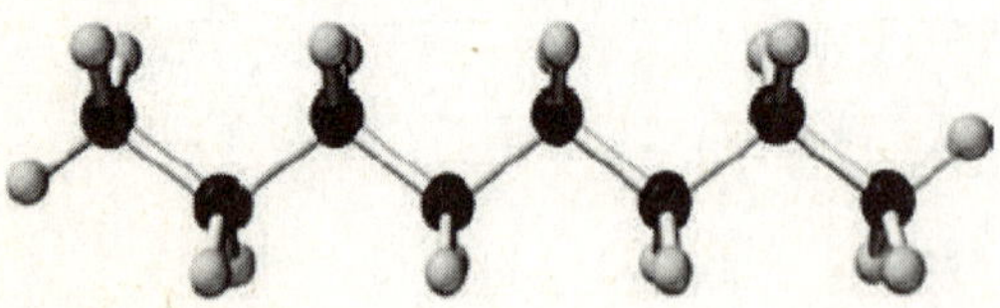

18. Which of the following molecules has the strongest London forces with other molecules of the same type?
A.

B.

C.

D.

19. Compare the boiling points of the following substances. In which of the pairs of substances would the compound listed on the **left** have the **higher** boiling point?
A. $CH_3CH_2CH_2CH_3$ and $CH_3CH_2CH_2OH$
B. $CH_3CH_2CH_2CH_3$ and $CH_3CH_2NHCH_2CH_3$
C. $CH_3CH_2CH_2CH_3$ and $CH_3CH_2CH_3$
D. $CH_3CH_2CH_2CH_3$ and $CH_3CH_2OCH_2CH_3$

20. Which of the following does not characterize both boiling and melting?
A. heat must be added
B. phase transition
C. involve the liquid state of matter
D. All characterize melting and boiling.

21. Fats:
A. contain mostly unsaturated fatty acids.
B. are carboxylate ions of carboxylic acids.
C. are liquids at room temperature.
D. have melting points near body temperature.

22. Which of the following contains a trans fatty acid?
A.

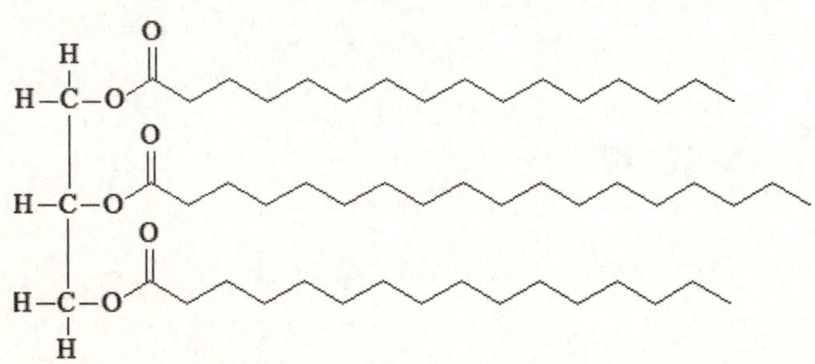

B.

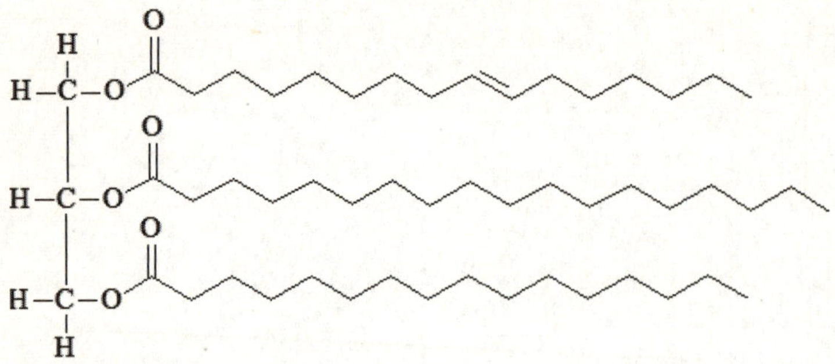

C.

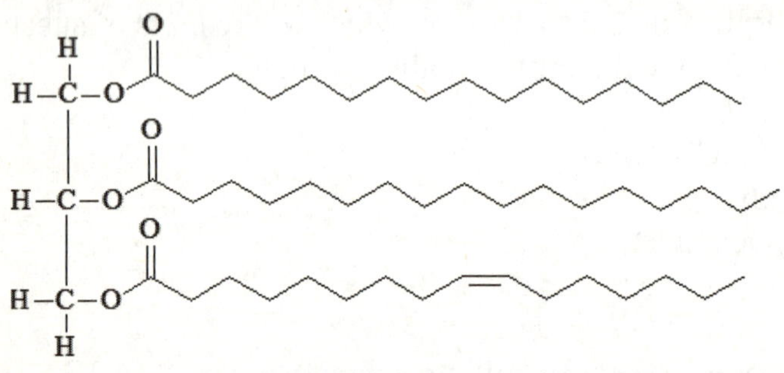

D.

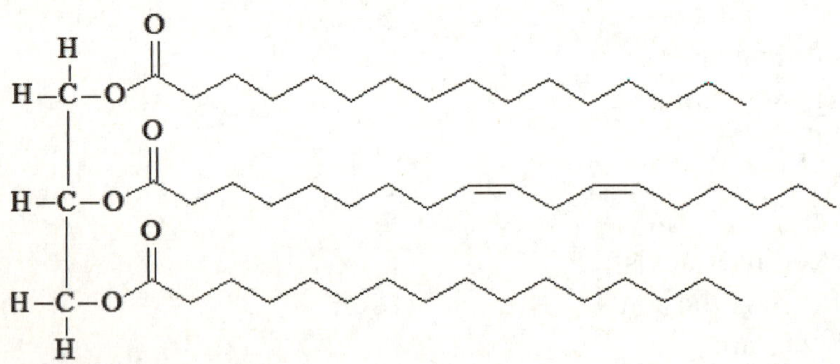

23. How many molecules of hydrogen would be required to completely saturate the following substance?

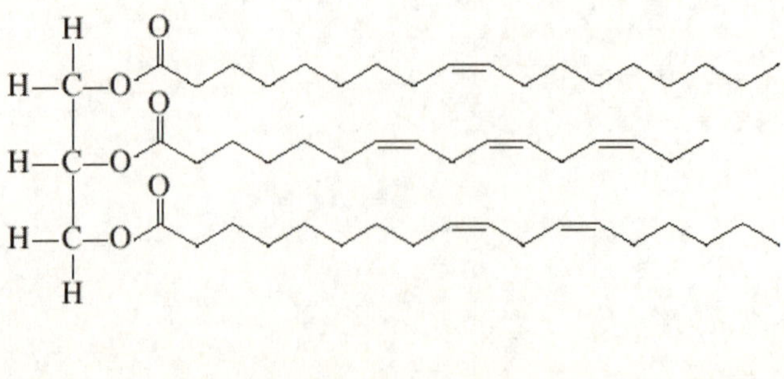

A. 6
B. 9
C. 12
D. 18

24. The following molecule was formed from the reaction of:

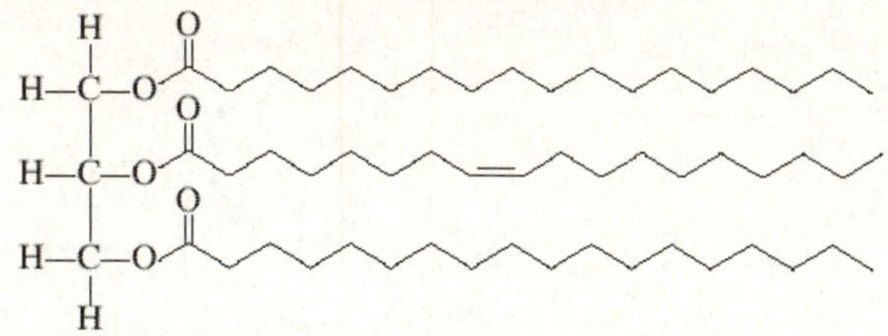

A. glycerol and a saturated fatty acid and an unsaturated fatty acid.
B. glycerol and two different saturated fatty acids and an unsaturated fatty acid.
C. glycerol and a saturated fatty acid and two different unsaturated fatty acids.
D. glycerol and three different saturated fatty acids.

25. An oil was completely hydrogenated with hydrogen gas (H_2) in the presence of a platinum catalyst. The triglycerides comprising the oil contained only one type of fatty acid. If it required 3 molecules of hydrogen for each oil molecule, which of the following fatty acids could be found in the triglycerides?

A.

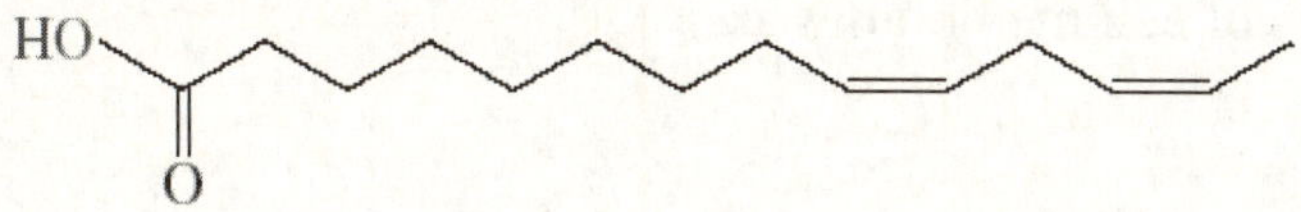

B.

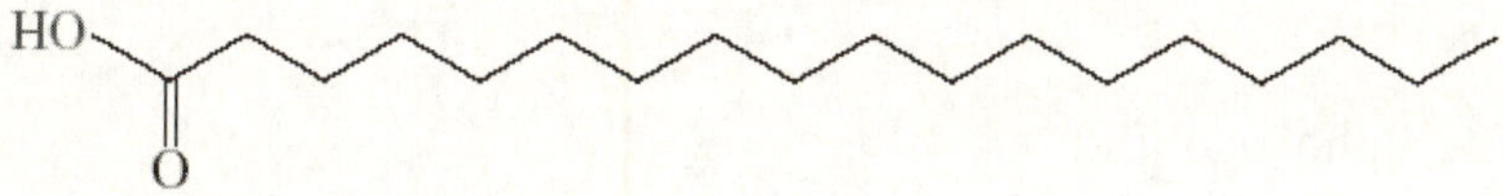

C.

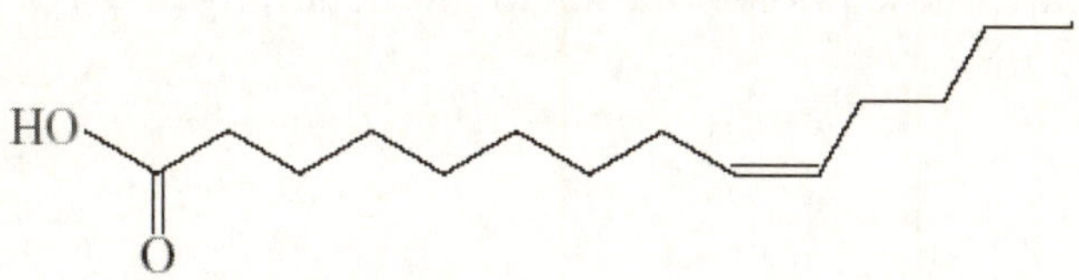

D. The structure of the fatty acid is not represented by any of the above.

26. Which of the following improves the spreadability of margarine?
A. hydrogenation
B. partial hydrogenation
C. condensation
D. esterification

27. The following is a cartoon of a soap molecule.

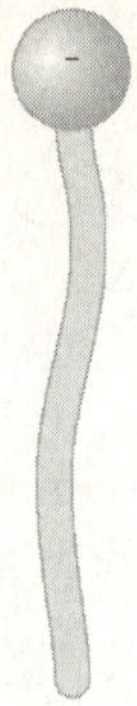

Water would be the most strongly attracted to:
A. the upper charged end.
B. the lower tail.
C. either end.
D. Water is not attracted to the molecule.

28. The primary structural components of cell membranes are:
A. fatty acids.
B. triglycerides.
C. phospholipids.
D. cholesterol.

29. If the following represents a phospholipid, 

between the layers of the cell membrane these phospholipids might be arranged as:
A.

B.

C.

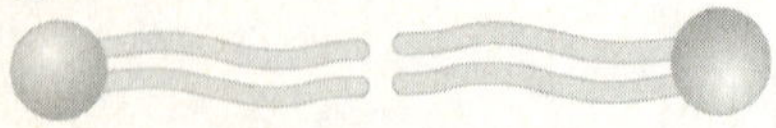

D. The order of orientation is random.

30. Which of the following modulates the flexibility of a cell membrane?
A. phospholipids
B. cholesterol
C. proteins
D. triglycerides

Answers

1. A	**2.** D	**3.** D	**4.** D	**5.** B	**6.** C	**7.** C	**8.** D	**9.** A	**10.** A
11. A	**12.** C	**13.** B	**14.** A	**15.** C	**16.** D	**17.** B	**18.** A	**19.** C	**20.** D
21. D	**22.** B	**23.** A	**24.** A	**25.** C	**26.** B	**27.** A	**28.** C	**29.** C	**30.** B

Chapter 6 – Solutions to Odd-Numbered Problems

Practice Problems

6.1 Explain the difference between a covalent bond and an intermolecular force.

Covalent bonds are intramolecular and involve the actual sharing of electrons between atoms *within* the same molecule; intermolecular forces are attractions *between* molecules. Intermolecular forces are weaker than even the weakest covalent bond. Intermolecular forces arise mainly from the attraction of an electron-rich area of one molecule (where electrons spend more of their time) to an electron-poor area of another molecule (where electrons spend less time). In other words, the partially negative (δ−) part of one molecule is attracted to the partially positive (δ+) part of a second molecule. This attraction between two molecules is an intermolecular force.

6.3 Polar molecules can contain more than one type of attractive force. Fill in the table with a yes or no answer to the following question: "Does this intermolecular force operate between molecules of the given compound?"

Molecule	H-Bond	Dipole-Dipole	London
CH_3OH			
NH_3			
CH_3CH_3			

Hydrogen bonding requires the interaction of two players, a donor and acceptor, as described in the following table:

REQUIREMENTS FOR HYDROGEN BONDING	
Name	**Description**
Hydrogen-bond donor (δ^+)	A molecule with a hydrogen atom covalently bonded to an oxygen, nitrogen, or fluorine (O, N, or F)
Hydrogen-bond acceptor (δ^-)	A molecule with a nonbonding pair of electrons on an oxygen, nitrogen, or fluorine (O, N, or F)

Some molecules—polar molecules—have a permanent, uneven distribution of electrons caused by electronegativity differences in the atoms that make up the molecules. This type of attraction involves the interaction of two dipoles and is thus called a dipole-dipole attraction.

London forces occur momentarily in all molecules when electrons become unevenly distributed over a molecule's surface. If the distribution of electrons over a molecule's surface becomes temporarily unequal, a temporary unequal charge, or induced dipole, is created.

CH_3OH: All molecules experience London forces, contains a polar —OH bond and is a polar molecule which also meets the criteria (in the table above) for hydrogen bonding.

NH_3: All molecules experience London forces, contains polar —NH bonds and is a polar molecule which also meets the criteria (in the table above) for hydrogen bonding.

CH_3CH_3: All molecules experience London forces, contains only nonpolar —CH bonds which do not meet the criteria for either dipole-dipole or H-bond intermolecular forces.

Molecule	H Bond	Dipole–Dipole	London
CH_3OH	Yes	Yes	Yes
NH_3	Yes	Yes	Yes
CH_3CH_3	No	No	Yes

6.5 Explain the difference between the dipole in London forces and the dipole in dipole-dipole attractions?

The dipole involved in London forces is temporary and therefore weak. The dipole in dipole-dipole attractions is permanent and therefore stronger than those involved in London forces.

6.7 Explain the requirements for a molecule to be a hydrogen bond acceptor.

A molecule with a nonbonding pair of electrons on an oxygen, nitrogen, or fluorine (O, N, or F).

6.9 Why is the ion-dipole attraction not classified as an intermolecular force?

Because a salt is made up of ions, a cation-anion combination, and not molecules, this force is technically not an inter*molecular* force.

6.11 For each of the following pairs of compounds, identify the strongest attractive force that would exist between the two: London forces, dipole-dipole attractions, hydrogen bonding, or ion-dipole attractions.

a. CH_3CH_3 and $CH_3CH_2CH_3$

b. $CH_3CH_2NH_2$ and KCl

c. $CH_3CH_2OCH_3$ and CH_3OH

d. 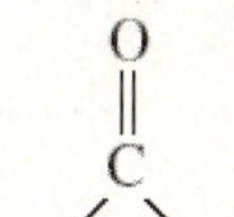 and NH_4Cl

Order of strength: London forces < dipole-dipole attractions < hydrogen bonds < ion-dipole attractions

All molecules experience London force.

Polar molecules experience dipole-dipole attractions.

Hydrogen bonding requires an electron-pair donor and an electron-pair acceptor. A hydrogen-bond donor is a molecule with a hydrogen atom covalently bonded to an oxygen, nitrogen, or fluorine (O, N, or F). A hydrogen-bond acceptor is a molecule with a nonbonding pair of electrons on an oxygen, nitrogen, or fluorine (O, N, or F).

a. CH_3CH_3 and $CH_3CH_2CH_3$: Both contain only nonpolar CH bonds and is are nonpolar molecules, experience only London forces

b. $CH_3CH_2NH_2$ and KCl: $CH_3CH_2NH_2$ experiences dipole-dipole attractions and hydrogen bonding, KCl is an ionic compound, strongest force is ion-dipole attractions

c. $CH_3CH_2OCH_3$ and CH_3OH: $CH_3CH_2OCH_3$: The oxygen of $CH_3CH_2OCH_3$ can act as a hydrogen bond acceptor and the –OH of CH_3OH can act as a hydrogen bond donor, strongest force is hydrogen bonding

d. 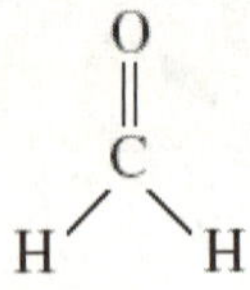 and NH_4Cl: 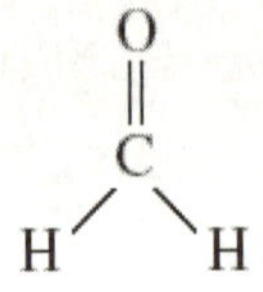 is a polar molecule which experiences dipole-dipole attractions and NH_4Cl is an ionic compound, strongest force is ion-dipole attractions.

6.13 For the pairs in problem 6.11 for which you listed ion-dipole attractions as a possibility, show how each of the ions would interact with the polar molecule. (See Figure 6.4.)

Anions (–) will be attracted to the δ+ end of the polar molecule and cations (+) will be attracted to the δ– end of the polar molecule. The δ– end of the polar molecule will be near the atoms with higher electronegativities and the δ+ end those with lower electronegativities.

b. $CH_3CH_2NH_2$ and KCl d. 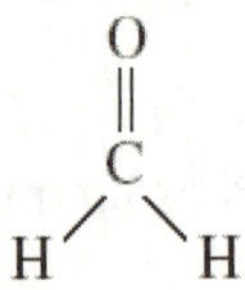 and NH_4Cl (Ion-dipole interactions are shown within the circles in the following - not in red.)

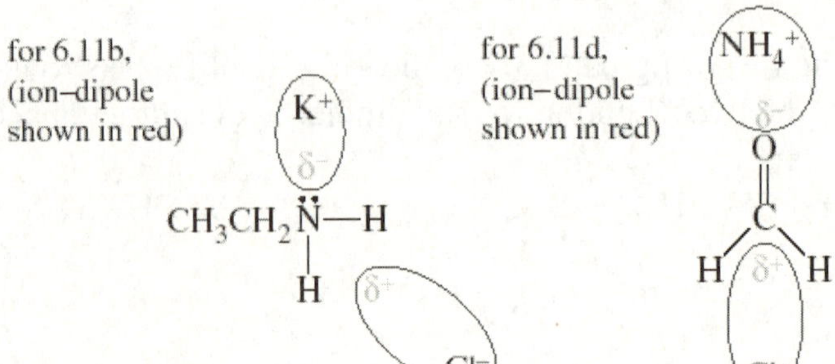

6.15 For each of the following pairs of compounds, predict whether they are likely to be *soluble* or *not soluble* in each other. If soluble, identify the attractive forces involved in their interaction. If not soluble, explain why not.

a. NaCl and 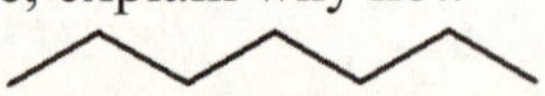

b. CH_3OH and CH_3OCH_3

The golden rule of solubility, "like dissolves like,"means that molecules that are similar will dissolve in each other. The "similarity" mentioned here relates to the polar character of the molecules. Specifically, molecules that have similar polarity and participate in the same types of intermolecular forces will dissolve each other.

a. NaCl and 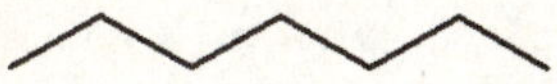: Not soluble, NaCl is ionic and  is nonpolar. Strong forces of attraction cannot develop between the two species.

b. CH_3OH and CH_3OCH_3: Soluble, CH_3OH is a hydrogen-bond donor and/or acceptor. CH_3OCH_3 is a hydrogen-bond acceptor. Hydrogen bonds can form between the two species.

6.17 Predict which member of each of the following pairs of compounds will be more soluble in water.

a. CH_3CH_2OH or CH_3CH_3 **b.** CH_3NH_2 or CS_2

The golden rule of solubility, "like dissolves like," means that molecules that are similar will dissolve in each other. The "similarity" mentioned here relates to the polar character of the molecules. Specifically, molecules that have similar polarity and participate in the same types of intermolecular forces will dissolve each other.

Water (H_2O or H—O—H) is an excellent hydrogen-bond acceptor and/or donor.

a. CH_3CH_2OH or CH_3CH_3: CH_3CH_2OH will be more soluble in water since it can form hydrogen bonds with water. CH_3CH_3 is nonpolar and cannot develop strong attraction with water.

b. CH_3NH_2 or CS_2: CH_3NH_2 will be more soluble in water since it can form hydrogen bonds with water. CS_2 is nonpolar and cannot develop strong attraction with water.

6.19 In the following pairs of molecules, predict which one of the molecules in the pair would have the lower boiling point. Name the strongest intermolecular force present in each molecule.

a. CH_4 and H_2O **b.** NH_3 and CO_2

c. 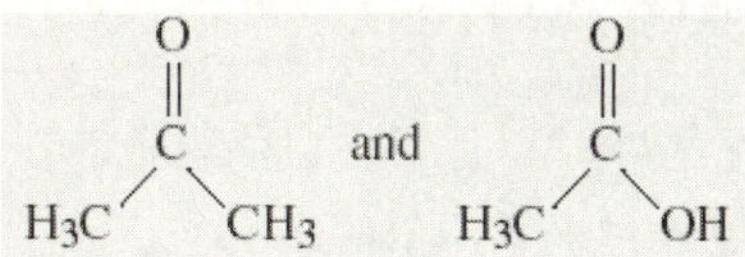

d. $CH_3CH_2CH_2CH_3$ and CH_3OCH_3

The stronger the type of intermolecular force, the higher will be the boiling point. For molecules with the same type of intermolecular force, the size of the molecule and the geometry must also be considered. These trends are summarized in Table 6.2.

TABLE 6.2 BOILING POINT TRENDS

Nonpolar Molecules	Similar Size Molecules, Different Forces	Molecules with Same Forces
The greater the surface area, the higher the boiling point.	The stronger the forces, the higher the boiling point.	The stronger and more numerous the forces, the higher the boiling point.

a. CH_4 and H_2O: CH_4 will have the lower boiling point since it is a nonpolar molecule experiencing only weak London forces while H_2O experiences strong hydrogen bonding.

b. NH_3 and CO_2: CO_2 will have the lower boiling point since it is a nonpolar molecule experiencing only weak London forces while NH_3 experiences strong hydrogen bonding.

c. 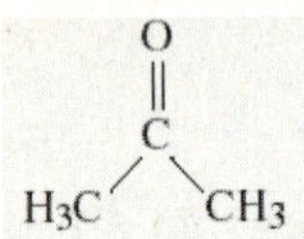 and 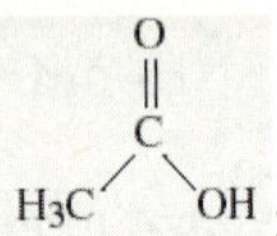: 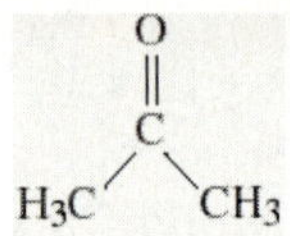 will have the lower boiling point since it a polar molecule but can only be a hydrogen-bond acceptor not a hydrogen-bond donor and will experience only dipole-dipole attractions, not hydrogen bonding.

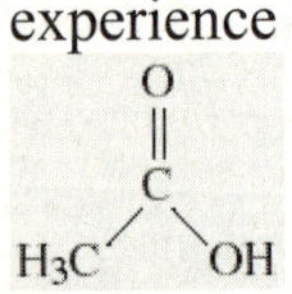 can be both a hydrogen-bond acceptor and donor and will experience hydrogen bonding.

d. $CH_3CH_2CH_2CH_3$ and CH_3OCH_3: $CH_3CH_2CH_2CH_3$ will have the lower boiling point since it is a nonpolar molecule experiencing only weak London forces while polar CH_3OCH_3 experiences stronger dipole-dipole attractions.

6.21 The formulas and boiling points of several similar compounds are listed in the following table. Determine the strongest intermolecular force present in each compound and explain the difference in their boiling points.

Formula	Boiling Point °C
H | H—C—H | H	−164
H | H—C—F | H	−78
H | H—C—OH | H	65

The highest boiling point is found in the compound with the strongest intermolecular forces (hydrogen bonding), CH_3OH. The lowest boiling point is found in the compound with the weakest intermolecular forces (London force), CH_4. CF_4 is intermediate with dipole-dipole forces that are stronger than London forces but weaker than hydrogen bonding.

6.23 Rank the following compounds in order of increasing boiling point (lowest to highest) and explain your ranking.

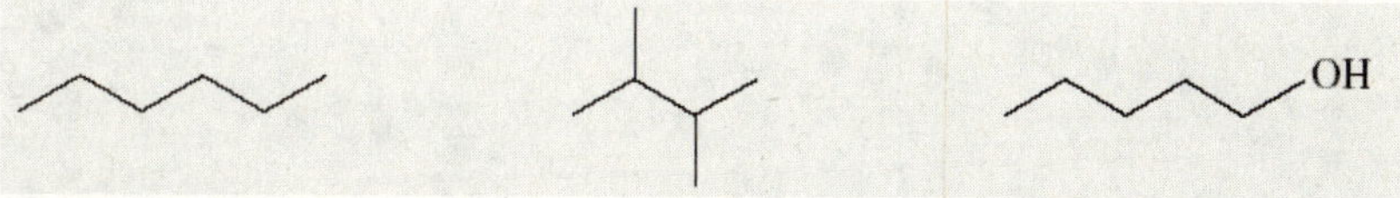

The molecule on the right is an alcohol and has hydrogen bonding and dipole-dipole attractions. The two remaining compounds are alkanes, which are nonpolar experiencing only London forces. The one with the more branching is the lower boiling because of its decreased surface area.

Lowest Boiling Point | Highest Boiling Point

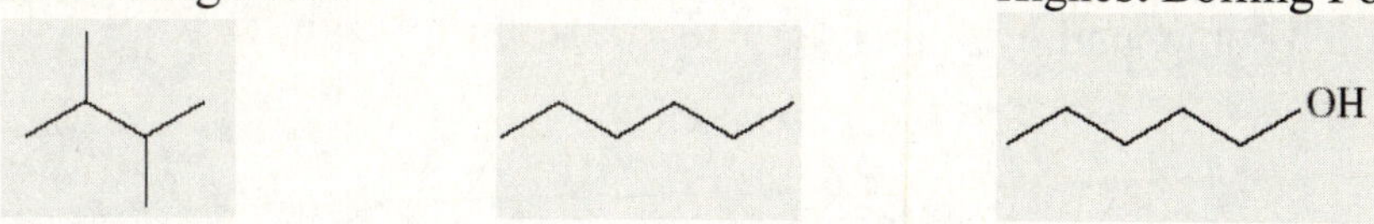

6.25 What is a triglyceride? Draw the structural formula for a simple triglyceride whose hydrocarbon tails contain six saturated carbons.

A triglyceride is the product of a condensation reaction of three fatty acids and glycerol; also known as a fat or an oil.

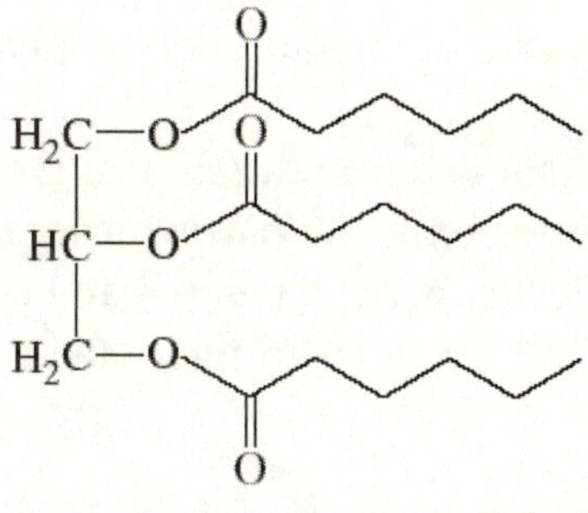

The glycerol "backbone" is shown on the left (by the vertical chain of H_2C—CH—H_2C) and the three fatty acid "residues" on the right.

6.27 Explain why fats are solids, but oils are liquids under normal conditions, despite the fact that both are triglycerides.

Even though fats and oils are both triglycerides, the large number of cis double bonds in the hydrocarbons tails of oils prevents them from interacting with each other as much as the saturated tails of the fats. (See structure below (a) fat and (b) oil.) More interactions mean that the tails move more slowly, becoming more like a solid, forming a fat.

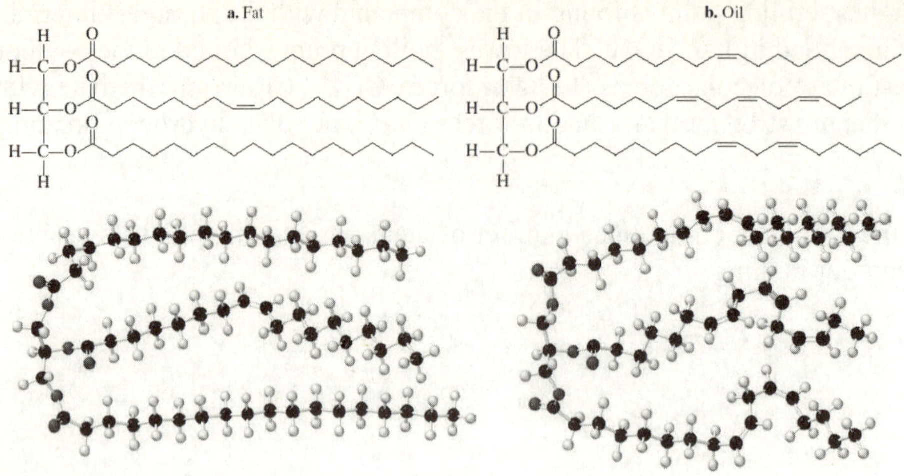

Tails are closer together in fat

Tails are farther apart in oil

6.29 Balance the following reaction equation by providing the missing number of hydrogen molecules.

One hydrogen molecule is required to hydrogenate each carbon-to-carbon double bond in an unsaturated compound.

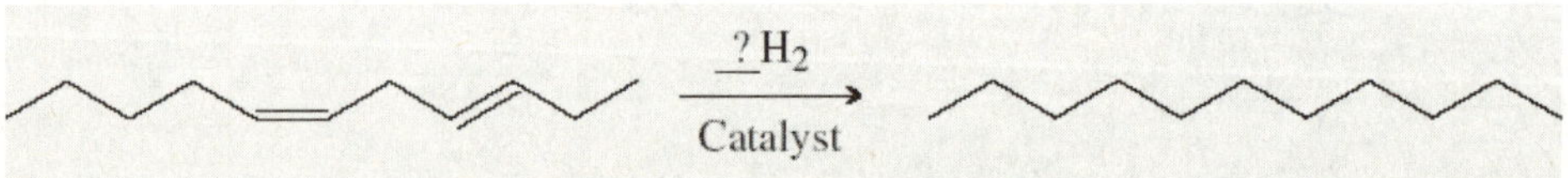

The alkene reactant contains 2 carbon-to-carbon double bonds and thus 2 H_2 molecules are needed to form the alkane product.

Additional Problems

6.31 Compare and contrast a dipole-dipole attraction and an ion-dipole attraction.

Both are attractive forces that occur between positive and negative charges. A dipole-dipole attraction occurs between two opposite partial charges that permanently exist in a molecule. An ion-dipole attraction occurs between an ion and a partial charge of opposite sign in a molecule. An ion-dipole attraction is stronger than a dipole-dipole attraction.

6.33 Show a possible hydrogen bond between methanol (CH_3OH) and water. Label the donor and the acceptor. Is there more than one possible hydrogen bond between these two molecules? Explain.

These molecules can form more than one hydrogen bond. They can form three, with methanol (CH_3OH) acting as an acceptor twice (once for each of the two lone pairs on oxygen) and a donor once (from the —OH bond). One such interaction (a-----d) is shown below.

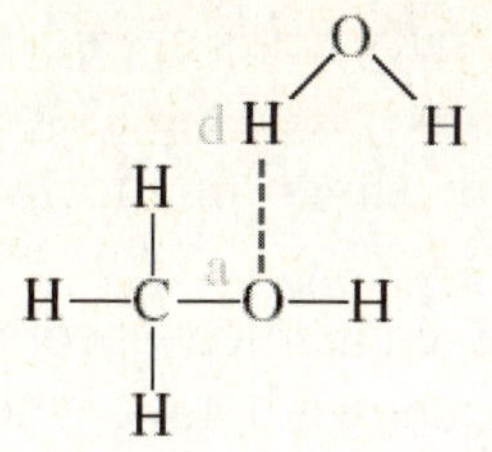

Another hydrogen bond could form between the oxygen of methanol and the positive end of the water molecule. The third would be between the H of the —OH bond in methanol and the negative (oxygen) end of the water molecule.

6.35 List all intermolecular forces found in each of the pure substances below.

a. 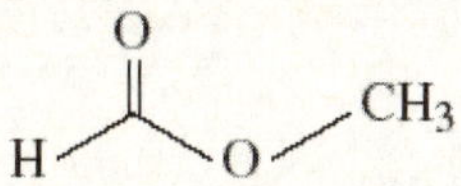

d. 

b. 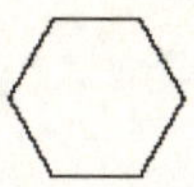

e. CH_2Cl_2

c. 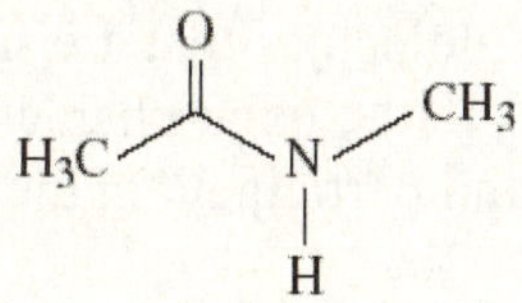

f. 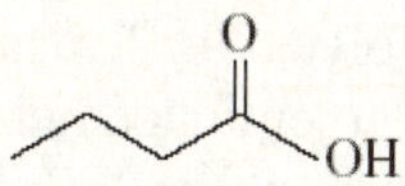

All molecules experience London forces.

Polar molecules experience dipole-dipole attractions.

Hydrogen bonding requires an electron-pair donor and an electron-pair acceptor. A hydrogen-bond donor is a molecule with a hydrogen atom covalently bonded to an oxygen, nitrogen, or fluorine (O, N, or F). A hydrogen-bond acceptor is a molecule with a nonbonding pair of electrons on an oxygen, nitrogen, or fluorine (O, N, or F).

a. London forces, dipole–dipole attractions
b. London forces
c. London forces, dipole–dipole attractions, hydrogen bonding
d. London forces
e. London forces, dipole–dipole attractions
f. London forces, dipole–dipole attractions, hydrogen bonding

6.37 In Chapter 4, we learned that fatty acids are not soluble in water because of their long hydrocarbon tails. However, fatty acid salts (soap molecules) are soluble and form micelles in water. Explain why fatty acid salts form micelles, but fatty acids do not.

Fatty acid salts are not the same chemical species as fatty acids. Fatty acid salts contain the carboxylate functional group, shown below, which can undergo ion-dipole attractions with water,

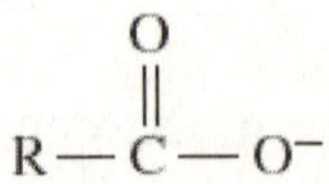

that are stronger than hydrogen bonding between water and fatty acids. These ion-dipole attractions allow the micelles to form.

6.39 Predict which member of each of the following pairs of compounds will be more soluble in vegetable oil.
a. fatty acid or fatty acid salt
b. fat or water
c. CH_3NH_2 or CS_2
d. KCl or $CH_3CH_2CH_2CH_2CH_2CH_3$

The golden rule of solubility, "like dissolves like," means that molecules that are similar will dissolve in each other. The "similarity" mentioned here relates to the polar character of the molecules. Specifically, molecules that have similar polarity and participate in the same types of intermolecular forces will dissolve each other.

Vegetable oil is a nonpolar hydrocarbon and will dissolve other nonpolar compounds.

a. fatty acid or fatty acid salt: fatty acid
A fatty acid is less polar overall than a fatty acid salt which is ionic.
b. fat or water: fat
A fat is much less polar overall than water.
c. CH_3NH_2 or CS_2: CS_2
CS_2 is much less polar than CH_3NH_2.
d. KCl or $CH_3CH_2CH_2CH_2CH_2CH_3$: $CH_3CH_2CH_2CH_2CH_2CH_3$
$CH_3CH_2CH_2CH_2CH_2CH_3$ which is much less polar than ionic KCl.

6.41 A stain on your shirt will not come out when you wipe it with a wet cloth. What clue does this give you to the nature of the stain?

Golden rule of solubility: like dissolves like. The stain must be nonpolar (hydrophobic) in nature since it does not dissolve in polar water.

6.43 Rank the four compounds in the figures below in decreasing order of boiling point. Explain the reasoning behind your ranking.

a. 

b. 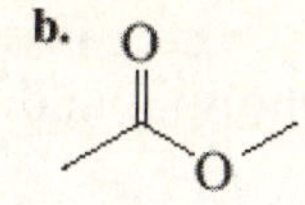

c. 

d. 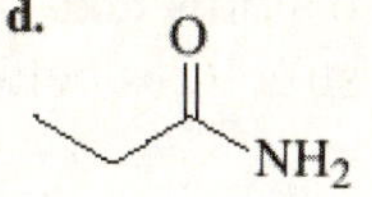

The stronger the type of intermolecular force the greater will be the boiling point. For molecules with the same type of intermolecular force, the size of the molecule and the geometry must also be considered. These trends are summarized in Table 6.2.

Highest to lowest boiling point:
d. 5 possible H-bonds per molecule (strongest)
c. 3 possible H-bonds per molecule
b. polar molecule
a. nonpolar molecule (weakest)

6.45 The boiling point of octane is much greater than that of hexane (see Table 6.1). Explain the difference in their boiling points.

Octane  Hexane 

Because the hydrocarbon chain is longer in octane (meaning the molecule has a greater surface area), the London forces between molecules are stronger, which makes the boiling point higher for octane.

6.47 Explain why a straight-chain alkane has a higher boiling point than a branched-chain alkane with the same molecular formula.

A branched alkane has less surface contact with neighboring molecules than does a straight-chain alkane. The intermolecular forces are therefore stronger between the straight-chain alkane molecules, raising the boiling point.

6.49 Draw the structure of the product of the following hydrogenation reaction.

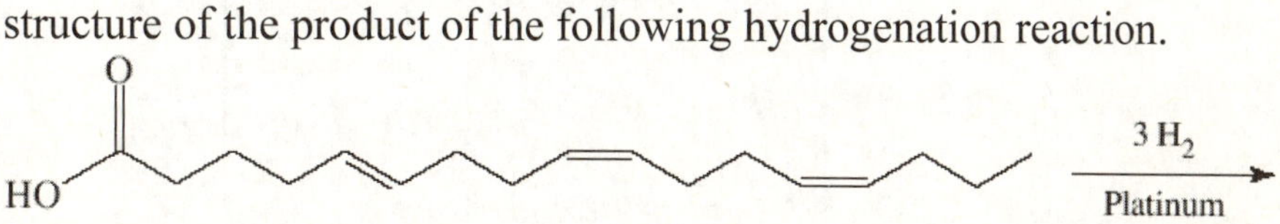

Hydrogenation converts each carbon-to-carbon double bond to a single bond.

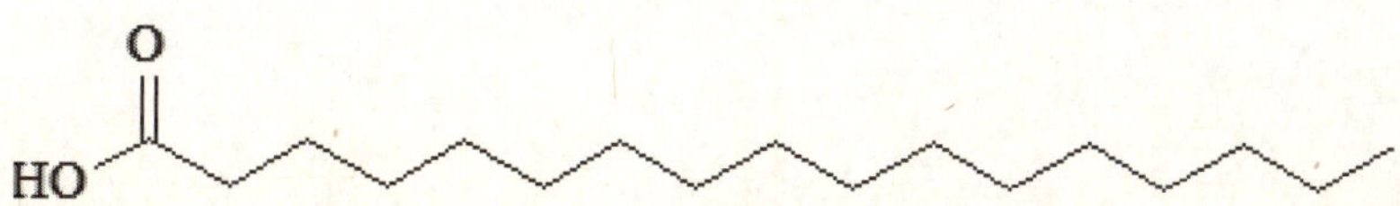

6.51 Compare the structure of a soap molecule to a phospholipid and explain why a soap's polar head is smaller than that of a phospholipid.

The phospholipid head group contains many more atoms than the three atoms in the carboxylate head group of a soap. Compare the structures below.

Soap:

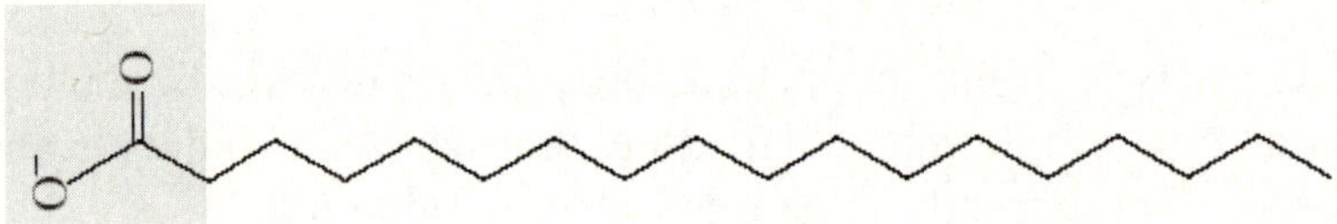

Phospholipid:

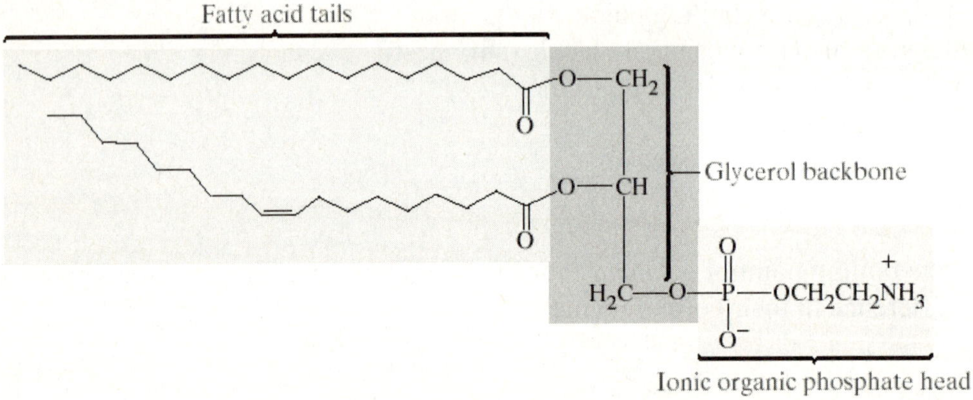

Challenge Problems

6.53 If very small amounts of soap are dissolved in water, the soap molecules can form a structure known as a monolayer on the surface of the water with part of the soap molecule dissolved in the water and part sticking into the air. Using the cartoon figures for soap molecules, draw a small part of a monolayer showing the water, the air, and the soap molecules as they would be arranged.

Hydrocarbon tails interact with air, which is basically nonpolar, and the polar heads interact with polar water as shown in the diagram below.

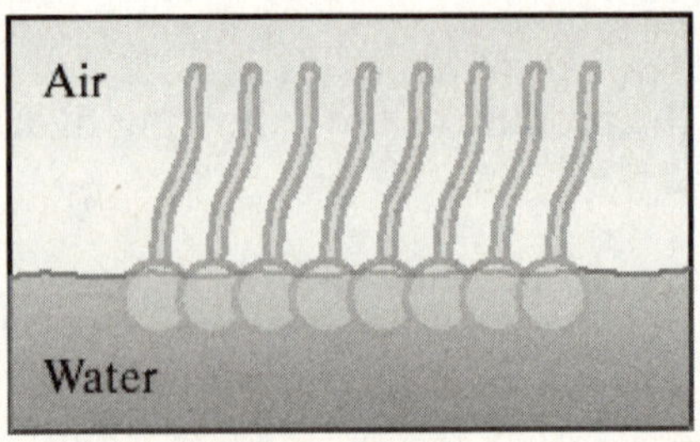

6.55 Water and hydrogen sulfide are structurally similar and geometrically identical (both have a bent shape); yet, H_2O is a liquid at room temperature and H_2S is a gas (rotten egg odor) at room temperature. In other words, water has a higher boiling point than hydrogen sulfide. Explain why two molecules that appear to be so similar have such different properties, especially in terms of their boiling points.

Sulfur cannot hydrogen bond, so H_2S molecules interact strongest through dipole–dipole interactions, which are weaker than hydrogen bonding. Boiling point increases with increasing strength of intermolecular attractions, resulting in a higher boiling point for water. Since this boiling point is above room temperature, water is a liquid.

6.57 Some other physical properties of liquids include viscosity, the resistance to flow, which can be thought of as the "thickness of a liquid," and surface tension, which measures the forces at a liquid surface. Consider the following liquids: *honey* (mainly glucose and fructose) versus *water*.

a. In your experience, which liquid has a higher viscosity?

b. What intermolecular forces are present? How many hydrogen bonding interactions are possible in a molecule of glucose and of water?

c. Explain viscosity in terms of intermolecular forces.

d. It is possible to float a small paper clip on the surface of a cup of water. Try it. What happens if you add soap to the water? Can you balance the same paper clip?

a. Honey which is much "thicker" than water and flows much more slowly.

b. Glucose and fructose are both polar with multiple —OH groups (See below for α-D-glucose.): Hydrogen bonding, dipole–dipole attractions, and London forces. A glucose in ring form could make up to 17 possible hydrogen bonds; water can make 4 possible hydrogen bonds per molecule.

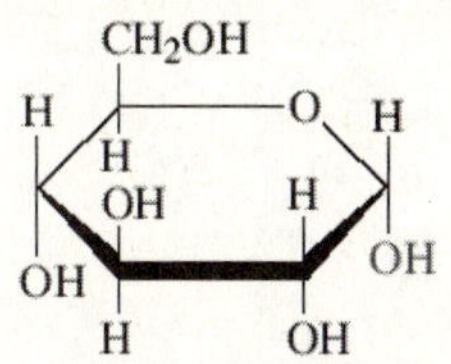

c. Because there are so many more hydrogen bonds possible between sugar molecules in honey, this increases the force of attraction across the molecules on the surface making the honey more viscous than water.

d. By adding soap to the water, the hydrogen bonding network between the water molecules is disrupted, and the paper clip can no longer float on the surface.

Chapter

7

Solution Chemistry: How Sweet *is* Your Tea?

7.1 Solutions Are Mixtures

Learning Objectives

Upon completion of this material, a student should be able to do the following:

A. define the following key terms:

solution	solute	solvent
aqueous solution	colloid	suspension

B. describe the properties of a solution
C. distinguish the solute and solvent from the description of a solution
D. distinguish solution and colloid
E. distinguish colloid and suspension

7.2 Formation of Solutions

Learning Objectives

Upon completion of this material, a student should be able to do the following:

A. define the following key terms:

hydration	solvation	solubility
unsaturated solution	saturated solution	Henry's law

B. distinguish saturated and unsaturated solution
C. describe the formation of a saturated solution in terms of equilibrium
D. demonstrate the effect of temperature on solubility
E. demonstrate the effect of pressure on the solubility of a gas (Henry's law)

7.3 Chemical Equations for Solution Formation

Learning Objectives

Upon completion of this material, a student should be able to do the following:

A. define the following key terms:

electrolyte	strong electrolyte	nonelectrolyte
ionize	weak electrolyte	law of conservation of mass
coefficient	equivalent (Eq)	

B. distinguish the solutes as electrolyte, weak electrolyte, or nonelectrolyte
C. relate the degree of ionization to the type of electrolyte
C. write chemical equations for the solvation of electrolytes, weak electrolytes, and nonelectrolytes
D. determine the number of equivalents in ionic solutions
E. relate the number of equivalents to the number of moles of solute

7.4 Concentrations

Learning Objectives

Upon completion of this material, a student should be able to do the following:

A. define the following key terms:

concentration	molarity

B. calculate the concentration of a solution in the following units:
   1. Milliequivalents per liter
   2. Molarity
   3. Percent concentration: % (m/m), % (m/v), and % (v/v)
   4. Parts per million (ppm) and parts per billion (ppb)

C. calculate how to prepare a solution from a solid solute

D. convert from one unit of concentration to another using the definitions of the concentration scales

7.5 Dilution

Learning Objectives

Upon completion of this material, a student should be able to do the following:

A. describe what is meant by dilution of a solution

B. state the dilution equation

C. calculate how to prepare a solution from a concentrated stock solution

7.6 Osmosis and Diffusion

Learning Objectives

Upon completion of this material, a student should be able to do the following:

A. define the following key terms:

semipermeable membrane	isotonic solution	hypotonic solution
osmosis	osmotic pressure	hypertonic solution
crenation	physiological solution	diffusion
dialysis		

B. classify solutions as isotonic, hypotonic, and hypertonic in general

C. determine direction of water flow across a membrane in osmosis and the direction of solute flow in diffusion

D. qualitatively relate osmotic pressure to solution concentration

E. predict whether a cell will swell or crenate based on concentration

F. compare and contrast osmosis and diffusion

G. classify solutions as isotonic, hypotonic, and hypertonic with physiological solutions

H. describe the process of dialysis and compare to diffusion

7.7 Transport Across Cell Membranes

Learning Objectives

Upon completion of this material, a student should be able to do the following:

A. define the following key terms:
   passive diffusion   facilitated transport   active transport

B. relate the structure of a cell membrane to the methods of transporting substances across the cell membrane

C. distinguish passive diffusion, facilitated transport, and active transport

Practice Test for Chapter 7

1. In which of the following would ethanol be classified the solvent?
   A. 35 mL of ethanol is mixed with 65 mL of methanol.
   B. 100 mL of ethanol is mixed with enough water to form 115 mL of solution.
   C. 35 g of ethanol is mixed with enough isopropyl alcohol to produce 150 g of solution.
   D. Ethanol is not the solvent in any of the solutions described.

2. Honey is mixed in hot tea. The resulting combination of honey and hot tea would be classified as:
   A. suspension.
   B. colloid.
   C. solution.
   D. solvent.

3. Blood is a ____________________ of plasma and blood cells that can be separated by centrifugation. Fill the blank with one of the choices.
   A. suspension
   B. colloid
   C. solution
   D. None of the above correctly classifies blood.

4. What happens to the solubility of oxygen in blood when a person drives from Kona, Hawaii at sea level to the top of Mauna Kea at almost 15,000 ft elevation?
   A. decreases
   B. increases
   C. remains constant
   D. cannot predict from the information given

5. Consider two identical bottles of soft drink. One is stored in the refrigerator and one is stored at room temperature. When opened, which one will release more gas from the liquid?

   A. the one stored at room temperature.
   B. the one stored in the refrigerator
   C. neither, they will release the same amount of gas regardless of temperature
   D. cannot predict based on the given information

6. The picture below was taken after table salt was added to water after the mixture was thoroughly stirred for 5 minutes. The pictured mixture probably represents a(n):

   A. heterogeneous mixture.
   B. saturated solution.
   C. equilibrium state.
   D. all of the above

7. Consider the following image of containers of the same carbonated beverage.

   In which container is the pressure of CO_2 likely to be the smallest?
   A. 2 L bottle
   B. beaker
   C. 12 oz can
   D. impossible to predict

8. For **both** solid and gaseous solutes, which of the following is true?
   A. Solubility increases with increasing temperature.
   B. Solubility decreases with increasing temperature.
   C. Solubility increases with increasing pressure.
   D. Solubility decreases with increasing pressure.
   E. None of the above is true for both solid and gaseous solutes.

9. Which of the following is most likely to be a weak electrolyte?
   A. $C_6H_{12}O_6(s)$
   B. $NH_4NO_3(s)$
   C. $CH_3COOH(l)$
   D. $CH_3OH(l)$

10. When HF(l) dissolves in water, the following reaction occurs. HF is classified as:

$$HF(l) \underset{H_2O}{\rightleftharpoons} H^+(aq) + F^-(aq)$$

A. weak electrolyte.
B. nonelectrolyte.
C. strong electrolyte.
D. The reaction alone cannot be used to predict.

11. The following reaction indicates:

$$MgCl_2(s) \xrightarrow[H_2O]{} Mg^{2+}(aq) + 2\ Cl^-(aq)$$

A. the solubility of the solute will decrease as the temperature is increased.
B. that the solute is a weak electrolyte.
C. the reaction is a hydration.
D. all of the above

12. The balanced equation for the hydration of $AlCl_3$ would contain how many total ions or molecules in the products?
A. 1
B. 2
C. 3
D. 4
E. 5

13. The products of the correctly balanced equation for the hydration of $BaCl_2$ would be:
A. $Ba^{2+}(aq) + Cl_2^{2-}(aq)$.
B. $Ba^{2+}(aq) + 2Cl^-(aq)$.
C. $BaCl_2(aq)$.
D. $Ba^{1+}(aq) + 2Cl^-(aq)$.

14. Which of the following concentration units is used to represent the amount of electrolytes in body fluid?
A. M
B. %(m/v)
C. %(m/m)
D. mEq/L

15. Which of the following conversion factors could be used to represent the relationship between equivalents and moles of PO_4^{3-}?

A. $\frac{1\,\text{Eq}\,PO_4^{3-}}{1\,\text{mol}\,PO_4^{3-}}$

B. $\frac{1\,\text{Eq}\,PO_4^{3-}}{3\,\text{mol}\,PO_4^{3-}}$

C. $\frac{3\,\text{Eq}\,PO_4^{3-}}{1\,\text{mol}\,PO_4^{3-}}$

D. $\frac{3\,\text{Eq}\,PO_4^{3-}}{3\,\text{mol}\,PO_4^{3-}}$

16. How many equivalents of CO_3^{2-} are present in a solution that contains 0.750 moles of CO_3^{2-}?
A. 1.50 Eq
B. 0.750 Eq
C. 0.250 Eq
D. 2.25 Eq

17. How many mmol of Ca^{2+} are present in a solution that contains 2.30 mEq of CO_3^{2-}?
A. 2.30 mmol
B. 4.60 mmol
C. 1.15 mmol
D. 6.90 mmol

18. A Ringer's solution for intravenous fluid replacement typically has a concentration of 77.5 mEq Cl^- per 0.500 L of solution. If a patient receives 0.750 L of Ringer's solution, how many equivalents of Cl^- were given?

A. 116 Eq
B. 0.116 Eq
C. 51. 7 Eq
D. 0.0775 Eq

19. What is the molarity of a solution prepared by dissolving 2.50 moles of NaCl in enough water to produce 3.00 L solution?
A. 0.833 M
B. 2.50 M
C. 7.50 M
D. 1.20 M

20. How many moles of KCl are present in 574 mL of a 0.660 M KCl solution?
A. 0.660 mol
B. 379 mol
C. 1.15 mol
D. 0.379 mol

21. Express the fraction shown below as a percent. Use two significant figures.

$$\frac{5}{8}$$

A. 1.6%
B. 63%
C. 0.63%
D. 52%
E. 160%

22. What is 45% of 250?
A. 45
B. 5.5
C. 140
D. 110

23. What is the molarity of a solution prepared by dissolving 50.0 g of KCl in enough water to yield 1.50 L of solution?
A. 33.3 M
B. 0.0300 M
C. 0.447 M
D. 0.671 M

24. Calculate the concentration in %(m/m) for a solution prepared by dissolving 58.5 g of Na_2SO_4 in 200.0 g of distilled water.
A. 22.6%
B. 29.2%
C. 3.41%
D. 11.7%

25. 0.125 L of a 55.0 mg/mL hydrocortisone solution is needed. What volume of a 210.0 mg/ml solution should be diluted to obtain this solution?
A. 32.7 mL
B. 477 mL
C. 0.0327 mL
D. 6.88 mL
E. 26.2 mL

26. How many grams of glucose would be needed to prepare 2.00 L of a 5.10% (m/v) solution?
A. 10.2 g
B. 102 g
C. 2.55 g
D. 25.5 g

27. How many liters of a 5.00% (m/v) solution of NaCl can be prepared from 575 mL of a stock 12.5% (m/v) solution?
    A. 230 L
    B. 1440 L
    C. 0.230 L
    D. 1.44 L

28. Consider the following image which depicts a cell in a solution.

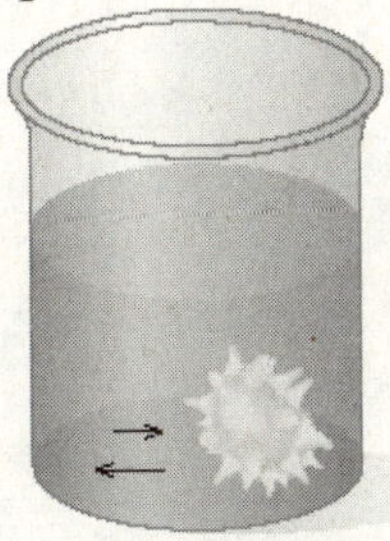

    Based on the effect on the cell, the solution would be classified as:
    A. hypertonic and the water is flowing out of the cell.
    B. hypotonic and the water is flowing into the cell.
    C. hypertonic and the water is flowing into the cell.
    D. hypotonic and the water is flowing out of the cell.
    E. isotonic and there is no net water flow.

29. Consider the diagram showing methods of transport across a cell membrane.

    A Na^+ ion is to be transported across the membrane with an input of energy. Which method of transport will be used?
    A. passive diffusion
    B. facilitated transport
    C. active transport
    D. Not enough information is given to determine the method.

30. A red blood cell that is isotonic with 0.90% (m/v) NaCl and 5% (m/v) glucose will undergo:
    A. lysis if placed in a solution of distilled water.
    B. crenation if placed in a 1.90% (m/v) NaCl and 5% (m/v) glucose solution.
    C. no change if placed in a 0.90% (m/v) NaCl and 5% (m/v) glucose solution.
    D. all of the above

Answers

1. B	**2.** C	**3.** A	**4.** A	**5.** D	**6.** D	**7.** B	**8.** E	**9.** C	**10.** A
11. C	**12.** D	**13.** B	**14.** D	**15.** C	**16.** A	**17.** C	**18.** B	**19.** A	**20.** D
21. B	**22.** D	**23.** C	**24.** A	**25.** A	**26.** B	**27.** D	**28.** A	**29.** C	**30.** D

Chapter 7 – Solutions to Odd-Numbered Problems

Practice Problems

7.1 Identify the solute and solvent in solutions composed of the following:

a. 80% nitrogen and about 20% oxygen in the air
b. 100.0 g of copper and 20.0 g of zinc
c. 500.0 mL of ethanol and a few drops of blue food coloring

The solute is the substance in a solution present in the smaller amount, and the solvent is the substance present in the larger amount.

a. 80% nitrogen and about 20% oxygen in the air: solute, oxygen; solvent, nitrogen
b. 100.0 g of copper and 20.0 g of zinc: solute, zinc; solvent, copper
c. 500.0 mL of ethanol and a few drops of blue food coloring: solute, blue food coloring; solvent ethanol

7.3 Identify the following as solutions or colloids:
a. yogurt **b.** cola drink **c.** gasoline

A colloid, like a solution, is a homogenous mixture. The difference lies in the size of the particles that each contains. By definition, the particles in a colloid must be between 1 nanometer (1×10^{-9} meter) and 1000 nanometers in diameter. Particles of this size remain suspended in solution, and so a colloid does not separate over time.

a. yogurt: colloid, does not separate, but like "homogenized" milk contains particles larger than a solution.
b. cola drink: a solution of a gas, sugar, etc. in water all less than 1 nanometer
c. gasoline: a solution of many additives in several organic liquids all less than 1 nanometer

7.5 Which of the following represents a saturated solution?
a. Throwing a pinch of salt into a pot of boiling water.
b. A layer of sugar forms at the bottom of a glass of tea as ice is added.

If a solution does *not* contain the maximum amount of the solute that the solvent can hold, it is referred to as an unsaturated solution. If more solute is added, it will dissolve. If a solution contains all the solute that can possibly dissolve, the solution is referred to as a saturated solution. If more solute is added, it will not dissolve.

a. unsaturated solution: the added salt will dissolve
b. saturated solution: all of the solute does not dissolve

7.7 Fill in the blank in the statements below with the words : "increase," "decrease," or "stay the same."
a. If the temperature of a solution increases, the solubility of most solid solutes will __________.
b. If the temperature of a solution increases, the solubility of a gaseous solute will __________.
c. If the pressure above a solution increases, the solubility of a gaseous solute will __________.

The solubility of most solids dissolved in water *increases* with temperature, while the solubility of a gas dissolved in water *decreases* with a rise in temperature.

Henry's law states that the solubility of a gas in a liquid is directly related to the pressure of that gas over the liquid. That is, the amount of a gas that can dissolve in a liquid increases as the pressure of the gas in the space above the liquid increases.

a. increase
b. decrease
c. increase

7.9 Explain what is happening in the following situations in terms of solubility.
a. An opened bottle of soda goes flat more slowly when recapped and stored in the refrigerator.
b. A bottle of club soda keeps its fizz longer if the lid is screwed on tightly.
c. Less sugar dissolves in cold iced tea than in hot tea.

The solubility of most solids dissolved in water *increases* with temperature, while the solubility of a gas dissolved in water *decreases* with a rise in temperature.

Henry's law states that the solubility of a gas in a liquid is directly related to the pressure of that gas over the liquid. That is, the amount of a gas that can dissolve in a liquid increases as the pressure of the gas in the space above the liquid increases.

a. An opened bottle of soda goes flat more slowly when recapped and stored in the refrigerator.
   The solubility of the gas (CO_2) which causes the fizz is increased at a lower temperature by refrigeration and by capping, the pressure of the gas over the liquid remains higher than if uncapped. Both factors keep more CO_2 dissolved.
b. A bottle of club soda keeps its fizz longer if the lid is screwed on tightly.
   Capping tightly keeps the pressure of the gas over the liquid higher than if uncapped or loosely capped, which increases the solubility of CO_2 (which causes the fizz) in the solution.
c. Less sugar dissolves in cold iced tea than in hot tea.
   The solubility of sugar (a solid) increases with increasing temperature.

7.11 Predict whether the following will fully dissociate, partially dissociate, or not dissociate when dissolved in solution:
a. potassium fluoride, KF (*s*), a strong electrolyte
b. hydrogen cyanide, HCN (*g*), a weak electrolyte
c. fructose, $C_6H_{12}O_6$ (*s*), a nonelectrolyte

Strong electrolytes are ionic compounds that completely dissociate in water, whereas weak electrolytes only partially dissociate into ions. Nonelectrolytes are substances (usually covalent compounds) that dissolve in water but do not dissociate and therefore do not conduct electrical current.
a. potassium fluoride, KF (*s*), a strong electrolyte: fully dissociates
b. hydrogen cyanide, HCN (*g*), a weak electrolyte: partially dissociates
c. fructose, $C_6H_{12}O_6$ (*s*), a nonelectrolyte: does not dissociate

7.13 Provide a balanced equation for the hydration of each compound in problem 7.11.
a. potassium fluoride, KF (*s*), a strong electrolyte
b. hydrogen cyanide, HCN (*g*), a weak electrolyte
c. fructose, $C_6H_{12}O_6$ (*s*), a nonelectrolyte

In writing the equation for a hydration reaction (or any reaction):
1. The number of atoms of each element must be the same on each side of the equation. To balance atoms, coefficients are used in front of the formula.
2. All hydration reactions show water under the reaction arrow.
3. The physical state of each component is shown with an abbreviation for the state enclosed in parentheses following the formula. (s, l, g, aq)
4. The equation for the hydration of a strong electrolyte shows the formula for the solute on the reactant side and the ions formed on the product side separated by a right arrow.
5. The equation for the hydration of a nonelectrolyte shows the formula solute on the reactant and product side of the equation separated by a right arrow.
4. The equation for the hydration of a weak electrolyte shows the formula for the solute on the reactant side and the ions formed on the product side separated by an equilibrium arrow.

a. potassium fluoride, KF (*s*), a strong electrolyte

$$KF(s) \xrightarrow[H_2O]{} K^+(aq) + F^-(aq)$$

b. hydrogen cyanide, HCN (*g*), a weak electrolyte

$$HCN(g) \underset{H_2O}{\rightleftharpoons} H^+(aq) + CN^-(aq)$$

c. fructose, $C_6H_{12}O_6$ (*s*), a nonelectrolyte

$$C_6H_{12}O_6(s) \xrightarrow[H_2O]{} C_6H_{12}O_6(aq)$$

7.15 For the following ionic compounds, write a balanced equation for their hydration in water:
a. $CaCl_2$ **b.** NaOH **c.** KBr **d.** $Fe(NO_3)_3$

In writing the equation for a hydration reaction (or any reaction):
1. The number of atoms of each element must be the same on each side of the equation. To balance atoms, coefficients are used in front of the formula.
2. All hydration reactions show water under the reaction arrow.
3. The physical state of each component is shown with an abbreviation for the state enclosed in parentheses following the formula. (s, l, g, aq)
4. The equation for the hydration of a strong electrolyte shows the formula for the solute on the reactant side and the ions formed on the product side separated by a right arrow.
5. The equation for the hydration of a nonelectrolyte shows the formula solute on the reactant and product side of the equation separated by a right arrow.
4. The equation for the hydration of a weak electrolyte shows the formula for the solute on the reactant side and the ions formed on the product side separated by an equilibrium arrow.

a. $CaCl_2$: Note that 2 Cl^- ions form.

$$CaCl_2(s) \xrightarrow[H_2O]{} Ca^{2+}(aq) + 2Cl^-(aq)$$

b. NaOH

$$NaOH(s) \xrightarrow[H_2O]{} Na^+(aq) + OH^-(aq)$$

c. KBr

$$KBr(s) \xrightarrow[H_2O]{} K^+(aq) + Br^-(aq)$$

d. $Fe(NO_3)_3$: Note that 3 NO_3^- ions form.

$$Fe(NO_3)_3(s) \xrightarrow[H_2O]{} Fe^{3+}(aq) + 3NO_3^-(aq)$$

7.17 How many equivalents of K^+ are present in a solution that contains 4.25 mol of K^+?

The number of equivalents present per mole of an ion equals the charge on that ion. The conversion factor for K^+ is:

$$\frac{1\,\text{Eq}\,K^+}{1\,\text{mol}\,K^+}$$

$$4.25\,\text{mol}\,K^+ \times \frac{1\,\text{Eq}\,K^+}{1\,\text{mol}\,K^+} = 4.25\,\text{Eq}\,K^+$$

7.19 A sodium chloride intravenous (IV) fluid contains 154 mEq of Na^+ per liter of solution. If a patient receives exactly 500 mL of the IV solution, how many moles of Na^+ were given?

The units on the numbers can be used as conversion factors to solve the problem, including some conversion factors given in previous chapters. One way to solve the problem is:

1.) The concentration can be written in the form of a conversion factor:

$$\frac{154\,\text{mEq}\,Na^+}{1\,\text{L}}$$

2.) mEq can be converted to Eq using the conversion factor

$$\frac{1\,\text{Eq}\,Na^+}{1000\,\text{mEq}\,Na^+}$$

3.) The number of equivalents present per mole of an ion equals the charge on that ion. The conversion factor for Na^+ is:

$$\frac{1\,\text{mol}\,Na^+}{1\,\text{Eq}\,Na^+}$$

4.) The concentration and the volume of the solution can be used to calculate the number of moles of solute. Remember to convert L to mL.

$$\frac{1\,\text{L}}{1000\,\text{mL}}$$

These factors are used to obtain the final answer as shown below and each is indicated by the boxed number.

$$\underset{\boxed{1}}{\frac{154\,\text{mEq}\,Na^+}{1\,\text{L}}} \times \underset{\boxed{2}}{\frac{1\,\text{Eq}\,Na^+}{1000\,\text{mEq}\,Na^+}} \times \underset{\boxed{3}}{\frac{1\,\text{mol}\,Na^+}{1\,\text{Eq}\,Na^+}} \times \underset{\boxed{4}}{\frac{1\,\text{L}}{1000\,\text{mL}}} \times 500\,\text{mL} = 0.077\,\text{mol}\,Na^+ \approx 0.08\,\text{mol}\,Na^+$$

7.21 What is the concentration in mmol/L of a Ca^{2+} solution that is 2.50 mEq/L?

The number of equivalents (or milliequivalents) present per mole (or millimole) of an ion equals the charge on that ion. Since the charge is +2, the conversion factor for Ca^{2+} is:

$$\frac{1\,\text{mmol}\,Ca^{2+}}{2\,\text{mEq}\,Ca^{2+}}$$

$$\frac{2.50\ \text{mEq}\,Ca^{2+}}{1\,\text{L}} \times \frac{1\ \text{mmol}\,Ca^{2+}}{2\ \text{mEq}\,Ca^{2+}} = \frac{1.25\ \text{mmol}\,Ca^{2+}}{1\,\text{L}} = 1.25\ \text{mmol}\,Ca^{2+}/\text{L}$$

7.23 What is the molarity of 250 mL of solution containing 4.20 mol of $ZnCl_2$?

1.) Molarity is defined as moles of solute per liter of solution, which can be written as:

$$M = \frac{\text{mole solute}}{\text{L solution}}$$

Molarity can then be calculated by dividing the number of moles of solute by the volume of the solution in liters.

2.) Here 250 mL must be converted to L.

$$\frac{1\ \text{L}}{1000\ \text{mL}}$$

These factors are used to obtain the final answer as shown below and each is indicated by the boxed number.

$$\underset{\boxed{1}}{\frac{4.20\ \text{mol}\ ZnCl_2}{250\ \text{mL}}} \times \underset{\boxed{2}}{\frac{1000\ \text{mL}}{1\ \text{L}}} = 17\ \text{M}$$

7.25 What is the molarity of 1.0 L of solution containing 25 g of NaCl?

1.) Molarity is defined as moles of solute per liter of solution, which can be written as:

$$M = \frac{\text{mole solute}}{\text{L solution}}$$

Molarity can then be calculated by dividing the number of moles of solute by the volume of the solution in liters.

2.) Mass must be converted to moles using the molar mass of NaCl, which can be written as:

$$\frac{1\ \text{mol}}{58.44\,\text{g}}$$

These factors are used to obtain the final answer as shown below and each is indicated by the boxed number.

$$\frac{25\ \text{g NaCl}}{1\ \text{L}} \times \frac{1\ \text{mol NaCl}}{58.44\ \text{g NaCl}} = 0.43\ \text{M}$$

$\boxed{1}$ $\boxed{2}$

7.27 How many grams of KBr are present in 300.0 mL of a 1.00 M solution?

1.) The concentration and the volume of the solution can be used to calculate the number of moles of solute. Remember to convert mL to L.

$$\frac{1\ \text{L}}{1000\ \text{mL}}$$

2.) The number of moles can be converted to a mass using the molar mass of KBr which can be written as:

$$\frac{119.0\ \text{g KBr}}{1\ \text{mol KBr}}$$

These factors are used to obtain the final answer as shown below, and each is indicated by the boxed number.

$$300\ \text{ml} \times \frac{1\ \text{L}}{1000\ \text{mL}} \times \frac{1.00\ \text{mol KBr}}{1\ \text{L}} \times \frac{119.0\ \text{g KBr}}{1\ \text{mol KBr}} = 35.7\ \text{g KBr}$$

$\boxed{1}$ $\boxed{2}$

7.29 What is the concentration in % (m/v) of a solution containing 0.30 g of glucose and 60 mL of solution?

% (m/v) is defined as:

$$\%(\text{m/v}) = \frac{\text{g solute}}{\text{mL solution}} \times 100\%$$

% (m/v) can then be calculated by dividing the number of grams of solute by the volume of the solution in milliliters.

$$\frac{0.30\ \text{g glucose}}{60\ \text{mL}} \times 100\% = 0.5\%$$

7.31 Calculate the mass/volume percent, % (m/v), for the solute in each of the following solutions:

a. 2.50g of KCl in 50.0 mL of solution

b. 7.5 g of casein in 120 mL of low-fat milk

% (m/v) is defined as:

$$\%(\text{m/v}) = \frac{\text{g solute}}{\text{mL solution}} \times 100\%$$

% (m/v) can then be calculated by dividing the number of grams of solute by the volume of the solution in milliliters.

a. $\dfrac{2.50 \text{ g KCl}}{50.0 \text{ mL}} \times 100\% = 5.00\,\%$

b. $\dfrac{7.5 \text{ g glucose}}{120 \text{ mL}} \times 100\% = 6.3\%$

7.33 How many grams of insulin are present in 26.0 g of a 0.450 % (m/m) solution?

To solve this problem, the definition of this concentration unit must be remembered. A 0.450% (m/m) solution means that there are 0.450 g of glucose per 100 g of solution, or

$$\frac{0.450 \text{ g insulin}}{100 \text{ g solution}}$$

A simple ratio can be set up and the number of grams in 26.0 g of solution determined:

$$\frac{0.450 \text{ g insulin}}{100 \text{ g solution}} = \frac{?\text{ g insulin}}{26.0 \text{ g solution}}$$

and ? g insulin can be calculated.

$$\frac{0.450 \text{ g insulin}}{100 \text{ g solution}} \times 26.0 \text{ g solution} = 0.117 \text{ g insulin}$$

7.35 What is the concentration in ppm of a solution containing 0.30 mg of fluoride and 60 mL of tap water? What is the concentration in ppb?

To solve this problem, the definition of this concentration unit must be remembered. A ppm solution is defined as:

$$\frac{\text{g solute}}{\text{mL solution}} \times (1\times10^6)$$

Convert mg to g and apply the definition.

$$\frac{0.30 \text{ mg fluoride}}{60 \text{ mL solution}} \times \frac{1 \text{ g}}{1000 \text{ mg}} \times (1\times10^6) = 5 \text{ ppm fluoride}$$

To solve this problem, the definition of this concentration unit must be remembered. A ppb solution is defined as:

$$\frac{\text{g solute}}{\text{mL solution}} \times (1\times10^9)$$

Convert mg to g and apply the definition.

$$\frac{0.30 \text{ mg fluoride}}{60 \text{ mL solution}} \times \frac{1 \text{ g}}{1000 \text{ mg}} \times (1\times10^9) = 5000 \text{ ppb fluoride}$$

7.37 How many liters of a 0.90% (m/v) NaCl solution can be prepared from 600 mL of a 9.0% (m/v) stock solution?

Establish the three quantities given and solve for the fourth unknown quantity.
$C_{initial}$ = 9.0% (m/v)
$V_{initial}$ = 600 mL
C_{final} = 0.90% (m/v)
V_{final} = ?

$$C_{initial} \times V_{initial} = C_{final} \times V_{final}$$

$$\frac{C_{initial} \times V_{initial}}{C_{final}} = V_{final}$$

$$\frac{9.0\% \times 600 \text{ mL}}{0.90\%} = 6000 \text{ mL}$$

Convert mL to L

$$6000 \text{ mL} \times \frac{1 \text{ L}}{1000 \text{ mL}} = 6 \text{ L}$$

7.39 What would the concentration of the resulting glucose solution be if 250 mL of a 12% (m/v) glucose solution were diluted to a final volume of 1 L with distilled water?

Establish the three quantities given and solve for the fourth unknown quantity.
$C_{initial}$ = 12% (m/v)
$V_{initial}$ = 250 mL
V_{final} = 1 L or 1000 mL
C_{final} = ?

$$C_{initial} \times V_{initial} = C_{final} \times V_{final}$$

$$\frac{C_{initial} \times V_{initial}}{V_{final}} = C_{final}$$

$$\frac{12\% \times 250 \text{ mL}}{1000 \text{ mL}} = 3\% \text{ (m/v)}$$

7.41 How would you prepare 250 mL of a 0.225% (m/v) NaCl solution from a 0.90% (m/v) NaCl?

Establish the three quantities given and solve for the fourth unknown quantity.
$C_{initial}$ = 0.90% (m/v)
C_{final} = 0.225% (m/v)
V_{final} = 250 mL
$V_{initial}$ = ?

$$C_{initial} \times V_{initial} = C_{final} \times V_{final}$$

$$\frac{C_{final} \times V_{final}}{C_{initial}} = V_{initial}$$

$$\frac{0.225\% \times 250 \text{ mL}}{0.90\%} = 62.5 \text{ mL} \approx 63 \text{ mL}$$

Add 62.5 mL of the 0 .90% NaCl stock solution to enough distilled water for a total volume of 250 mL solution.

7.43 A cucumber placed in a briny salt water solution makes a pickle.
a. Does water leave or enter the cucumber's cells?
b. Does the cucumber swell or crenate?
c. Is the salt water solution hypertonic or hypotonic to the cucumber?

When the concentration of salt outside the cell is not the same as inside, water molecules will travel across the cell membrane to equalize concentrations.

a. The salt concentration is higher outside the cell (hypertonic) so water will leave.
b. The cucumber cells will crenate (shrink).
c. The salt water is hypertonic to the cucumber.

7.45 A patient is undergoing hemodialysis. As the blood leaves the patient, its solute concentrations are ___________ (hypertonic, hypotonic, isotonic) to the dialyzing solution. As the blood re-enters the patient, its solute concentrations are __________(hypertonic, hypotonic, isotonic) to the dialyzing solution.

Hemodialysis is used to "cleanse the blood." In this process, blood is removed from the patient and passes through one side of a semipermeable membrane in contact on the opposite side with a dialyzing solution that is isotonic with normal blood solute concentrations. Urea and small waste molecules are present in greater concentration in the blood (hypertonic) and diffuse out of the passing blood and into the dialyzing solution, and the dialyzed blood which is now isotonic returns to the patient.

hypertonic
isotonic

7.47 Are the following solutions isotonic, hypotonic, or hypertonic with physiological solutions?
a. 10% (m/v) NaCl **b.** 0.90% (m/v) glucose
c. 5% (m/v) NaCl **d.** 5% (m/v) glucose

Common isotonic IV solutions used in hospitals include 0.90% (m/v) NaCl (saline) and a 5% (m/v) D-glucose (dextrose) solution commonly referred to as D5W (Dextrose 5% in Water). These solutions are called physiological solutions because they exert the same osmotic pressure as the cells and are isotonic with cells.

a. 10% (m/v) NaCl: $10\% > 0.90\%$, hypertonic
b. 0.90% (m/v) glucose: $0.90\% < 5\%$, hypotonic
c. 5% (m/v) NaCl: $5\% > 0.90\%$, hypertonic
d. 5% (m/v) glucose: $5\% = 5\%$, isotonic

7.49 Identify the type of transport (passive diffusion, facilitated transport, or active transport) that will occur for the following molecules:

a. oxygen
b. glucose, no energy required
c. Na^+, energy required
d. K^+, no energy required

Review the definitions of each type of transport:

active transport – The transport of ions or small molecules across the cell membrane in a direction opposite to diffusion. This process involves an integral membrane protein and requires energy.

facilitated transport – The diffusion of ions or small polar molecules across the cell membrane with the assistance of a protein; does not require energy.

passive diffusion – Diffusion of solutes across the cell membrane to equalize their concentration; does not require energy.

a. oxygen: small molecule moving to equalize concentrations, passive diffusion
b. glucose, no energy required: polar molecule across a membrane without energy expenditure, facilitated transport using glucose transporter proteins
c. Na^+, energy required: ion and requires energy, active transport
d. K^+, no energy required: ion across a membrane without energy expenditure, facilitated transport

Additional Problems

7.51 Identify the following as homogeneous or heterogeneous mixtures:

a. whipped cream **b.** wine **c.** coffee with cream and sugar

See Section 1.3 for a review of homogeneous (uniform composition throughout) and heterogeneous (composition varies throughout) mixtures.

a. whipped cream: homogeneous
b. wine: homogeneous
c. coffee with cream and sugar: homogeneous

7.53 Identify the mixtures in 7.51 as a solution, colloid, or suspension.

A colloid, like a solution, is a homogenous mixture. The difference lies in the size of the particles that each contains. By definition, the particles in a colloid must be between 1 nanometer (1×10^{-9} meter) and 1000 nanometers in diameter. Particles of this size remain suspended in solution, and so a colloid does not separate over time. In suspensions, the particles are larger and will separate with time.

a. whipped cream: colloid
b. wine: solution
c. coffee with cream and sugar: colloid (cream portion may separate with time)

7.55 Does the solubility of the solute increase or decrease in each of the following situations:
a. Sugar is dissolved in iced tea instead of hot tea.
b. A bottle of soda (solute is CO_2 gas) is placed in the refrigerator instead of in a pantry at room temperature.
c. An opened bottle of champagne (solute is CO_2 gas) is allowed to sit at the same temperature open to the air for hours.

The solubility of most solids dissolved in water *increases* with temperature, while the solubility of a gas dissolved in water *decreases* with a rise in temperature.

Henry's law states that the solubility of a gas in a liquid is directly related to the pressure of that gas over the liquid. That is, the amount of a gas that can dissolve in a liquid increases as the pressure of the gas in the space above the liquid increases.

a. If sugar is dissolved in iced tea instead of hot tea. Decrease: Decreasing temperature decreases the solubility of solids.
b. If a bottle of soda (solute is CO_2 gas) is placed in the refrigerator instead of in a pantry at room temperature. Increase: Decreasing temperature (refrigeration) increases the solubility of a gas.
c. If an opened bottle of champagne (solute is CO_2 gas) is allowed to sit at the same temperature open to the air for hours. Decrease: the open container decreases the pressure of the gas (CO_2) and decreases solubility.

7.57 Predict whether the following will fully dissociate, partially dissociate, or not dissociate when dissolved in solution:
a. sodium iodide, NaI (s), a strong electrolyte
b. formic acid, HCOOH (*l*), a weak electrolyte
c. glucose, $C_6H_{12}O_6$ (s), a nonelectrolyte

Strong electrolytes are ionic compounds that completely dissociate in water, whereas weak electrolytes only partially dissociate into ions. Nonelectrolytes are substances (usually covalent compounds) that dissolve in water but do not dissociate and therefore do not conduct electrical current.

a. sodium iodide, NaI (s), a strong electrolyte: fully dissociate
b. formic acid, HCOOH (*l*), a weak electrolyte: partially dissociate
c. glucose, $C_6H_{12}O_6$ (s), a nonelectrolyte: does not dissociate

7.59 Provide a balanced equation for the hydration of each compound in problem 7.57.
a. sodium iodide, NaI (s), a strong electrolyte
b. formic acid, HCOOH (*l*), a weak electrolyte
c. glucose, $C_6H_{12}O_6$ (s), a nonelectrolyte

In writing the equation for a hydration reaction (or any reaction):
1. The number of atoms of each element must be the same on each side of the equation. To balance atoms, coefficients are used in front of the formula.
2. All hydration reactions show water under the reaction arrow.

3. The physical state of each component is shown with an abbreviation for the state enclosed in parentheses following the formula. (s, l, g, aq)
4. The equation for the hydration of a strong electrolyte shows the formula for the solute on the reactant side and the ions formed on the product side separated by a right arrow.
5. The equation for the hydration of a nonelectrolyte shows the formula for the solute on the reactant and product side of the equation separated by a right arrow.
4. The equation for the hydration of a weak electrolyte shows the formula for the solute on the reactant side and the ions formed on the product side separated by an equilibrium arrow.

a. sodium iodide, NaI (s), a strong electrolyte

$$NaI(s) \xrightarrow[H_2O]{} Na^+(aq) + I^-(aq)$$

b. formic acid, HCOOH (*l*), a weak electrolyte

$$HCOOH(l) \underset{H_2O}{\rightleftharpoons} HCOO^-(aq) + H^+(aq)$$

c. glucose, $C_6H_{12}O_6$ (s), a nonelectrolyte

$$C_6H_{12}O_6(s) \xrightarrow[H_2O]{} C_6H_{12}O_6(aq)$$

7.61 How many equivalents of Ca^{2+} are present in a solution that contains 1.34 mol of Ca^{2+}?

The number of equivalents present per mole of an ion equals the charge on that ion. The conversion factor for Ca^{2+} for this problem is:

$$\frac{2\ \text{Eq}\ Ca^{2+}}{1\ \text{mol}\ Ca^{2+}}$$

$$1.34\ \text{mol}\ K^+ \times \frac{2\ \text{Eq}\ Ca^{2+}}{1\ \text{mol}\ Ca^{2+}} = 2.68\ \text{Eq}\ Ca^{2+}$$

7.63 A Ringer's solution for intravenous fluid replacement contains 4 mEq Ca^{2+} per liter of solution. If a patient receives 1750 mL of Ringer's solution, how many milliequivalents of Ca^{2+} were given?

1.) The concentration can be written in the form of a conversion factor:

$$\frac{4\ \text{mEq}\ Ca^{2+}}{1\ \text{L}}$$

2.) The concentration and the volume of the solution can be used to calculate the number of moles of solute. Remember to convert mL to L.

$$\frac{1\ \text{L}}{1000\ \text{mL}}$$

These factors are used to obtain the final answer as shown below and each is indicated by the boxed number.

$$\underset{\boxed{1}}{\frac{4\ \text{mEq}\ Na^+}{1\ \text{L}}} \times 1750\ \text{mL} \times \underset{\boxed{2}}{\frac{1\ \text{L}}{1000\ \text{mL}}} = 7\ \text{mEq}\ Ca^{2+}$$

7.65 What is the concentration in mmol/L of a SO_4^{2-} solution that is 25.0 mEq/L?

1.) The concentration can be written in the form of a conversion factor:

$$\frac{25.0\,\text{mEq}\,SO_4^{2-}}{1\text{ L}}$$

2.) The number of equivalents present per mole of an ion equals the charge on that ion. The conversion factor for Ca^{2+} for this problem is:

$$\frac{1\text{ mmol }SO_4^{2-}}{2\text{ mEq }SO_4^{2-}}$$

These factors are used to obtain the final answer as shown below and each is indicated by the boxed number.

$$\underset{\boxed{1}}{\frac{25.0\text{ mEq }Na^{+}}{1\text{ L}}} \times \underset{\boxed{2}}{\frac{1\text{ mmol }SO_4^{2-}}{2\text{ mEq }SO_4^{2-}}} = \frac{12.5\text{ mmol }SO_4^{2-}}{1\text{ L}}$$

7.67 What is the molarity of 1.50 L of solution, containing 2.80 mol KCl?

Molarity is defined as moles of solute per liter of solution which can be written as:

$$M = \frac{\text{mole solute}}{\text{L solution}}$$

Molarity can then be calculated by dividing the number of moles of solute by the volume of the solution in liters.

$$\frac{2.80\text{ mol KCl}}{1.50\text{ L}} = 1.87\text{ M}$$

7.69 What is the molarity of 0.50 L of solution containing 40.0 g of $CaCO_3$?

1.) Molarity is defined as moles of solute per liter of solution, which can be written as:

$$M = \frac{\text{mole solute}}{\text{L solution}}$$

Molarity can then be calculated by dividing the number of moles of solute by the volume of the solution in liters.

2.) The mass can be converted to number of moles using the molar mass of $CaCO_3$ which can be written as:

$$\frac{100.09\text{ g }CaCO_3}{1\text{ mol }CaCO_3}$$

$$\frac{40.0 \text{ g CaCO}_3}{0.50 \text{ L}} \times \frac{1 \text{ mol CaCO}_3}{100.09 \text{ g CaCO}_3} = 0.80 \text{ mol/L CaCO}_3$$

$\boxed{1}$ $\boxed{2}$

7.71 A 750 mL bottle of wine contains 12% (v/v) ethanol. How many milliliters of ethanol are in the bottle of wine?

To solve this problem, the definition of this concentration unit must be remembered. A 12% (v/v) solution means that there are 12 mL of ethanol per 100 mL of solution, or

$$\frac{12 \text{ mL ethanol}}{100 \text{ mL solution}}$$

A simple ratio can be set up and the number of mL of ethanol in 750 mL of solution can be determined:

$$\frac{12 \text{ mL ethanol}}{100 \text{ mL solution}} = \frac{? \text{ mL ethanol}}{750 \text{ mL solution}}$$

and ? mL ethanol can be calculated.

$$\frac{12 \text{ mL ethanol}}{100 \text{ mL solution}} \times 750 \text{ mL solution} = 90 \text{ mL ethanol}$$

7.73 How many grams of dextrose must be combined with water to prepare 500 mL of a 5.0% (m/v) dextrose maintenance solution? (Note: In the following calculation it will be assumed that the volume of the solution (500 mL) will be measured in a piece of glassware in which the volume is known to three significant figures.)

To solve this problem, the definition of this concentration unit must be remembered. A 5.0% (m/v) solution means that there are 5.0 g of dextrose per 100 mL of solution, or

$$\frac{5.0 \text{ g dextrose}}{100 \text{ mL solution}}$$

A simple ratio can be set up and the number of grams of dextrose in 500 mL of solution can be determined:

$$\frac{5.0 \text{ g dextrose}}{100 \text{ mL solution}} = \frac{? \text{ g dextrose}}{500 \text{ mL solution}}$$

and ? g dextrose can be calculated.

$$\frac{5.0 \text{ g dextrose}}{100 \text{ mL solution}} \times 500 \text{ mL solution} = 25 \text{ g dextrose}$$

7.75 The normal level of urea nitrogen in adults is 7–18 mg/dL. What is this value in ppm?

To solve this problem, the definition of this concentration unit must be remembered. A ppm solution is defined as:

$$\frac{\text{g solute}}{\text{mL solution}} \times (1\times10^6) = \frac{\text{1 mg solute}}{\text{1 L solution}}$$

One way to solve the problem is to convert dL to L, mg to g, and apply the definition.

For the lower limit (7 mg/dl):

$$\frac{\text{7 mg urea nitrogen}}{\text{1 dL}} \times \frac{\text{1 g}}{\text{1000 mg}} \times \frac{\text{1 dL}}{\text{100 mL}} \times (1\times10^6) = 70 \text{ ppm urea nitrogen}$$

For the upper limit (18 mg/dl):

$$\frac{\text{18 mg urea nitrogen}}{\text{1 dL}} \times \frac{\text{1 g}}{\text{1000 mg}} \times \frac{\text{1 dL}}{\text{100 mL}} \times (1\times10^6) = 180 \text{ ppm urea nitrogen}$$

Range: 70–180 ppm urea nitrogen

7.77 Express the following numbers as percents.
a. ¼ **b.** 3/8 **c.** 66/100 **d.** 0.58 **e.** 0.36 **f.** 0.125

A fraction can be converted to a percent by dividing the numerator by the denominator and multiplying by 100 and adding a percent sign. It is assumed in the following that the numerator and denominator values are exact and the answer will be rounded to two significant figures for simplicity.

a. ¼

$$\frac{1}{4} = 0.25\times100 = 25\%$$

b. 3/8

$$\frac{3}{8} = 0.375\times100 = 37.5\%$$

c. 66/100

$$\frac{66}{100} = 0.66\times100 = 66\%$$

A decimal number can be converted to a percent by multiplying by 100 and adding a percent sign.

d. 0.58

$0.58\times100 = 58\%$

e. 0.36
$0.36 \times 100 = 36\%$

f. 0.125
$0.125 \times 100 = 12.5\%$

7.79 Express the following numbers as a fraction
a. 20% **b.** 75% **c.** 40% **d.** 12%

A percent is defined as parts per one hundred parts or expressed as a fraction:

$$\frac{\text{parts}}{100\ \text{parts}}$$

a. 20%
$\frac{20}{100} = \frac{1}{5}$ (obtained by dividing each part of the fraction by 20)

b. 75%
$\frac{75}{100} = \frac{3}{4}$ (obtained by dividing each part of the fraction by 25)

c. 40%
$\frac{40}{100} = \frac{2}{5}$ (obtained by dividing each part of the fraction by 20)

d. 12%
$\frac{12}{100} = \frac{3}{25}$ (obtained by dividing each part of the fraction by 4)

7.81 Determine the percentage from two numbers given
a. 50 is what percent of 125?
b. 6 is what percent of 600?
c. What percent of 300 is 15?
d. What percent of 400 is 30?

A percent is defined as parts per one hundred parts or expressed as a fraction:

$$\frac{\text{parts}}{100\ \text{parts}}$$

The fraction can be converted to a percent by dividing the numerator by the denominator and multiplying by 100 and adding a percent sign. It is assumed in the following that the numerator and denominator values are exact and the answer will be rounded to one significant figure for simplicity.

a. 50 is what percent of 125?
$\frac{50}{125} = 0.40 \times 100 = 40\%$

b. 6 is what percent of 600?

$$\frac{6}{600} = 0.010 \times 100 = 1\%$$

c. What percent of 300 is 15?

$$\frac{15}{300} = 0.050 \times 100 = 5\%$$

d. What percent of 400 is 30?

$$\frac{30}{400} = 0.075 \times 100 = 7.5\%$$

7.83 Calculate the final concentration of each of the following diluted solutions:
a. 2.0 L of a 3.0 M HNO_3 solution is added to water so that the final volume is 6.0 L.
b. Water is added to 0.50 L of a 6.0 M KOH solution to make 4.0 L of a diluted KOH solution.

a. 2.0 L of a 3.0 M HNO_3 solution is added to water so that the final volume is 6.0 L.

Establish the three quantities given and solve for the fourth unknown quantity.
$C_{initial} = 3.0$ M
$V_{initial} = 2.0$ L
$V_{final} = 6.0$ L
$C_{final} = ?$

$$C_{initial} \times V_{initial} = C_{final} \times V_{final}$$

$$\frac{C_{initial} \times V_{initial}}{V_{final}} = C_{final}$$

$$\frac{3.0 \text{ M} \times 2.0 \text{ L}}{6.0 \text{ L}} = 1.0 \text{ M } HNO_3$$

b. Water is added to 0.50 L of a 6.0 M KOH solution to make 4.0 L of a diluted KOH solution.

Establish the three quantities given and solve for the fourth unknown quantity.
$C_{initial} = 6.0$ M
$V_{initial} = 0.50$ L
$V_{final} = 4.0$ L
$C_{final} = ?$

$$C_{initial} \times V_{initial} = C_{final} \times V_{final}$$

$$\frac{C_{initial} \times V_{initial}}{V_{final}} = C_{final}$$

$$\frac{6.0 \text{ M} \times 0.50 \text{ L}}{4.0 \text{ L}} = 0.75 \text{ M KOH}$$

7.85 What is the final volume in liters that can be prepared from each of the following concentrated solutions?

a. A 1.00 M HCl solution prepared from 150.0 mL of a 6.0 M HCl solution.

b. A 4.00% (m/v) $NaHCO_3$ solution prepared from 250 mL of a 20.0% (m/v) $NaHCO_3$.

a. A 1.00 M HCl solution prepared from 150.0 mL of a 6.0 M HCl solution.

Establish the three quantities given and solve for the fourth unknown quantity.

$C_{initial} = 6.0$ M

$V_{initial} = 150.0$ mL

$C_{final} = 1.00$ M

$V_{final} = ?$

$$C_{initial} \times V_{initial} = C_{final} \times V_{final}$$

$$\frac{C_{initial} \times V_{initial}}{C_{final}} = V_{final}$$

$$\frac{6.0\ \text{M} \times 150\ \text{mL}}{1.00\ \text{M}} = 900\ \text{mL}$$

Convert mL to L

$$900\ \text{mL} \times \frac{1\ \text{L}}{1000\ \text{mL}} = 0.90\ \text{L}$$

b. A 4.00% (m/v) $NaHCO_3$ solution prepared from 250 mL of a 20.0% (m/v) $NaHCO_3$.

Establish the three quantities given and solve for the fourth unknown quantity.

$C_{initial} = 20.0\%$ (m/v)

$V_{initial} = 250$ mL

$C_{final} = 4.00\%$ (m/v)

$V_{final} = ?$

$$C_{initial} \times V_{initial} = C_{final} \times V_{final}$$

$$\frac{C_{initial} \times V_{initial}}{C_{final}} = V_{final}$$

$$\frac{20.0\% \times 250\ \text{mL}}{4.00\%} = 1250\ \text{mL}$$

Convert mL to L

$$1250\ \text{mL} \times \frac{1\ \text{L}}{1000\ \text{mL}} = 1.25\ \text{L} \approx 1.3\ \text{L}$$

7.87 Consider a cell placed in solution shown below.

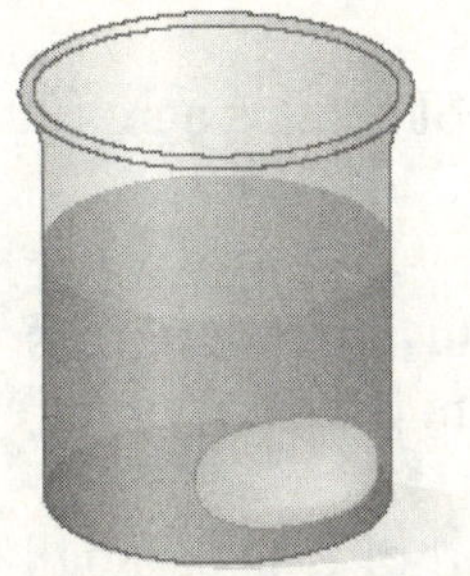

If the inside of the cell has a concentration equivalent to 0.9% (m/v) NaCl, is the cell in a hypertonic, hypotonic, or isotonic solution if the solution outside of the cell (O) is the following:
a. 5% NaCl
b. 0.09% NaCl
c. 0.9% NaCl
Classification of solutions:
hypertonic solution – A solution outside of a cell having a higher concentration of solutes than the solution inside the cell.
hypotonic solution – A solution outside of a cell having a lower concentration of solutes than the solution inside the cell.
isotonic solution – A solution outside of a cell having the identical concentration of solutes as the solution on the inside of a cell.

a. 5% NaCl: $5\% > 0.9\%$, hypertonic
b. 0.09% NaCl: $0.09\% < 0.9\%$, hypotonic
c. 0.9% NaCl: $0.9\% = 0.9\%$, isotonic

7.89 Edema, more commonly referred to as water retention, is characterized by a swelling of the tissues. Individuals with kidney disease will not excrete normal amounts of Na^+ leading to higher levels of Na^+ in the tissues. In terms of osmosis, explain how high levels of Na^+ in the tissues can lead to edema.

Osmosis is water moving from a solution with lower concentration to a solution with higher concentration, balancing out the concentrations. The cells will attempt to dilute the higher than normal concentration of Na^+ in the tissues by moving more water into the tissues, causing fluid retention.

7.91 A process called reverse osmosis purifies tap water by pushing water through a set of semipermeable membrane filters. Explain why this process is called *reverse* osmosis.

If osmosis is water moving from a solution with lower concentration to a solution with higher concentration, balancing out the concentrations, then reverse osmosis would require water moving in the opposite direction. Less pure water (higher concentration of solutes) is forced (requires energy) through a filter (semipermeable membrane) removing more impurities, making the water even purer for drinking.

7.93 Do the following processes require energy when transporting molecules across the cell membrane?
a. passive diffusion
b. facilitated transport
c. active transport

Review the definitions of each type of transport:

passive diffusion – Diffusion of solutes across the cell membrane to equalize their concentration; does not require energy.

facilitated transport – The diffusion of ions or small polar molecules across the cell membrane with the assistance of a protein; does not require energy.

active transport – The transport of ions or small molecules across the cell membrane in a direction opposite to diffusion. This process involves an integral membrane protein and requires energy.

a. passive diffusion: does not require energy
b. facilitated transport: does not require energy
c. active transport: does require energy

Challenge Problems

7.95 A scuba diver diving 100 m down in the ocean experiences twenty times as much pressure on their body than a person swimming at sea level experiences from the atmosphere. The high pressure of the air in a scuba diver's air tank affects the solubility of gases in the bloodstream. A condition called "the bends," which is related to nitrogen gas solubility, can occur in a scuba diver who ascends too quickly from a deep dive to the surface. Can you provide an explanation for this condition?

The explanation lies in the drastic decrease in pressure that occurs when the diver comes back to the surface. According to Henry's law, more nitrogen would dissolve in the blood at lower depths (higher pressure) than at the surface (lower pressure). When a diver ascends to the surface quickly, the pressure of the air on the lungs decreases rapidly. As a result, the nitrogen in the blood can no longer remain dissolved (Henry's law). Nitrogen gas bubbles begin forming in the bloodstream, causing the condition known as "the bends."

7.97 Two containers of equal volume are separated by a membrane that allows free passage of water, but completely restricts passage of solute molecules. Solution A has 3 molecules of the protein albumin (molecular weight 66,000) and Solution B contains 15 molecules of glucose (molecular weight 180). Into which compartment will water flow, or will there be no net movement of water? Which solution is at the higher concentration? Which chamber is exerting the higher osmotic pressure?

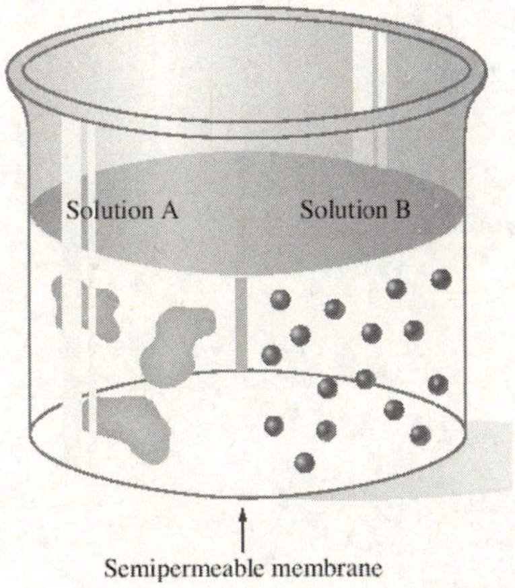

Water will move from solution A to solution B because solution B is a higher concentrated solution (more solute/solution volume). It does not matter that the albumin particles are huge, osmotic flow depends on the concentration, not the size of the particles.

Chapter

8

Acids, Bases, and Buffers in the Body

8.1 Acids and Bases–Definitions

Learning Objectives

Upon completion of this material, a student should be able to do the following:

A. define the following key terms:

acid	hydronium ion	base

B. describe the general characteristics of acids and bases

C. recognize that a proton and a hydrogen ion represent the same thing

D. realize the relationship between a hydrogen ion and a hydronium ion in aqueous solution

E. identify an acid and a base using the Arrhenius and Brønsted–Lowry definitions

8.2 Strong Acids and Bases

Learning Objectives

Upon completion of this material, a student should be able to do the following:

A. define the following key terms:

strong acid	weak acid	strong base
weak base	exothermic reaction	salt
neutralization		

B. recognize the common strong acids (See Table 8.1.) and strong bases

C. relate strong acids and weak acids to strong electrolytes and weak electrolytes, respectively

D. write a balanced neutralization reaction

E. identify a reaction as exothermic given information about the heat involved

8.3 Chemical Equilibrium

Learning Objectives

Upon completion of this material, a student should be able to do the following:

A. define the following key terms:

chemical equilibrium	equilibrium constant	Le Châtelier's principle
endothermic reaction		

B. describe what is meant by a reversible reaction

C. understand the concept of equilibrium as applied to a chemical reaction

D. write an equilibrium expression for K

E. predict the relative amounts of reactants and products in an equilibrium reaction from its K

F. apply LeChâtelier's principle to predict the effect of changes on equilibrium equations including

1. changes in temperature
2. changes in the amount of reactants and products

8.4 Weak Acids and Bases

Learning Objectives

Upon completion of this material, a student should be able to do the following:

A. define the following key terms:

acid dissociation constant	conjugate base	conjugate acid
conjugate acid-base pair	lone pair	

B. write an equilibrium expression for K_a
C. predict the strength of an acid from its K_a
D. predict the formula for the conjugate base of a weak acid and the conjugate acid of a weak base
E. write balanced conjugate acid-base reactions
F. identify conjugate acid-base pairs in a given reaction

8.5 pH and the pH Scale

Learning Objectives

Upon completion of this material, a student should be able to do the following:

A. define the following key terms:

autoionization of water	acidic	basic
pH		

B. describe the relationship between $[H_3O^+]$ and $[OH^-]$ in neutral, acidic, and basic solutions.
C. determine the acidity of a solution from the pH scale
E. appreciate a logarithmic (log) scale of measurement
F. use the pH equation to calculate the pH of a solution from its hydronium ion concentration and vice versa

8.6 pK_a

Learning Objectives

Upon completion of this material, a student should be able to do the following:

A. define the following key terms:

pK_a

B. predict the strength of a weak acid from its pK_a
C. describe the relationship between pH and pK_a
D. use pK_as to establish the predominant species present at a given pH

8.7 Amino Acids: Common Biological Weak Acids

Learning Objectives

Upon completion of this material, a student should be able to do the following:

A. define the following key terms:

amino acids	zwitterion	isoelectric point

B. describe the structure of amino acids
C. write the zwitterion forms of amino acids
D. apply the concepts of acid-base conjugate pairs to amino acids
E. describe the relationships among zwitterions, pI and pH

F. establish the predominant species present in aqueous solution at a given pH for the functional groups amine and carboxylic acid found in amino acids.

8.8 Buffers: An Important Property of Weak Acids and Bases

Learning Objectives

Upon completion of this material, a student should be able to do the following:

A. define the following key terms:

buffer	homeostasis	hypoventilation
respiratory acidosis	hyperventilation	respiratory alkalosis
metabolic acidosis	metabolic alkalosis	

B. describe the H_2CO_3/HCO_3^- buffer system in the blood
C. apply Le Châtelier's principle to the bicarbonate buffer system.
D. predict the effect of changes in ventilation rate on blood pH
E. distinguish between acidosis and alkalosis
F. distinguish among respiratory acidosis and alkalosis and metabolic acidosis and alkalosis

Practice Test for Chapter 8

1. Which of the following describes a base?
   A. slippery feel
   B. sour taste
   C. corrodes metal
   D. donates a proton

2. An acid can be described as producing which of the following?
   A. proton
   B. H^+ ion
   C. H_3O^+ ion
   D. all of these

3. For the following reaction:
   $$NH_3(aq) + H_3O^+(aq) \rightarrow H_2O(l) + NH_4^+(aq)$$
   which of the following is a correct statement?
   A. NH_3 is the acid and H_3O^+ is the base.
   B. NH_3 is the base and H_3O^+ is the acid.
   C. H_2O and H_3O^+ are both acids.
   D. NH_3 and NH_4^+ are both bases.

4. In which of the following reactions does water act as a proton donor?
   A. $HBr(aq) + H_2O(l) \rightarrow H_3O^+(aq) + Br^-(aq)$

   B. $H_2O(l) + CN^-(aq) \rightarrow HCN(aq) + OH^-(aq)$

   C. $HCl(g) + H_2O(l) \rightarrow H_3O^+(aq) + Cl^-(aq)$

   D. none of the above

5. Which of the following would be classified as a strong acid?
   A. HF
   B. HNO_2
   C. HCl
   D. H_3PO_4
   E. All of the above are strong acids.

6. Which of the following compounds will completely ionize in water?
   A. NH_3
   B. H_2CO_3
   C. CH_3COOH
   D. KOH

7. When the equation is balanced, the products of the following reaction are:

$$H_2SO_4 + Ca(OH)_2 \rightarrow$$

A. $CaSO_4 + 2\ H_2O$
B. $CaSO_4 + H_2O$
C. $SO_4(OH)_2 + CaH_2$
D. $2SO_4OH + 2CaH$

8. What is the equilibrium constant expression for the reaction given below?

$$CaCO_3(s) + 2\ HI(aq) \rightleftharpoons CaI_2(aq) + CO_2(g) + H_2O(l)$$

A. $K = \dfrac{[CaI_2][CO_2][H_2O]}{[CaCO_3][HI]}$

B. $K = \dfrac{[CaI_2][CO_2][H_2O]}{[CaCO_3][HI]^2}$

C. $K = \dfrac{[CaI_2][CO_2]}{[HI]}$

D. $K = \dfrac{[CaI_2][CO_2]}{[HI]^2}$

9. At a certain temperature, *K* for the following reaction is 54.

$$H_2(g) + I_2(g) \rightleftharpoons 2HI(g)$$

This means at equilibrium
A. the products are present in greater amounts than the reactants.
B. the reactants are present in greater amounts than the products.
C. the products and reactants are present in equal amounts.
D. the amounts of reactants and products will vary depending on the starting amount of reactant.

10. Consider the following reaction.

$$2\ SO_2(g) + O_2(g) \rightleftharpoons 2\ SO_3(g) + \text{heat}$$

Which of the following will increase the amount of SO_3?
A. increasing the temperature
B. removing O_2
C. adding SO_2
D. All increase the amount of SO_3.

11. Which of the following changes will increase the amount of product at equilibrium for an endothermic reaction?

A. remove the product as it forms
B. increase the concentration of a reactant
C. increase the temperature of the reaction
D. A and B
E. All of the above will increase the amount of product.

12. Rank the following acids in order of increasing strength, weakest to strongest.

HCN HCl HNO_2 NH_4^+

TABLE 8.5 K_a VALUES FOR SUBSTANCES ACTING AS WEAK ACIDS (25 °C)

Name	Formula	K_a
Hydrogen sulfate ion	HSO_4^-	1.0×10^{-2}
Phosphoric acid	H_3PO_4	7.5×10^{-3}
Hydrofluoric acid	HF	6.5×10^{-4}
Nitrous acid	HNO_2	4.5×10^{-4}
Formic acid	$HCOOH$	1.8×10^{-4}
Acetic acid	CH_3COOH	1.75×10^{-5}
Carbonic acid	H_2CO_3	4.5×10^{-7}
Dihydrogen phosphate ion	$H_2PO_4^-$	6.6×10^{-8}
Ammonium ion	NH_4^+	6.3×10^{-10}
Hydrocyanic acid	HCN	6.2×10^{-10}
Bicarbonate ion	HCO_3^-	4.8×10^{-11}
Hydrogen phosphate ion	HPO_4^{2-}	1×10^{-12}

A. $HCN < HNO_2 < NH_4^+ < HCl$
B. $HCN < NH_4^+ < HNO_2 < HCl$
C. $HCl < HNO_2 < NH_4^+ < HCN$
D. $HCl \approx HNO_2 < NH_4^+ < HCN$

13. What is the conjugate base of HCN in the following reaction?

$$H_2O(l) + CN^-(aq) \rightleftharpoons HCN(aq) + OH^-(aq)$$

A. H_2CN^+
B. OH^-
C. H_2O
D. CN^-

14. Which of the following is true of HCO_3^-?
A. CO_3^{2-} is the conjugate base.
B. H_2CO_3 is the conjugate acid.
C. H_2O is the conjugate base.
D. H_3O^+ is the conjugate acid.
E. A and B

15. The products of the following reaction would be:

$$F^-(aq) + H_2O(l) \rightleftharpoons$$

A. $HF(aq) + H_3O^+(aq)$
B. $H_2F^+(aq) + OH^-(aq)$
C. $HF(aq) + OH^-(aq)$
D. $H_2F^+(aq) + H_3O^+(aq)$

16. What is the log of 1.75?
    A. 0.243
    B. 0.75
    C. 56.2
    D. 0.560

17. A sample of seawater has a pH of 8.23. This solution is:
    A. basic.
    B. acidic.
    C. neutral.

18. A grapefruit has a $[H_3O^+]$ of 3.35×10^{-3}. This solution is:
    A. basic.
    B. acidic.
    C. neutral.

19. What is the pH of solution of coffee with $[H_3O^+]$ of 6.3×10^{-5}?
    A. 9.67
    B. −4.20
    C. 1.30
    D. 4.20

20. What is the $[H_3O^+]$ of a baking soda solution with a pH of 9.22?
    A. 2.2×10^{-9}
    B. 9.65×10^{-1}
    C. 6.03×10^{-10}
    D. 1.01×10^{4}

21. As the pK_a of an acid increases, the strength of the acid:
    A. increases.
    B. decreases.
    C. remains constant.

22. The pK_a of HF is 3.19. At a pH of 5.25,
    A. $[HF] = [F^-]$
    B. $[HF] > [F^-]$
    C. $[HF] < [F^-]$
    D. Relative concentrations cannot be determined from the given information.

23. The pK_a of HCN is 9.12. If $[HCN] = [CN^-]$, the pH must be:
    A. 9.12.
    B. > 9.12.
    C. < 9.12.
    D. between 7.00 and 9.12.

24. Which of the following has the strongest acid of the pair listed on the left?

TABLE 8.7 pK_a AND K_a VALUES FOR SUBSTANCES ACTING AS WEAK ACIDS (25 °C)

Name	Formula	pK_a	K_a
Hydrogen sulfate ion	HSO_4^-	2.00	1.0×10^{-2}
Phosphoric acid	H_3PO_4	2.12	7.5×10^{-3}
Hydrofluoric acid	HF	3.19	6.5×10^{-4}
Nitrous acid	HNO_2	3.35	4.5×10^{-4}
Formic acid	$HCOOH$	3.74	1.8×10^{-4}
Acetic acid	CH_3COOH	4.76	1.75×10^{-5}
Carbonic acid	H_2CO_3	6.35	4.5×10^{-7}
Dihydrogen phosphate ion	$H_2PO_4^-$	7.18	6.6×10^{-8}
Ammonium ion	NH_4^+	9.20	6.3×10^{-10}
Hydrocyanic acid	HCN	9.21	6.2×10^{-10}
Bicarbonate ion	HCO_3^-	10.32	4.8×10^{-11}
Hydrogen phosphate ion	HPO_4^{2-}	12.00	1×10^{-12}

A. HCO_3^- and H_2CO_3
B. HPO_4^{2-} and $H_2PO_4^-$
C. $H_2PO_4^-$ and H_3PO_4
D. All pairs have the stronger acid listed on the right.

25. Using the pK_a values in the table below, predict the products of the following reaction.

$$HCO_3^-(aq) + H_2PO_4^{2-}(aq) \rightleftharpoons$$

TABLE 8.7 pK_a AND K_a VALUES FOR SUBSTANCES ACTING AS WEAK ACIDS (25 °C)

Name	Formula	pK_a	K_a
Hydrogen sulfate ion	HSO_4^-	2.00	1.0×10^{-2}
Phosphoric acid	H_3PO_4	2.12	7.5×10^{-3}
Hydrofluoric acid	HF	3.19	6.5×10^{-4}
Nitrous acid	HNO_2	3.35	4.5×10^{-4}
Formic acid	$HCOOH$	3.74	1.8×10^{-4}
Acetic acid	CH_3COOH	4.76	1.75×10^{-5}
Carbonic acid	H_2CO_3	6.35	4.5×10^{-7}
Dihydrogen phosphate ion	$H_2PO_4^-$	7.18	6.6×10^{-8}
Ammonium ion	NH_4^+	9.20	6.3×10^{-10}
Hydrocyanic acid	HCN	9.21	6.2×10^{-10}
Bicarbonate ion	HCO_3^-	10.32	4.8×10^{-11}
Hydrogen phosphate ion	HPO_4^{2-}	12.00	1×10^{-12}

A. $CO_3^{2-}(aq) + H_3PO_4(aq)$
B. $H_2CO_3(aq) + HPO_4^{2-}(aq)$
C. $H_2CO_3(aq) + PO_4^{3-}(aq)$
D. The products cannot be predicted.

26. For an amino acid, the pK_a for the carboxylic acid is 4.50 and pK_a for the protonated amine is 11.00. What the pI for this amino acid?
A. 4.50
B. 11.00
C. 15.50
D. 6.50
E. 7.75

27. Which of the following represents the zwitterion of the amino acid valine?
A.

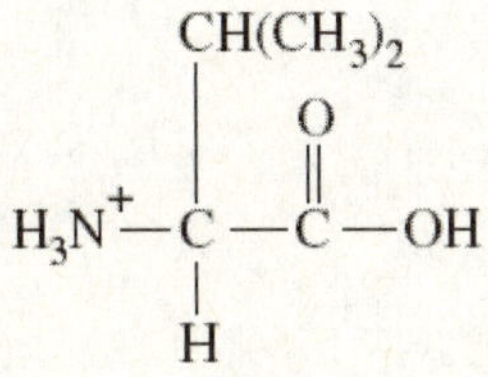

B.

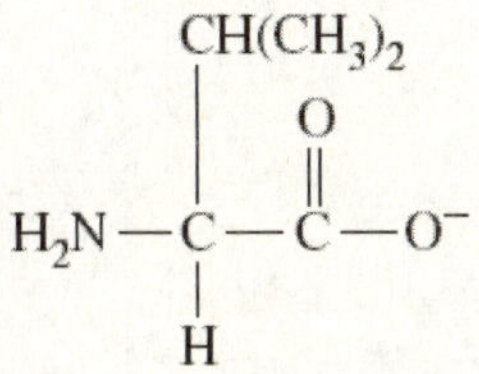

C.

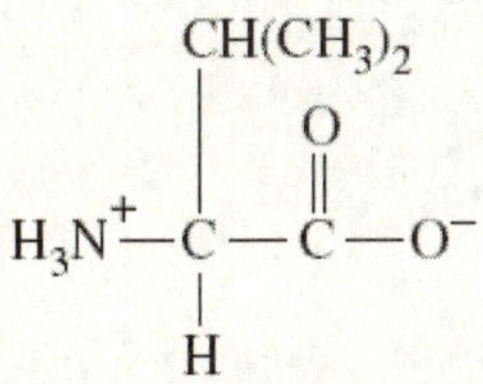

D. None of these represents the zwitterion.

28. Under what conditions would the following reaction occur?

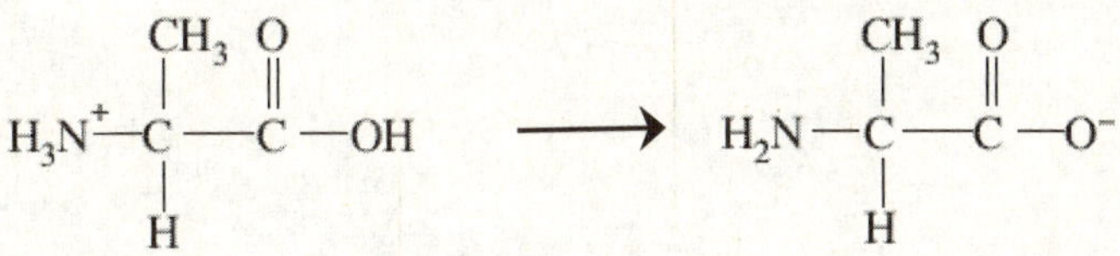

A. add an acid
B. decrease the pH
C. increase the hydronium ion concentration
D. add a proton acceptor

29. Hyperventilation:
A. is also known as metabolic acidosis.
B. results in decreased amounts of CO_2 in the blood.
C. increases the pH of the blood.
D. upsets the bicarbonate balance of the blood.
E. All of the above are characteristic of hyperventilation.

30. The blood buffer system can be represented as:

$$CO_2(g) + H_2O(l) \rightleftharpoons H_3O^+(aq) + HCO_3^-(aq)$$

Vomiting can cause a loss of the hydronium ion from blood. This causes:
A. metabolic alkalosis.
B. a shift in the reaction toward the formation of more CO_2.
C. a decrease in blood pH.
D. an imbalance which can be treated with bicarbonate solution.
E. all of the above

Answers

1. A	**2.** D	**3.** B	**4.** B	**5.** C	**6.** D	**7.** A	**8.** D	**9.** A	**10.** C
11. E	**12.** B	**13.** D	**14.** E	**15.** C	**16.** A	**17.** A	**18.** B	**19.**D	**20.** C
21. B	**22.** C	**23.** A	**24.** D	**25.** B	**26.** E	**27.** C	**28.** D	**29.** D	**30.** A

Chapter 8 – Solutions to Odd-Numbered Problems

Practice Problems

8.1 Indicate whether each of the following statements is characteristic of an acid or a base:

a. has a sour taste **b.** accepts a proton
c. produces H^+ ions in water **d.** is named potassium hydroxide

Review the general properties of acids and bases in Section 8.1.
a. has a sour taste: acid (bases are bitter)
b. accepts a proton: base (acids donate protons)
c. produces H^+ ions in water: acid (bases accept H^+ ions in water)
d. is named potassium hydroxide: base (metal hydroxides are generally basic)

8.3 In your own words, explain how a proton and the hydrogen ion represent the same thing.

Most hydrogen atoms contain one proton and one electron (no neutrons). Therefore, a positively charged hydrogen ion (H^+) contains just one proton, being the same as a proton.

8.5 In each of the following equations, identify the acid (proton donor) and base (proton acceptor) for the reactants:

a. $HI(aq) + H_2O(l) \rightarrow H_3O^+(aq) + I^-(aq)$
b. $F^-(aq) + H_2O(l) \rightarrow HF(aq) + OH^-(aq)$

Compare the reactant to its product form. If it has one less H it is an acid, one more H it is a base.
a. acid: $HI(aq)$
base: $H_2O(l)$
b. acid: $H_2O(l)$
base: $F^-(aq)$

8.7 Which of the following are strong acids?

a. H_2SO_4 **b.** HCl **c.** HF **d.** HNO_3 **e.** H_3PO_4

Refer to Table 8.1

TABLE 8.1
SIX COMMON STRONG ACIDS

Acid Name	Formula
Perchloric acid	$HClO_4$
Sulfuric acid*	H_2SO_4
Hydroiodic acid	HI
Hydrobromic acid	HBr
Hydrochloric acid	HCl
Nitric acid	HNO_3

a. H_2SO_4: strong
b. HCl: strong
c. HF: weak
d. HNO_3: strong
e. H_3PO_4: weak

8.9 Complete and balance the following neutralization reactions:
a. $HNO_3(aq)$ + $LiOH(s) \rightarrow$
b. $H_2SO_4(aq)$ + $Ca(OH)_2(s) \rightarrow$

When completing a neutralization reaction:
<u>STEP 1 – Form the Products</u>
The products will always be (a) a salt and (b) H_2O. The salt produced must be a neutral ionic compound (refer to Section 3.3 for a refresher on forming ionic compounds like salts).

a. $HNO_3(aq) + LiOH(s) \rightarrow LiNO_3(aq) + H_2O(l)$
b. $H_2SO_4(aq) + Ca(OH)_2(s) \rightarrow CaSO_4(aq) + H_2O(l)$

<u>STEP 2 – Balance the Chemical Equation</u>
The same number of each kind of atom must appear in both the reactants and products. After forming the salt, inspect the reaction to see that it is balanced. Balance the reaction by adding coefficients in front of product or reactant compounds where appropriate.

a. $HNO_3(aq) + LiOH(s) \rightarrow LiNO_3(aq) + H_2O(l)$ Note: already balanced
b. $H_2SO_4(aq) + Ca(OH)_2(s) \rightarrow CaSO_4(aq) + 2\ H_2O(l)$

8.11 Complete and balance the following neutralization reactions:
a. $HBr(aq)$ + $Al(OH)_3(s) \rightarrow$
b. $HI(aq)$ + $CaCO_3(s) \rightarrow$

When completing a neutralization reaction:

<u>STEP 1 – Form the Products</u>
The products will always be (a) a salt and (b) H_2O. The salt produced must be a neutral ionic compound (refer to Section 3.3 for a refresher on forming ionic compounds like salts).

a. $HBr(aq) + Al(OH)_3(s) \rightarrow AlBr_3(aq) + H_2O(l)$

When carbonates are used to neutralize acid, the reaction produces a salt, water, *and* carbon dioxide gas.

b. $HI(aq) + CaCO_3(s) \rightarrow CaI_2(aq) + H_2O(l) + CO_2(g)$

STEP 2 – Balance the Chemical Equation
The same number of each kind of atom must appear in both the reactants and products. After forming the salt, inspect the reaction to see that it is balanced. Balance the reaction by adding coefficients in front of product or reactant compounds where appropriate.

a. $3HBr(aq) + Al(OH)_3(s) \rightarrow AlBr_3(aq) + 3H_2O(l)$
b. $2HI(aq) + CaCO_3(s) \rightarrow CaI_2(aq) + H_2O(l) + CO_2(g)$

8.13 What is meant by the term "reversible reaction"?

Reversible reactions are reactions that proceed in both the forward and reverse directions. Reactants can form products but the products can also reform the reactants.

8.15 Write an equilibrium constant expression for the following reactions:

a. $CO(g) + H_2O(g) \rightleftharpoons H_2(g) + CO_2(g)$

b. $CH_3COOH(aq) + H_2O(l) \rightleftharpoons H_3O^+(aq) + CH_3COO^-(aq)$

For an equilibrium reaction of the form:

$$aA + bB \rightleftharpoons cC + dD$$

The general equilibrium expression is given as

$$K = \frac{[C]^c[D]^d}{[A]^a[B]^b}$$

The brackets, [ ], mean "concentration of." Only substances whose concentrations change appear in an equilibrium expression, *K*. Solids (*s*) and pure liquids (*l*) (as solvents) have constant concentrations and so do not appear in the equilibrium expression.

a. $K = \frac{[CO_2][H_2]}{[CO][H_2O]}$

All coefficients are 1, so all exponents are 1. There were no solids or liquids in reaction, so all reactants and products appear in the expression.

b. $K = \frac{[CH_3COO^-][H_3O^+]}{[CH_3COOH]}$

All coefficients are 1, so all exponents are 1. The water was a pure liquid and does not appear in the expression. The remaining reactant and the products appear in the expression.

8.17 For the following values of *K*, indicate whether the products or reactants are present in larger amounts:

a. 1×10^{-5}
b. 156
c. 1

Use Table 8.3 which summarizes the significance of the value of *K*.

TABLE 8.3

INTERPRETING VALUES OF *K*

Value of *K*	Predominant Species at Equilibrium
K = 1	Equal amounts of products and reactants
K > 1	Products
K < 1	Reactants

a. 1×10^{-5}: K< 1, reactants
b. 156: K> 1, products
c. 1: K=1, equal amounts of reactants and products

8.19 Hydrogen chloride can be made by reacting hydrogen gas and chlorine gas.

$$H_2(g) + Cl_2(g) \rightleftharpoons 2\,HCl(g) + \text{heat}$$

What effect does each of the following changes have on equilibrium?

a. add H_2 **b.** add heat (raise temperature) **c.** remove HCl **d.** remove Cl_2

Use Le Châtelier's principle, which states if we disturb an equilibrium (chemists refer to this as applying stress to the equilibrium) the reaction will change the rates of the forward and reverse reaction to offset the stress and regain its equilibrium.

a. add H_2: Increases a reactant and favors the forward reaction.
b. add heat (raise temperature): The reaction is exothermic (heat is a product). Increasing temperature increases "heat" and favors the reverse reaction.
c. remove HCl: Decreases a product concentration and favors the forward reaction.
d. remove Cl_2: Decreases a reactant concentration and favors the reverse reaction.

8.21 In the lower atmosphere, oxygen is converted to ozone (O_3) by the energy provided from lightning.

$$3\,O_2(g) + \text{heat} \rightleftharpoons 2\,O_3(g)$$

What effect does each of the following changes have on equilibrium?

a. add O_2 **b.** add O_3 **c.** raise temperature **d.** lower temperature

Use Le Châtelier's principle, which states if we disturb an equilibrium (chemists refer to this as applying stress to the equilibrium) the reaction will change the rates of the forward and reverse reaction to offset the stress and regain its equilibrium.

a. add O_2: Increases a reactant and favors the forward reaction.
b. add O_3: Increases a product concentration and favors the reverse reaction.
c. raise temperature: The reaction is endothermic (heat is a reactant). Increasing temperature increases "heat" and favors the forward reaction.
d. lower temperature: The reaction is endothermic (heat is a reactant). Decreasing temperature decreases "heat" and favors the reverse reaction.

8.23 When you open a bottle of a soft drink, the drink eventually goes flat. How can LeChâtelier's principle explain this?

$$H^+(aq) + HCO_3^-(aq) \rightleftharpoons CO_2(g) + H_2O(l)$$

Le Châtelier's principle states if we disturb an equilibrium (chemists refer to this as applying stress to the equilibrium) the reaction will change the rates of the forward and reverse reaction to offset the stress and regain its equilibrium.

Opening the can allows the escape of CO_2 gas, which removes CO_2 from the reaction, thereby reducing the concentration of a product. The reaction shifts toward products causes even more CO_2 to be lost. Eventually all of the CO_2 is lost and the soda no longer bubbles.

8.25 Identify the stronger acid in each pair:

a. $H_2PO_4^-$ or HPO_4^{2-}

b. HF or HCOOH

c. HBr or HNO_2

The relative strength of an acid can be determined from the size of the K_a, the smaller the K_a the weaker the acid. Refer to Tables 8.1 and 8.5

TABLE 8.1

SIX COMMON STRONG ACIDS

Acid Name	Formula
Perchloric acid	$HClO_4$
Sulfuric acid*	H_2SO_4
Hydroiodic acid	HI
Hydrobromic acid	HBr
Hydrochloric acid	HCl
Nitric acid	HNO_3

TABLE 8.5 K_a VALUES FOR SUBSTANCES ACTING AS WEAK ACIDS (25 °C)

Name	Formula	K_a
Hydrogen sulfate ion	HSO_4^-	1.0×10^{-2}
Phosphoric acid	H_3PO_4	7.5×10^{-3}
Hydrofluoric acid	HF	6.5×10^{-4}
Nitrous acid	HNO_2	4.5×10^{-4}
Formic acid	HCOOH	1.8×10^{-4}
Acetic acid	CH_3COOH	1.75×10^{-5}
Carbonic acid	H_2CO_3	4.5×10^{-7}
Dihydrogen phosphate ion	$H_2PO_4^-$	6.6×10^{-8}
Ammonium ion	NH_4^+	6.3×10^{-10}
Hydrocyanic acid	HCN	6.2×10^{-10}
Bicarbonate ion	HCO_3^-	4.8×10^{-11}
Hydrogen phosphate ion	HPO_4^{2-}	1×10^{-12}

a. $H_2PO_4^-$ or HPO_4^{2-}: K_a $H_2PO_4^-$> K_a HPO_4^{2-}, $H_2PO_4^-$ is the stronger acid.

b. HF or HCOOH: K_a HF> K_a HCOOH, HF is the stronger acid.

c. HBr or HNO_2: K_a HBr (a strong acid)> K_a HNO_2, HBr is the stronger acid.

8.27 Identify the acid and base on the left side of the following equations and identify their conjugate species on the right side.

a. $HSO_4^-(aq) + H_2O(l) \rightleftharpoons H_3O^+(aq) + SO_4^{2-}(aq)$

b. $NH_4^+(aq) + H_2O(l) \rightleftharpoons H_3O^+(aq) + NH_3(aq)$

c. $HCN(aq) + NO_2^-(aq) \rightleftharpoons CN^-(aq) + HNO_2(aq)$

Molecules or ions related by the loss or gain of one H^+ are referred to as conjugate acid-base pairs. The conjugate acid has one more H^+ than the conjugate base or, vice versa, the conjugate base has one less H^+ than the conjugate acid.

a. acid HSO_4^-; conjugate base SO_4^{2-}
   base H_2O; conjugate acid H_3O^+
b. acid NH_4^+; conjugate base NH_3
   base H_2O; conjugate acid H_3O^+
c. acid HCN; conjugate base CN^-
   base NO_2^-; conjugate acid HNO_2

8.29 Write the formula and name of the conjugate base formed from each of the following acids:
a. HF **b.** H_2O **c.** H_2CO_3 **d.** HSO_4^-

The conjugate base has one less H than the conjugate acid due to the donation of an H^+ ion. Remove an H^+ from the acid and write the formula for the remaining species with a charge −1 more than that of the acid.

a. HF: Removing H^+ from neutral HF produces the fluoride ion, F^-.
b. H_2O: Removing H^+ from neutral H_2O produces the hydroxide ion, OH^-.
c. H_2CO_3: Removing H^+ from neutral H_2CO_3 produces the bicarbonate ion, HCO_3^-.
d. HSO_4^-: Removing H^+ from charged HSO_4^- produces the sulfate ion, SO_4^{2-}.

8.31 Write the formula and name of the conjugate acid formed from each of the following bases:
a. CO_3^{2-} **b.** H_2O **c.** $H_2PO_4^-$ **d.** Br^-

The conjugate acid has one more H than the conjugate base due to the acceptance of an H^+ ion. Add an H^+ to the base and write the formula for the species with a charge +1 more than that of the base.

a. CO_3^{2-}: Adding an H^+ to CO_3^{2-} with a 2− charge produces the bicarbonate ion, HCO_3^-.
b. H_2O: Adding an H^+ to neutral H_2O produces the hydronium ion, H_3O^+.
c. $H_2PO_4^-$: Adding an H^+ to $H_2PO_4^-$ with a 1− charge produces neutral phosphoric acid, H_3PO_4.
d. Br^-: Adding an H^+ to Br^- with a 1− charge produces neutral hydrobromic acid, HBr.

8.33 Complete the reactions below and label the conjugate acid and conjugate base.

a. $HA(aq) + H_2O(l) \rightleftharpoons$

Acid Base

b. $H_2PO_4^-(aq) + H_2O(l) \rightleftharpoons$

Acid Base

c. $NH_3(aq) + H_2O(l) \rightleftharpoons$

Base Acid

Molecules or ions related by the loss or gain of one H^+ are referred to as conjugate acid-base pairs. The conjugate acid has one more H^+ than the conjugate base or, vice versa, the conjugate base has one less H^+ than the conjugate acid.

To form the conjugate acid: Add an H^+ to the base and write the formula for the new species with a charge +1 more than that of the base.

To form the conjugate base: Remove an H^+ from the acid and write the formula for the new species with a charge −1 more than that of the acid.

a. $HA(aq) + H_2O(l) \rightleftharpoons A^-(aq) + H_3O^+(aq)$

Acid Base Conjugate base Conjugate acid

b. $H_2PO_4^-(aq) + H_2O(l) \rightleftharpoons HPO_4^{2-}(aq) + H_3O^+(aq)$

Acid Base Conjugate base Conjugate acid

c. $NH_3(aq) + H_2O(l) \rightleftharpoons NH_4^+(aq) + OH^-(aq)$

Base Acid Conjugate acid Conjugate base

8.35 State whether each of the following solutions is acidic, basic, or neutral:

a. blood, pH 7.38
b. vinegar, pH 2.8
c. drain cleaner, pH 11.2
d. coffee, pH 5.5

Refer to Table 8.6.

TABLE 8.6 RELATIONSHIP BETWEEN ACIDITY, pH, AND $[H_3O^+]$

Solution Acidity	pH	$[H_3O^+]$
Basic	>7	$< 1.0 \times 10^{-7}$ M
Neutral	$=7$	$= 1.0 \times 10^{-7}$ M
Acidic	<7	$> 1.0 \times 10^{-7}$ M

a. blood, pH 7.38: pH > 7, basic
b. vinegar, pH 2.8: pH < 7, acidic
c. drain cleaner, pH 11.2: pH > 7, basic
d. coffee, pH 5.5: pH < 7, acidic

8.37 State whether each of the following solutions is acidic, basic, or neutral:
a. $[H_3O^+] = 1.2 \times 10^{-8}$ M **b.** $[H_3O^+] = 7.0 \times 10^{-3}$ M
c. $[H_3O^+] = 4.7 \times 10^{-11}$ M **d.** $[H_3O^+] = 1.0 \times 10^{-7}$ M

Refer to Table 8.6.

TABLE 8.6 RELATIONSHIP BETWEEN ACIDITY, pH, AND $[H_3O^+]$

Solution Acidity	pH	$[H_3O^+]$
Basic	>7	$< 1.0 \times 10^{-7}$ M
Neutral	$=7$	$= 1.0 \times 10^{-7}$ M
Acidic	<7	$> 1.0 \times 10^{-7}$ M

a. $[H_3O^+] = 1.2 \times 10^{-8}$ M: $[H_3O^+] < 1.0 \times 10^{-7}$ M, basic
b. $[H_3O^+] = 7.0 \times 10^{-3}$ M: $[H_3O^+] > 1.0 \times 10^{-7}$ M, acidic
c. $[H_3O^+] = 4.7 \times 10^{-11}$ M: $[H_3O^+] < 1.0 \times 10^{-7}$ M, basic
d. $[H_3O^+] = 1.0 \times 10^{-7}$ M: $[H_3O^+] = 1.0 \times 10^{-7}$ M, neutral

8.39 Calculate the pH of each of the solutions in problem 8.37.
a. $[H_3O^+] = 1.2 \times 10^{-8}$ M **b.** $[H_3O^+] = 7.0 \times 10^{-3}$ M
c. $[H_3O^+] = 4.7 \times 10^{-11}$ M **d.** $[H_3O^+] = 1.0 \times 10^{-7}$ M

By definition:

$$pH = -\log [H_3O^+]$$

a. $[H_3O^+] = 1.2 \times 10^{-8}$ M: $pH = -\log [H_3O^+] = -\log (1.2 \times 10^{-8}) = -(-7.92) = 7.92$
b. $[H_3O^+] = 7.0 \times 10^{-3}$ M: $pH = -\log [H_3O^+] = -\log (7.0 \times 10^{-3}) = -(-2.15) = 2.15$
c. $[H_3O^+] = 4.7 \times 10^{-11}$ M: $pH = -\log [H_3O^+] = -\log (4.7 \times 10^{-11}) = -(-10.33) = 10.33$
d. $[H_3O^+] = 1.0 \times 10^{-7}$ M: $pH = -\log [H_3O^+] = -\log (1.0 \times 10^{-7}) = -(-7.00) = 7.00$

Note: In pH calculations with logarithms, the number of significant figures in the $[H_3O^+]$ will be the number of decimal places in the pH value.

8.41 Calculate the $[H_3O^+]$ for each of the measured pH values below.
a. drain cleaner, pH 12.10 **b.** coffee, pH 5.5
c. gastric juice, pH 2.00 **d.** toothpaste, pH 8.3

The equation relating pH and $[H_3O^+]$ is:

$$\text{INV} \log -pH = [H_3O^+]$$

or alternatively,

$$10^{-pH} = [H_3O^+]$$

a. drain cleaner, pH 12.10: $[H_3O^+]$ = INV log –12.10 = 7.9×10^{-13} M
$[H_3O^+] = 10^{-12.10} = 7.9 \times 10^{-13}$ M

b. coffee, pH 5.5: $[H_3O^+]$ = INV log –5.5= 3.2×10^{-6} M
$[H_3O^+] = 10^{-5.5} = 3 \times 10^{-6}$ M

c. gastric juice, pH 2.00: $[H_3O^+]$ = INV log –2.00 = 1.00×10^{-2} M
$[H_3O^+] = 10^{-2.00} = 1.00 \times 10^{-2}$ M

d. toothpaste, pH 8.3: $[H_3O^+]$ = INV log –8.3 = 5.0×10^{-9} M
$[H_3O^+] = 10^{-8.3} = 5 \times 10^{-9}$ M

Note: In pH calculations with logarithms, the number of decimal places in the pH will be the number of significant figures in the $[H_3O^+]$,

8.43 Using Table 8.7, determine the stronger acid from the pairs of acids below.
a. $H_2PO_4^-$ or HPO_4^{2-}
b. H_2SO_4 or acetic acid
c. formic acid or carbonic acid
d. ammonium ion or hydrocyanic acid

TABLE 8.7 pK_a AND K_a VALUES FOR SUBSTANCES ACTING AS WEAK ACIDS (25 °C)

Name	Formula	pK_a	K_a
Hydrogen sulfate ion	HSO_4^-	2.00	1.0×10^{-2}
Phosphoric acid	H_3PO_4	2.12	7.5×10^{-3}
Hydrofluoric acid	HF	3.19	6.5×10^{-4}
Nitrous acid	HNO_2	3.35	4.5×10^{-4}
Formic acid	HCOOH	3.74	1.8×10^{-4}
Acetic acid	CH_3COOH	4.76	1.75×10^{-5}
Carbonic acid	H_2CO_3	6.35	4.5×10^{-7}
Dihydrogen phosphate ion	$H_2PO_4^-$	7.18	6.6×10^{-8}
Ammonium ion	NH_4^+	9.20	6.3×10^{-10}
Hydrocyanic acid	HCN	9.21	6.2×10^{-10}
Bicarbonate ion	HCO_3^-	10.32	4.8×10^{-11}
Hydrogen phosphate ion	HPO_4^{2-}	12.00	1×10^{-12}

The smaller the K_a (or the larger the pK_a) the weaker the acid, or the larger the K_a (or the smaller the pK_a) the stronger the acid. Using either value will give the same answer.

a. $H_2PO_4^-$ or HPO_4^{2-}: K_a $H_2PO_4^-$ > K_a HPO_4^{2-}, $H_2PO_4^-$ is the stronger acid or
pK_a $H_2PO_4^-$ < pK_a HPO_4^{2-}, $H_2PO_4^-$ is the stronger acid.

b. H_2SO_4 or acetic acid: K_a H_2SO_4 (strong acid Table 8.1) > K_a acetic acid, H_2SO_4 is the stronger acid or
pK_a H_2SO_4 < pK_a acetic acid, H_2SO_4 is the stronger acid.

c. Formic acid or carbonic acid: K_a formic acid > K_a carbonic acid, formic acid is the stronger acid or
pK_a formic acid < pK_a carbonic acid, formic acid is the stronger acid.

d. ammonium ion or hydrocyanic acid: K_a ammonium ion > K_a hydrocyanic acid, ammonium ion is the stronger acid or
pK_a ammonium ion < pK_a hydrocyanic, ammonium ion is the stronger acid.

8.45 Consider the weak acid , benzoic acid, shown below, whose pK_a value is 4.2. Will the acid or the conjugate base predominate at the following pHs?

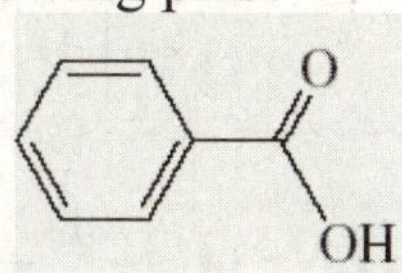

a. 3.0 **b.** 4.2 **c.** 11.3

Compare the pH and the pK_a, then use Table 8.8.

TABLE 8.8 RELATIONSHIP BETWEEN ACID (HA), CONJUGATE BASE (A^-), pH, AND pK_a

$pH < pK_a$	$pH = pK_a$	$pH > pK_a$
$[HA] > [A^-]$	$[HA] = [A^-]$	$[HA] < [A^-]$

a. 3.0: $pH < pK_a$ and $[HA] > [A^-]$, acid predominates
b. 4.2: $pH = pK_a$ and $[HA] = [A^-]$, neither form predominates, are present in equal concentrations
c. 11.3: $pH > pK_a$ and $[HA] < [A^-]$, conjugate base predominates

8.47 Valine has the zwitterion structure shown below. The pK_a and pI values are listed. Given this information, what form of the molecule will predominate at the indicated pH values?

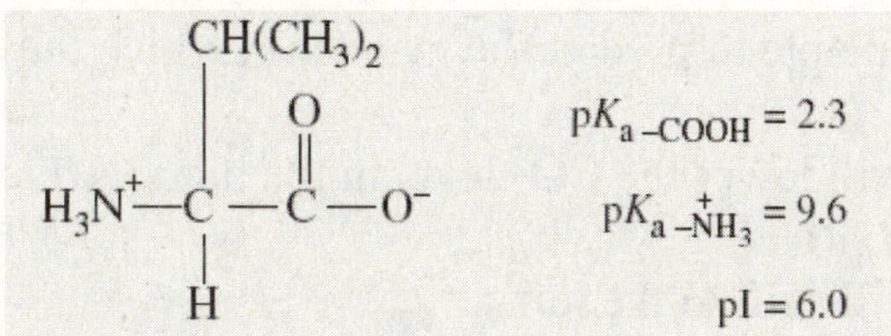

a. pH 1.0 **b.** pH 6.0 **c.** pH 11.0

At the pI, an amino acid exists in its zwitterion form. At any pH below the pK_a of the carboxylic acid, both the carboxyl group and amino group will be protonated. At any pH above the pK_a of the amino group, both the carboxyl group and the amino group will be deprotonated.

a. pH 1.0: pH below pK_a of the carboxyl group, both groups protonated.

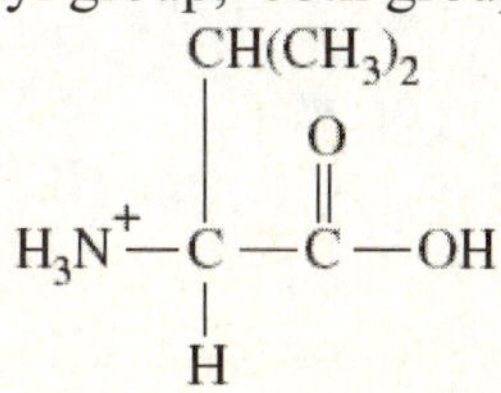

b. pH 6.0: pH equals pI, the carboxyl group deprotonated, amino group protonated (zwitterion).

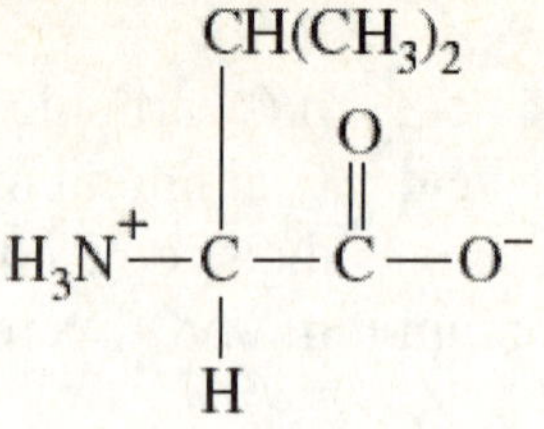

c. pH 11.0: pH above pK_a of the amino group, both groups deprotonated.

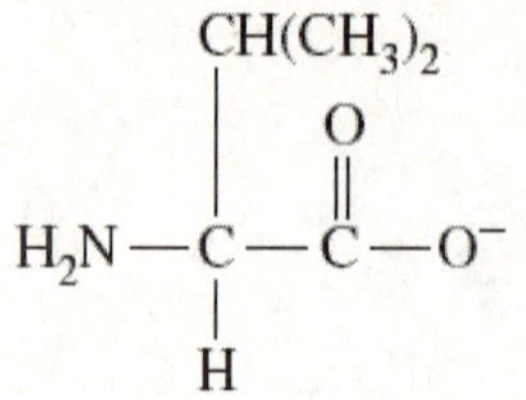

8.49 Sometimes during stress or trauma, a person can start to hyperventilate. The person may then be instructed to breathe into a paper bag to avoid fainting. The bicarbonate equilibrium below is affected.

$$H_2O(l) + CO_2(g) \rightleftharpoons H_3O^+(aq) + HCO_3^-(aq)$$

a. How does blood pH change during hyperventilation?
b. In which direction will the bicarbonate equilibrium shown above shift during hyperventilation?
c. What is this condition called?
d. How does breathing into a paper bag help return the blood pH to normal?
e. Which direction will the bicarbonate equilibrium shown above shift as the pH returns to normal?

Use LeChâtelier's principle to predict the manner in which the equilibrium shifts.

a. Hyperventilation will lower the CO_2 level in the blood, which decreases the H_3O^+ and increases the blood pH.
b. The equilibrium will shift to the left.
c. This condition is called respiratory alkalosis.
d. Breathing into a bag will increase the CO_2 level, which increases H_3O^+ , and lowers the blood pH.
e. The equilibrium will shift to the right.

Additional Problems

8.51 Identify the following acids as strong or weak:
a. H_2SO_4 **b.** H_2CO_3 **c.** HCl **d.** HF

Refer to Table 8.1.

TABLE 8.1

SIX COMMON STRONG ACIDS

Acid Name	Formula
Perchloric acid	$HClO_4$
Sulfuric acid*	H_2SO_4
Hydroiodic acid	HI
Hydrobromic acid	HBr
Hydrochloric acid	HCl
Nitric acid	HNO_3

a. H_2SO_4: strong
b. H_2CO_3: weak
c. HCl: strong
d. HF: weak

8.53 What are some similarities and differences between strong and weak acids?

Both strong and weak acids produce H_3O^+ in water. Weak acids are only slightly ionized, whereas a strong acid exists as completely as ions in solution. Per mole a strong acid produces more H_3O^+ in water than a weak acid.

8.55 For the reaction below:

$$PbI_2(s) \rightleftharpoons Pb^{2+}(aq) + 2\, I^-(aq)$$

a. Write an equilibrium constant expression.
b. If the value for $K = 7.1 \times 10^{-9}$, which substance predominates, the solid or the ions in solution?

For an equilibrium reaction of the form:

$$aA + bB \rightleftharpoons cC + dD$$

The general equilibrium expression is given as

$$K = \frac{[C]^c[D]^d}{[A]^a[B]^b}$$

The brackets, [ ], mean "concentration of." Only substances whose concentrations change appear in an equilibrium expression, *K*. Solids (*s*) and pure liquids (*l*) (as solvents) have constant concentrations and so do not appear in the equilibrium expression.

a. Write an equilibrium constant expression. Note PbI_2 is solid and is omitted from the expression.

$$K = [Pb^{2+}][I^-]^2$$

Use Table 8.3, which summarizes the significance of the value of *K*.

TABLE 8.3

INTERPRETING VALUES OF *K*

Value of *K*	Predominant Species at Equilibrium
K = 1	Equal amounts of products and reactants
K > 1	Products
K < 1	Reactants

b. If the value for $K = 7.1 \times 10^{-9}$, which substance predominates, the solid or the ions in solution?
Since $K < 1$, reactant (the solid) predominates at equilibrium.

8.57 Consider the reaction below:

$$CO(g) + 2\,H_2(g) \rightleftharpoons CH_3OH(g) + \text{heat}$$

a. Is the reaction above exothermic or endothermic as written?

Since heat is a product of the reaction, heat is released and the reaction is classified as exothermic.

b. Provide an equilibrium expression (K) for the reaction.

For an equilibrium reaction of the form:

$$aA + bB \rightleftharpoons cC + dD$$

The general equilibrium expression is given as

$$K = \frac{[C]^c[D]^d}{[A]^a[B]^b}$$

The brackets, [ ], mean "concentration of." Only substances whose concentrations change appear in an equilibrium expression, *K*. Solids (*s*) and pure liquids (*l*) (as solvents) have constant concentrations and so do not appear in the equilibrium expression.

$$K = \frac{[CH_3OH]}{[CO][H_2]^2}$$

Use Le Châtelier's principle, which states if we disturb an equilibrium (chemists refer to this as applying stress to the equilibrium) the reaction will change the rates of the forward and reverse reaction to offset the stress and regain its equilibrium.

c. How will the equilibrium above shift if hydrogen gas is added to the reaction?
Added H_2 increases a reactant concentration which favors the forward reaction.

d. How will the equilibrium above shift if carbon monoxide is removed from the reaction?
Removal of CO decreases the concentration of a reactant and favors the reverse reaction.

e. How will the equilibrium shift if the temperature is lowered?
The reaction is exothermic (heat is a product). Decreasing temperature decreases "heat" and favors the forward reaction.

8.59 Determine the pH for the following solutions:
a. $[H_3O^+] = 2.5 \times 10^{-8}$ M
b. $[H_3O^+] = 7.0 \times 10^{-2}$ M

By definition:

$$pH = -\log [H_3O^+]$$

a. $[H_3O^+] = 2.5 \times 10^{-8}$ M: $pH = -\log [H_3O^+] = -\log (2.4 \times 10^{-8}) = -(-7.61) = 7.61$
b. $[H_3O^+] = 7.0 \times 10^{-2}$ M: $pH = -\log [H_3O^+] = -\log (7.0 \times 10^{-2}) = -(-1.15) = 1.15$

Note: In pH calculations with logarithms, the number of significant figures in the $[H_3O^+]$ will be the number of decimal places in the pH value.

8.61 Calculate the H_3O^+ concentration for a solution with the following pH:
a. 3.52 **b.** 5.00 **c.** 9.25

The equation relating pH and $[H_3O^+]$ is:

$$\text{INV log -pH} = [H_3O^+]$$

or alternatively,

$$10^{-pH} = [H_3O^+]$$

a. 3.52: $H_3O^+] = \text{INV log} -3.52 = 3.0 \times 10^{-4}$ M
$[H_3O^+] = 10^{-3.52} = 3.0 \times 10^{-4}$ M
b. 5.00: $H_3O^+] = \text{INV log} -5.00 = 1.0 \times 10^{-5}$ M
$[H_3O^+] = 10^{-5.00} = 1.0 \times 10^{-5}$ M
c. 9.25: $H_3O^+] = \text{INV log} -9.25 = 5.6 \times 10^{-10}$ M
$[H_3O^+] = 10^{-9.25} = 5.6 \times 10^{-10}$ M

8.63 Consider the buffer system of nitrous acid, HNO_2, and its salt, $NaNO_2$.

$$HNO_2(aq) + H_2O(l) \rightleftharpoons H_3O^+ + NO_2^-(aq)$$

a. What is the purpose of the buffer system?
b. What is the purpose of $NaNO_2$ in the buffer?
c. How does the buffer react when some acid (H_3O^+) is added?
d. How does the buffer react when some base (OH^-) is added?

Addition of acid or base to a buffer system can be interpreted in terms of Le Châtelier's principle:

a. A buffer system keeps the pH relatively constant upon addition of an acid or base.
b. The conjugate base, NO_2^-, neutralizes any acid added by forming HNO_2 in the reverse reaction.
c. Added H_3O^+ reacts with NO_2^-, forming HNO_2 in the reverse reaction.
d. Added OH^- is neutralized by HNO_2, forming NO_2^- in the forward reaction.

8.65 In the blood plasma, pH is maintained by the carbonic acid-bicarbonate buffer equilibrium shown below:

$$H_2O(l) + CO_2(g) \rightleftharpoons H_3O^+(aq) + HCO_3^-(aq)$$

a. If excess acid is added to the bloodstream by a physiological condition other than breathing, what condition occurs?

b. Is the pH of the blood below or above normal?

c. How could such a condition be treated?

a. The condition is called metabolic acidosis. Respiratory acidosis is produced by a breathing condition.

b. Since this is an "acidosis" condition, the pH is below normal

c. The excess acid needs to be neutralized. The conjugate base that can accomplish this is bicarbonate ion, so administer bicarbonate (HCO_3^-).

8.67 Consider the amino acid valine shown in its zwitterion form below. The pI of valine is 6.0.

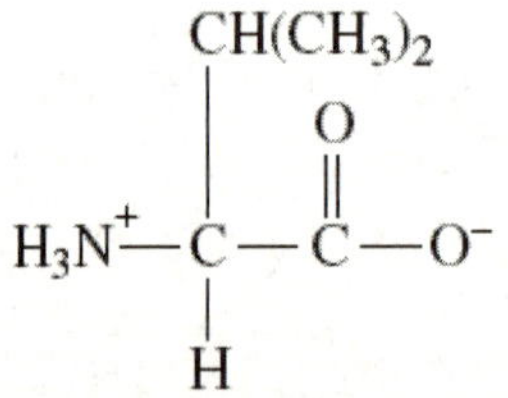

a. Draw the structure of the predominant form of valine at pH = 10.0

b. Draw the structure of the predominant form of valine at pH = 3.0

At the pI, an amino acid exists in it zwitterion form. At any pH below the pK_a of the carboxylic acid, both the carboxyl group and amino group will be protonated. At any pH above the pK_a of the amino group, both the carboxyl group and the amino group will be deprotonated.

a. Draw the structure of the predominant form of valine at pH = 10.0. Assuming that pH 10.0 is above the pK_a of the amino group, the amino acid will be fully deprotonated.

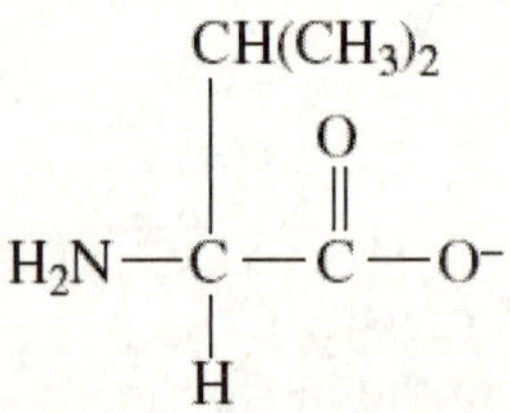

b. Draw the structure of the predominant form of valine at pH = 3.0. Assuming that pH 3.0 is below the pK_a of the carboxyl group, the amino acid will be fully protonated.

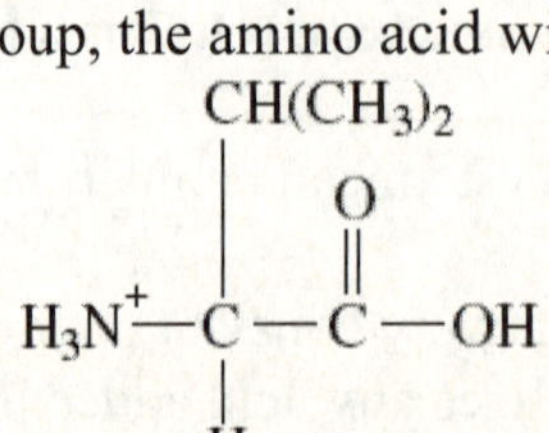

Challenge Problems

8.69 To determine the concentration of an unknown weak acid solution, a known amount of strong base can be titrated (added) to it until complete neutralization occurs (an endpoint is reached). If a 10.0 mL sample of vinegar (aqueous acetic acid, CH_3COOH) requires 16.5 mL of 0.500 M NaOH to reach the endpoint, what is the original molarity of the acetic acid solution? *Hint: 1 mole of NaOH will neutralize 1 mole of acetic acid.*

Since it is given that one mole of NaOH neutralizes (reacts with) one mole of acetic acid, M and volume can be used to calculate moles of NaOH added. Once moles of NaOH (and acetic acid) are known, the volume of acetic acid can be used to calculate its molarity.
\# moles of NaOH = # moles of acetic acid

$$\frac{0.500 \text{ moles NaOH}}{1 \cancel{\text{L}}} \times \frac{1 \cancel{\text{L}}}{1000 \cancel{\text{mL}}} \times 16.5 \cancel{\text{mL}} = 0.00825 \text{ moles NaOH}$$

$$\frac{0.00825 \text{ moles acetic acid}}{10.0 \cancel{\text{mL}}} \times \frac{1000 \cancel{\text{mL}}}{1\text{L}} = 0.825 \text{ M acetic acid}$$

Chapter

9

Proteins – Amides at Work

9.1 Amino Acids—A Second Look

Learning Objectives

Upon completion of this material, a student should be able to do the following:

A. define the following key terms:

alpha (α) carbon	alpha (α) amino group	alpha (α) carboxylate group
nonpolar amino acids	polar amino acids	

B. draw the general structure of an amino acid and recognize the common functional groups
C. identify the alpha (α) carbon in an amino acid
D. identify the chiral carbon atoms in amino acids
E. classify the 20 common amino acids by the polarity of their side chains (Table 9.1)
F. identify the organic functional groups present in 20 common amino acid side chains
G. give the one-letter and three-letter designation for the common amino acids (Table 9.1)

9.2 Protein Formation

Learning Objectives

Upon completion of this material, a student should be able to do the following:

A. define the following key terms:

amide	peptide bond	dipeptide
N-terminus	C-terminus	

A. provide the products of condensation of amino acids (formation of a peptide bond), or the reverse reaction (hydrolysis of a peptide bond)
B. identify and name the N-terminus amino acid in a peptide
C. identify and name the C-terminus amino acid in a peptide
D. identify the peptide bonds in a peptide
E. give the one-letter and three-letter designations for a peptide
F. draw the structural formula for a peptide based on the one-letter or three-letter name

9.3 The Three-Dimensional Structure of Proteins

Learning Objectives

Upon completion of this material, a student should be able to do the following:

A. define the following key terms:

primary structure	protein backbone	secondary structure
alpha (α) helix	beta (β) pleated sheet -	
tertiary structure	nonpolar interactions	hydrophobic effect
polar interactions	salt bridges (ionic attractions)	

disulfide bond globular protein fibrous protein
quaternary structure

B. describe the levels of protein structure through attractive forces present
C. realize that more than one primary structure can be formed by the bonding of individual amino acids
D. recognize and describe alpha helices and beta pleated sheets
E. describe the importance of hydrogen bonds in the secondary structure of proteins
F. relate the type of interactions in the tertiary structure of proteins to the specific types of amino acids present
G. describe the role of the various types of interactions (nonpolar interactions, polar interactions, salt bridges, and disulfide bonds) in determining the tertiary structure of proteins
H. apply the hydrophobic effect to the presence of specific amino acids
I. compare and contrast globular and fibrous proteins
J. relate the hydrophobic or hydrophilic nature of the component amino acids in a protein to their positions in the tertiary structure of proteins.

9.4 Denaturation of Proteins

Learning Objectives

Upon completion of this material, a student should be able to do the following:

A. define the following key terms:
denaturation
B. define protein denaturation and identify denaturing agents
C. give examples of the application of denaturation in medicine

9.5 Examples of Biologically Important Proteins

Learning Objectives

Upon completion of this material, a student should be able to do the following:

A. define the following key terms:
triple helix prosthetic group conformational change
antigen integral membrane protein
B. recognize the relationship between protein structure and function through the examination of sample proteins
C. describe the importance of the role of a prosthetic group in protein function
D. describe the importance and role of conformational changes in protein function
E. define the role of an antigen and antibody in the body's defense system
F. describe the role of integral membrane proteins in membrane transport
G. apply the hydrophilic nature and the hydrophobic nature of chemical substances to the functioning of proteins in the cell membrane

Practice Test for Chapter 9

1. The side chain in the following amino acid is a:

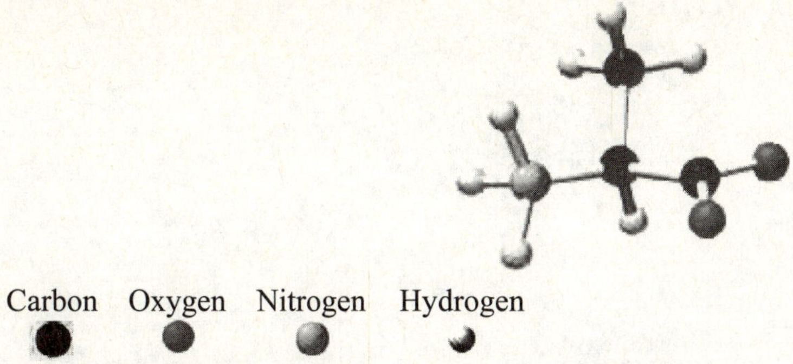

   A. hydrogen atom.
   B. methyl group.
   C. carboxylate group.
   D. amino group.
   E. none of the above

2. The following amino acid would be classified as:

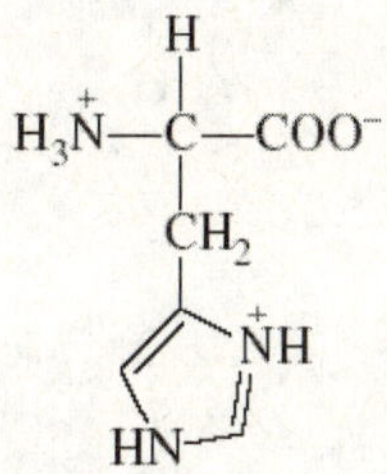

   A. nonpolar.
   B. polar (neutral).
   C. polar (acidic).
   D. polar (basic).

3. The amino acid shown below could be represented as:

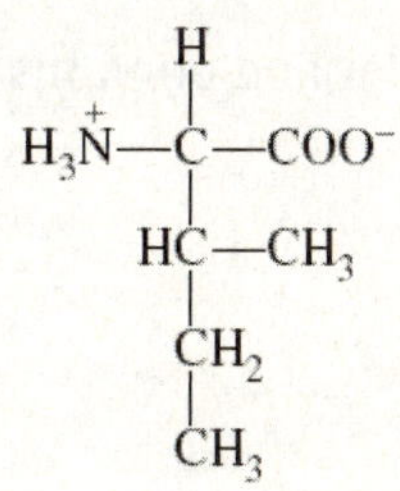

   A. isoleucine.
   B. Ile.
   C. I.
   D. all of the above

4. How many chiral carbon atoms are present in the following amino acid?

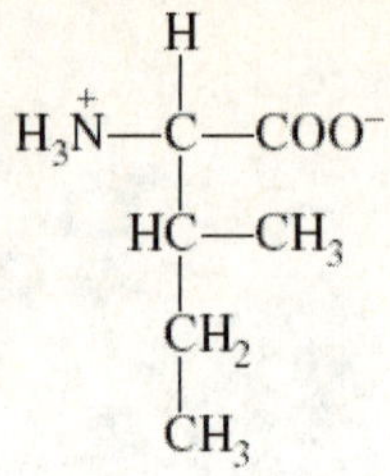

A. none
B. 1
C. 2
D. 3
E. 5

5. The side chain of the following amino acid would be:

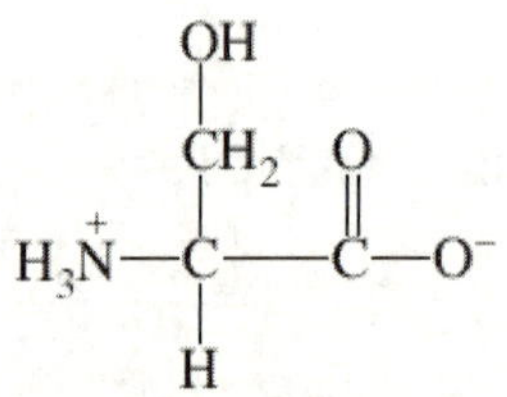

A. nonpolar and hydrophilic.
B. polar and hydrophilic.
C. nonpolar and hydrophobic.
D. polar and hydrophobic.
E. amphipathic.

6. Which amino acid contains a side chain that is hydrophilic and contains an aromatic ring?
A. phenylalanine
B. tryptophan
C. tyrosine
D. histidine

7. Which amino acid is classified as polar and contains an amide in the side chain?
A. R
B. Q
C. H
D. K

8. When amino acids react and form a peptide:
A. a condensation reaction occurs.
B. an amide bond forms.
C. the $-COO^-$ of one amino acid reacts with the $-NH_3^+$ of another amino acid.
D. both A and B
E. A, B, and C

9. Consider the following structure.

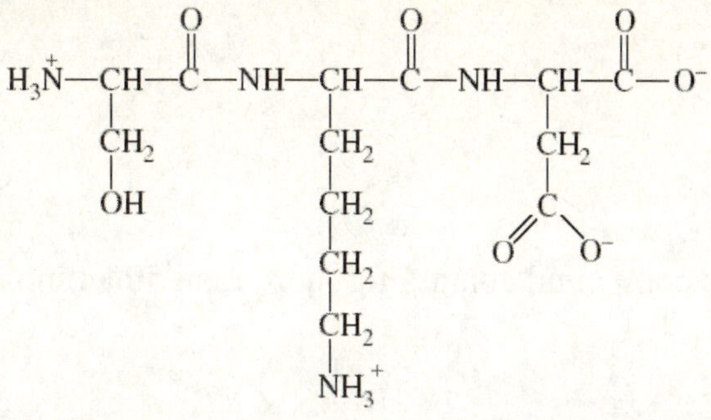

This represents a ________________ with____________ as the N-terminal amino acid.

Fill in the blanks with the correct choice.
A. tripeptide, serine
B. tripeptide, aspartate
C. dipeptide, Ser
D. dipeptide, D

10. Consider the following structure.

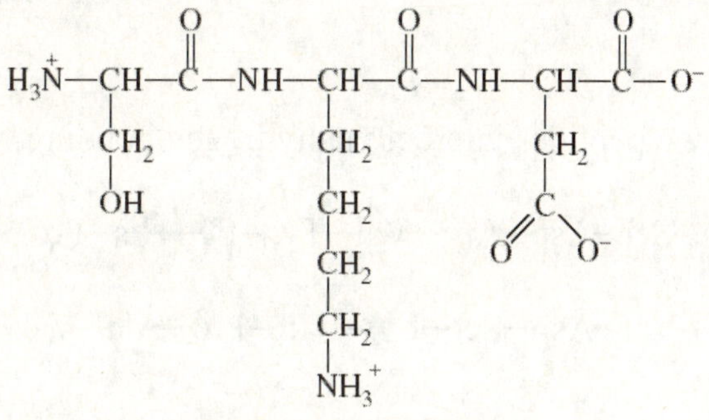

Which of the following correctly describes this structure?
A. contains three peptide bonds and all polar side chains
B. contains three peptide bonds and all hydrophilic side chains
C. contains two peptide bonds and all polar side chains
D. contains two peptide bonds and all hydrophobic side chains

11. The following peptide could be named as:

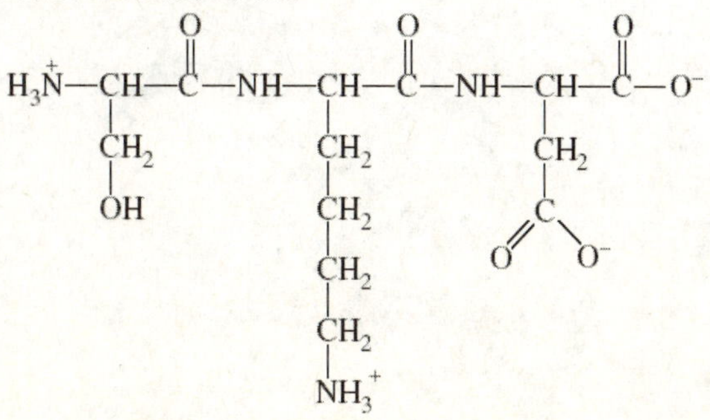

A. Ser—Lys—Asp.
B. S—L—A.
C. Asp—Lys—Ser.
D. D—R—S
E. either A or B
F. either C and D

12. A biologically active compound containing more than 50 amino acids is referred to as a:
    A. protein.
    B. peptide.
    C. polypeptide.
    D. oligopeptide.
    E. All of the above terms can be used.

13. Consider the following.

    Asp—Lys—Ser—Asp—Arg—Val—Tyr—Ile—His—Pro—Phe—Ser—Lys—Asp

    What level of protein structure is represented?
    A. primary
    B. secondary
    C. tertiary
    D. quaternary

14. When the following two peptides interact via hydrogen bonding, what level of protein structure is formed?

    Asp—Lys—Ser—Asp—Arg—Val—Tyr—Ile—His—Pro—Phe—Ser—Lys—Asp

    Tyr—Gln—Val—Asp—Ser—Lys—Thr—Pro—Ala—Val—Ile—Leu—Gln—Arg

    A. primary
    B. secondary
    C. tertiary
    D. quaternary

15. What level of protein structure is depicted in the following image?

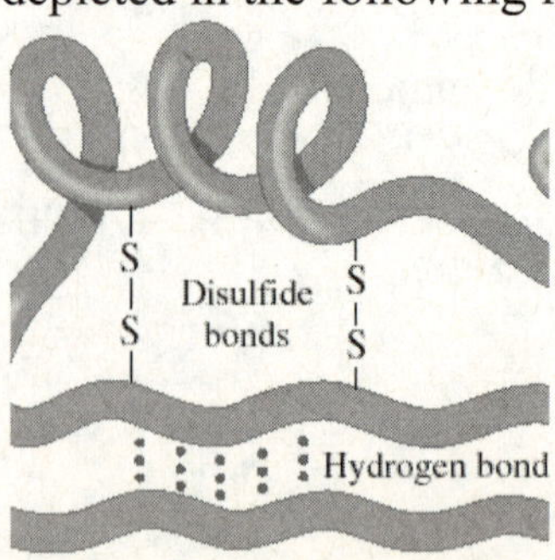

    A. primary
    B. secondary
    C. tertiary

D. quaternary

16. Which level of protein structure does not usually involve hydrogen bonding?
    A. primary
    B. secondary
    C. tertiary
    D. quaternary

17. Consider the following description of protein structure:

    "Open extended zigzag with amino acid side chains pointing upward and downward"

    This description corresponds to a:
    A. secondary structure called a β pleated sheet.
    B. tertiary structure called a β pleated sheet.
    C. secondary structure called an α helix.
    D. tertiary structure called an α helix.

18. How many different tripeptides could be produced containing one glycine, one cytosine, and one glutamine?
    A. 1
    B. 2
    C. 3
    D. 4
    E. 6
    F. 8

19. The amino acid shown would most likely be involved in what type of interaction in the tertiary structure of protein?

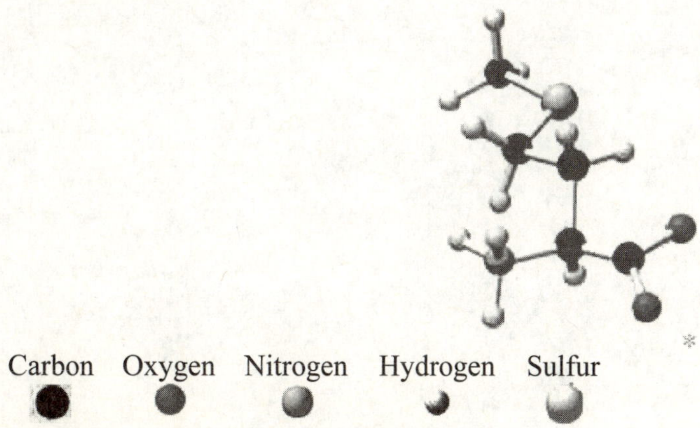

A. nonpolar interactions
B. polar interactions
C. salt bridges
D. disulfide bridges

20. What type(s) of interaction(s) could occur between the side chains of the following pairs of amino acids?

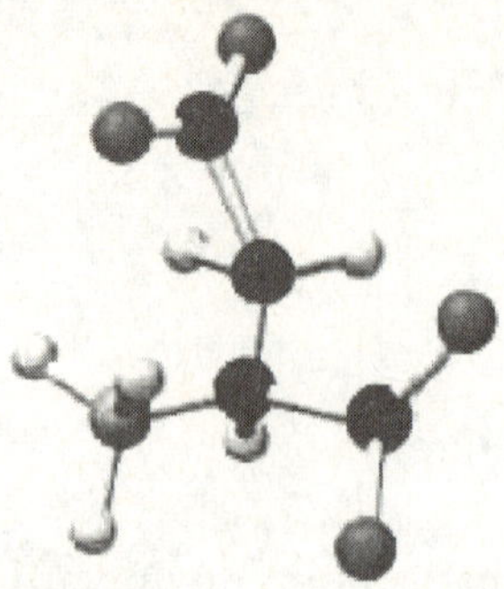

and

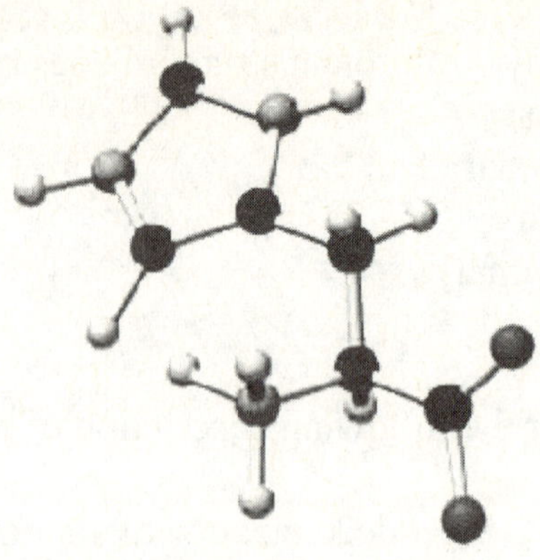

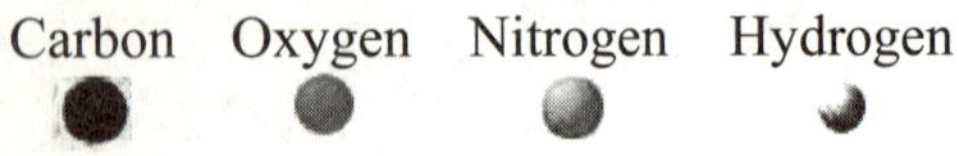

A. nonpolar interactions
B. polar interactions
C. salt bridges
D. disulfide bonds
E. all of the above

21. The stabilizing force in the secondary structure of proteins is:
A. nonpolar interactions.
B. polar interactions.
C. salt bridges.
D. disulfide bridges.
E. hydrogen bonding.

22. Consider the following image.

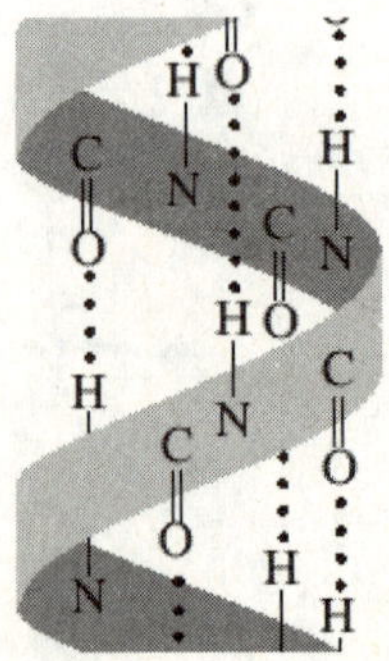

This excerpt represents a:
A. secondary protein structure stabilized by hydrogen bonds.
B. primary protein structure stabilized by hydrogen bonds.
C. secondary protein structure stabilized by salt bridges.
D. tertiary protein structure stabilized by hydrogen bonds.

23. The following represents an excerpt from a polypeptide chain:

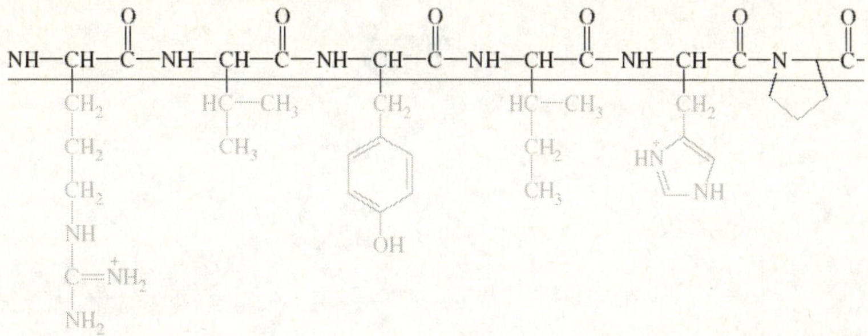

Which of the following correctly describes this image?

A. The underlined portion represents the protein backbone.
B. This is a representation of the primary structure of the protein.
C. The gray structures represent the protein's side chains.
D. All of the above are correct for this image.

24. When two —SH groups are brought close together when a protein folds:
A. reduction can occur, forming a disulfide bond.
B. oxidation can occur, forming an —S—S— link which stabilizes the tertiary structure of the protein.
C. oxidation can occur, forming a salt bridge.
D. reduction can occur, which stabilizes the quaternary structure of the protein.

25. Which of the following amino acids would be more likely to be on the interior of a protein after it has folded into its tertiary structure?
A. lysine
B. aspartate
C. tyrosine
D. phenylalanine

26. Denaturation of protein does **not** disrupt what level of protein structure?
A. primary
B. secondary
C. tertiary
D. quaternary

27. Which of the following is most likely to disrupt the hydrogen bonding in proteins?
A. application of heat
B. addition of an acid
C. addition of a base
D. presence of organic compounds

28. Which of the following denaturing agents would most likely disrupt the stabilizing force represented below?

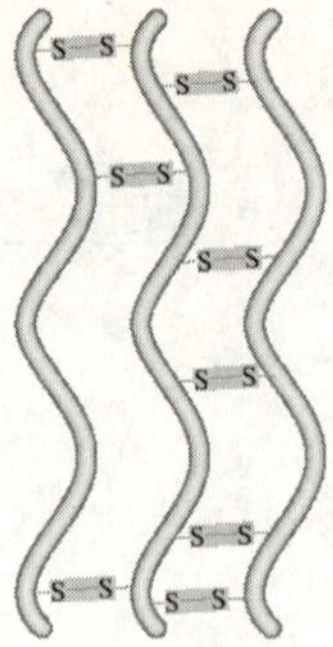

A. addition of an acid
B. heavy metals
C. mechanical agitation
D. both B and C

29. Consider the following depiction of a protein.

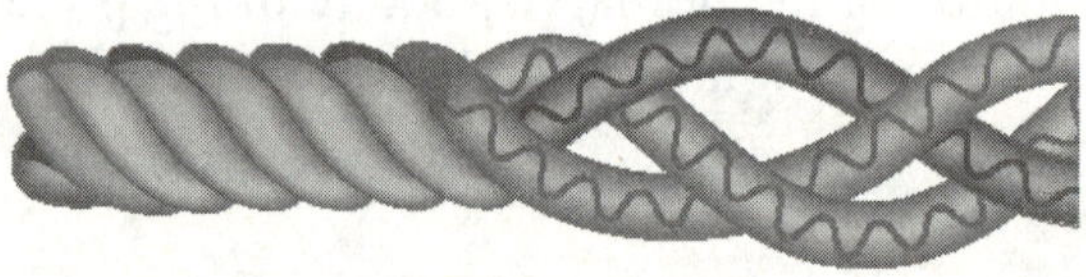

This image could represent:
A. hemoglobin.
B. an antibody.
C. collagen.
D. Na^+/K^+ ATPase.

30. In the following image showing a cell membrane, the structures designated α and β might represent:

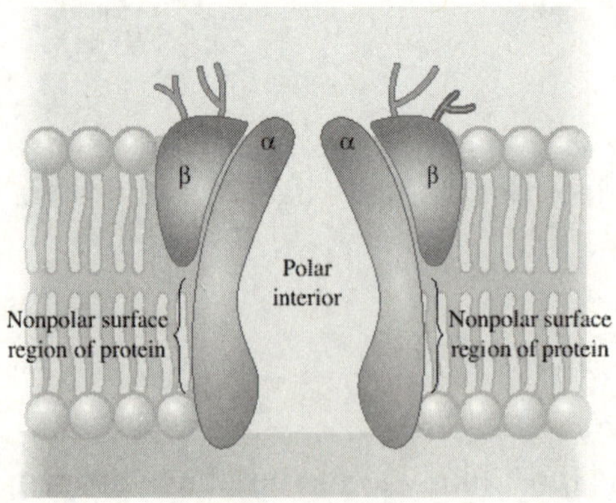

A. hemoglobin.
B. an antibody.
C. collagen.
D. Na^+/K^+ ATPase.

Answers

1. B	**2.** D	**3.** D	**4.** C	**5.** B	**6.** C	**7.** B	**8.** E	**9.** A	**10.** C
11. A	**12.** A	**13.** A	**14.** B	**15.** C	**16.** A	**17.** A	**18.** E	**19.** A	**20.** C
21. E	**22.** A	**23.** D	**24.** B	**25.** D	**26.** A	**27.** A	**28.** B	**29.** C	**30.** D

Chapter 9 – Solutions to Odd-Numbered Problems

Practice Problems

9.1 Draw the structural formula for each of the following amino acids and put an asterisk (*) next to any chiral carbon atoms in your structure:
a. alanine
b. lysine
c. tryptophan
d. aspartate

Refer to Table 9.1 for the structures of the amino acids.

To determine the chiral carbon atoms:

Locate the tetrahedral carbons in the molecule. (Any carbons with double or triple bonds cannot be chiral centers since they do not have four groups attached.) Determine if four different groups are attached to the tetrahedral carbons. If they are, the carbon atom is a chiral center.

a. alanine

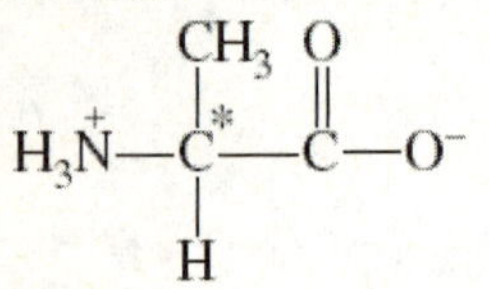

b. lysine

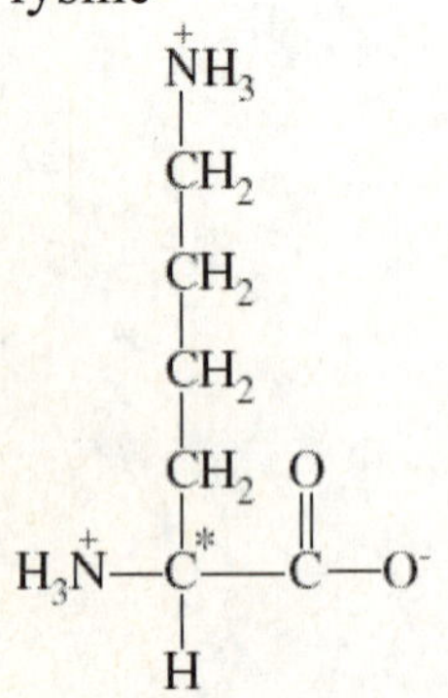

c. tryptophan

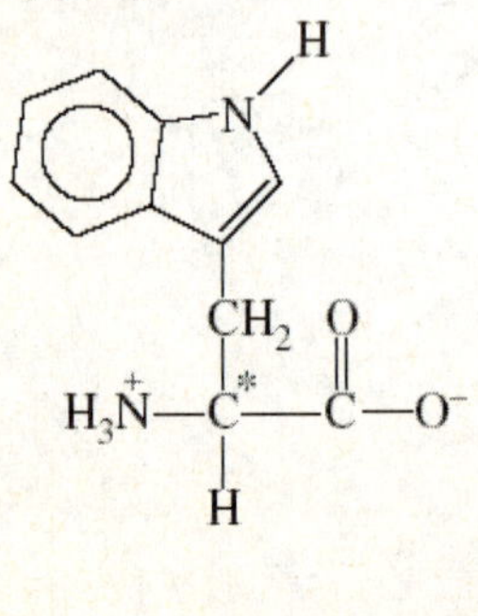

d. aspartate

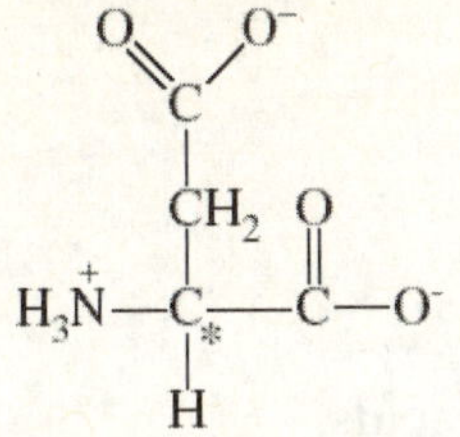

9.3 Give the three-letter and one-letter abbreviations and identify the functional group in the side chain for each of the amino acids in problem 9.1.
a. alanine
b. lysine
c. tryptophan
d. aspartate

Refer to Table 9.1 for the structures, names, and abbreviations of the amino acids and to Table 4.6 for the structural names of the functional groups.
a. alanine: Ala, A, hydrocarbon
b. lysine: Lys, K, protonated amine
c. tryptophan: Trp, W, aromatic
d. aspartate: Asp, D, carboxylate

9.5 Classify each of the amino acids in problem 9.4 as hydrophobic (nonpolar) or hydrophilic (polar).
a. leucine **b.** glutamate **c.** methionine **d.** threonine

The side chain of the amino acid is the R group attached to the alpha carbon.

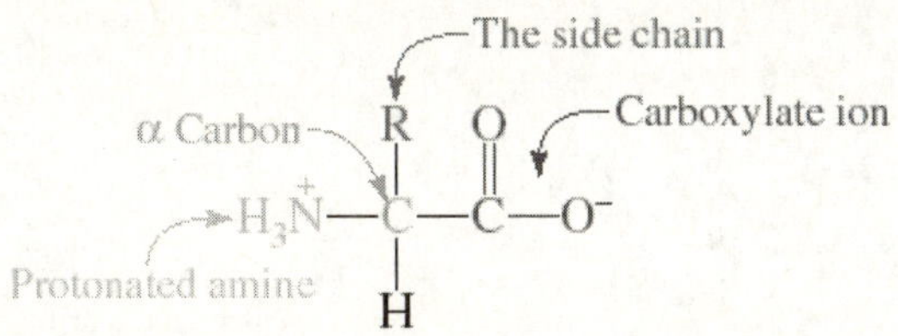

The amino acids are separated into two broad classifications based on the characteristics of the side chains: nonpolar amino acids and polar amino acids.
1.) Nonpolar: The nonpolar amino acids, with a few exceptions, have side chains that are composed entirely of carbon and hydrogen.
2.) Polar: The side chains of each of the amino acids classified as polar contain functional groups such as hydroxyl (–OH) and amide (–$CONH_2$) that create an uneven distribution of electrons in the side chain, making them polar.

a. leucine: : hydrocarbon side chain, nonpolar (hydrophobic)
b. glutamate: side chain contains a carboxylate group (—COO^-), polar (hydrophilic)
c. methionine: : side chain mostly hydrocarbon (no polar groups), nonpolar (hydrophobic)
d. threonine: side chain contains a hydroxyl (–OH) group, polar (hydrophilic)

9.7 Draw the structure for the amino acid represented by each of the following abbreviations:
a. G
b. His
c. Q
d. Ile

Refer to Table 9.1 for the structures, names, and abbreviations of the amino acids.

a. G: glycine

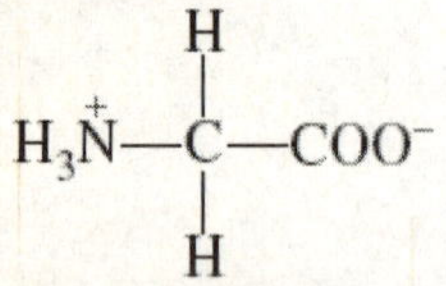

b. His: histidine

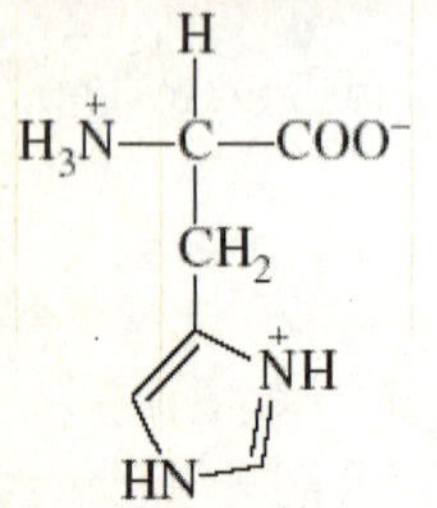

c. Q: glutamine

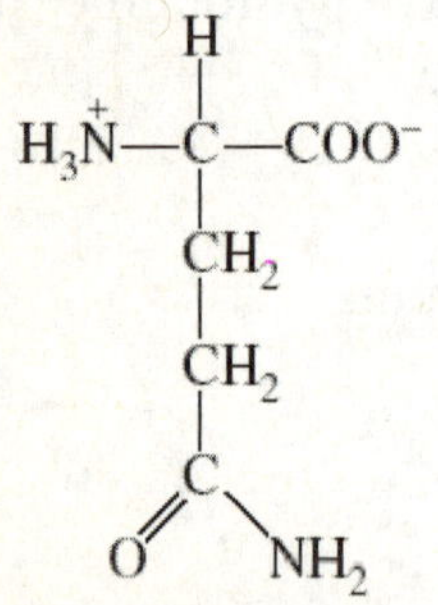

d. Ile: isoleucine

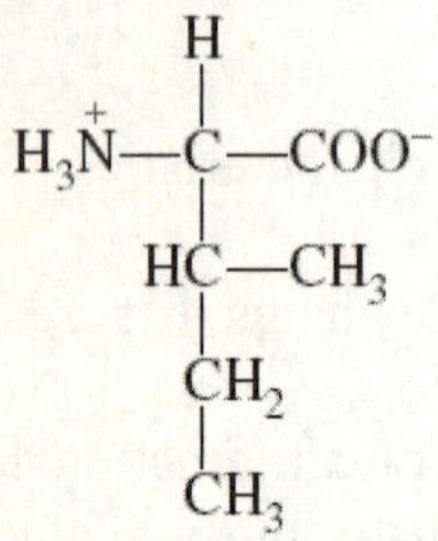

9.9 Give the name of the amino acids you drew in problem 9.7.
a. G
b. His
c. Q
d. Ile

Refer to Table 9.1 for the structures, names, and abbreviations of the amino acids.

a. G: glycine
b. His: histidine
c. Q: glutamine
d. Ile: isoleucine

9.11 Consider the following tripeptide:

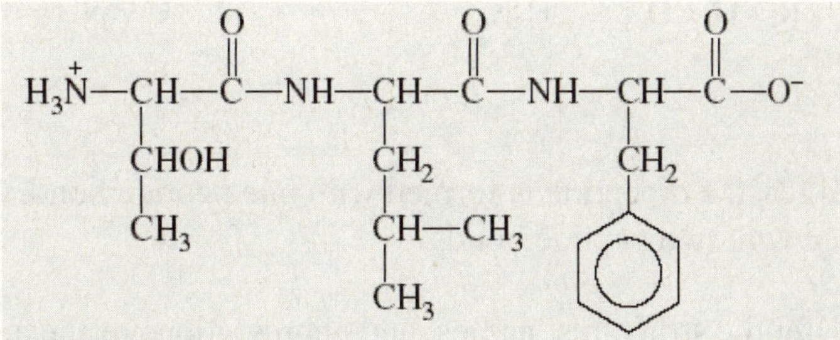

a. Draw a circle around the N-terminal amino acid and give its name. Draw a square around the C-terminal amino acid and give its name.
b. Give the one-letter and three-letter abbreviation of this tripeptide.

The N-terminus (or N-terminal amino acid) has an unreacted (free) α-amino group. The C-terminus (or C-terminal amino acid) has an unreacted α-carboxylate group. By convention, peptides are always written from the N-terminus to the C-terminus (left to right) whether using full structures or the one- or three-letter abbreviations.

Refer to Table 9.1 for the structures, names, and abbreviations of the amino acids.

a. Draw a circle around the N-terminal amino acid and give its name. Draw a square around the C-terminal amino acid and give its name.

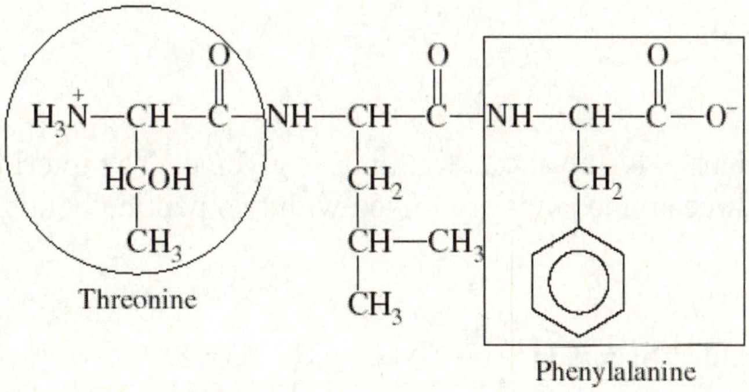

b. Give the one-letter and three-letter abbreviation of this tripeptide: TLF; Thr–Leu–Phe

9.13 Draw the structural formula for each of the following peptides:
a. Phe–Glu **b.** KCG
c. His–Met–Gln **d.** WY

Peptides form from the condensation reaction between amino acids. The carboxylate ion (—COO^-) of one amino acid molecule reacts with the protonated amine (—NH_3^+) of a second amino acid. A water molecule is removed, and an amide functional group (peptide bond) is formed.

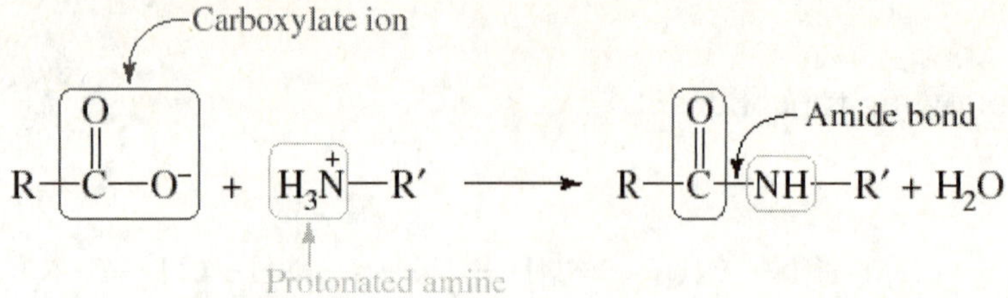

If two amino acids react, a dipeptide is formed with one peptide bond. If three amino acids react, a tripeptide is formed with two peptide bonds.

Refer to Table 9.1 for the structures, names, and abbreviations of the amino acids.

The N-terminus (or N-terminal amino acid) has an unreacted (free) α-amino group. The C-terminus (or C-terminal amino acid) has an unreacted α-carboxylate group. By convention, peptides are always written from the N-terminus to the C-terminus (left to right) whether using full structures or the one- or three-letter abbreviations.

a. Phe–Glu: N-terminus is phenylalanine and C-terminus is glutamate. The two amino acids are linked with one peptide bond.

a.

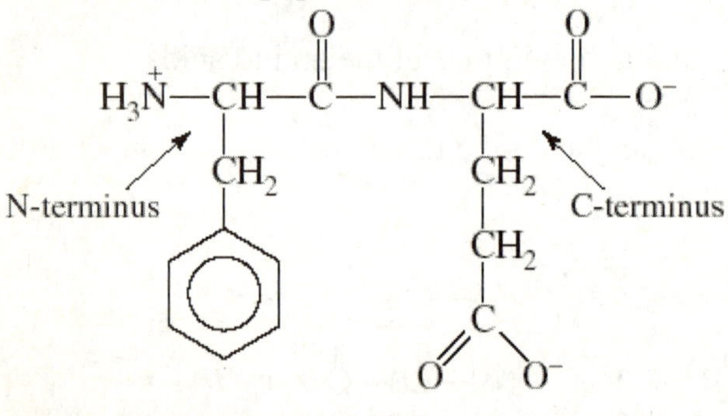

b. KCG: N-terminus is lysine and C-terminus is glycine. The interior amino acid residue is cysteine. The three amino acids are linked with two peptide bonds.

b.

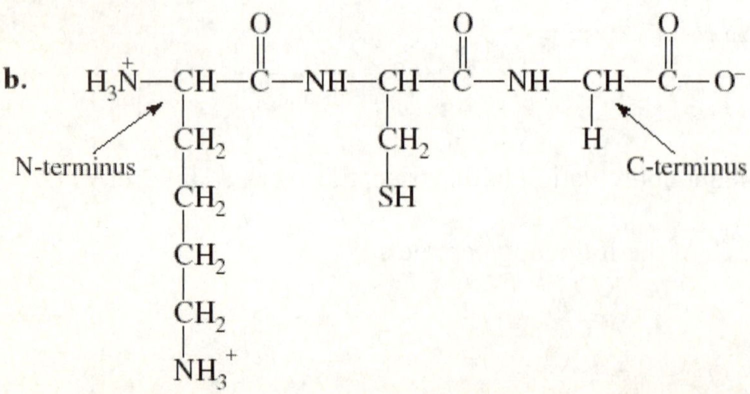

c. His–Met–Gln: N-terminus is histidine and C-terminus is glutamine. The interior amino acid residue is methionine. The three amino acids are linked with two peptide bonds.

c.

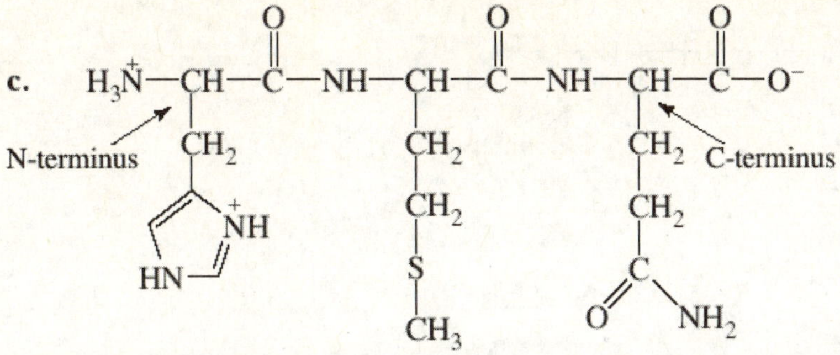

d. WY: N-terminus is tryptophan and C-terminus is tyrosine. The two amino acids are linked with one peptide bond.

d.

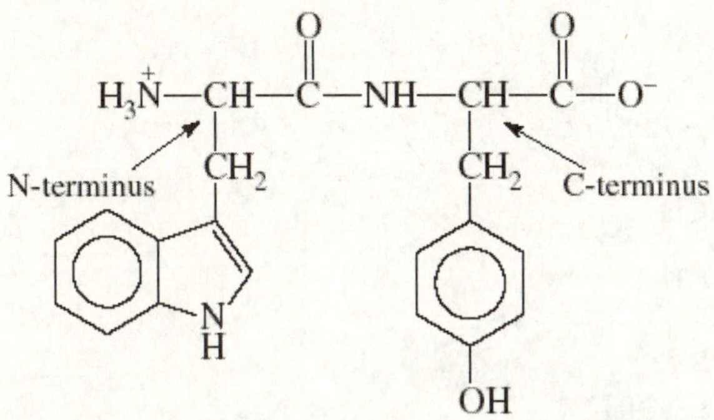

9.15 Label the N-terminus and the C-terminus for each of the peptides in problem 9.13.

a. Phe–Glu **c.** His–Met–Gln

b. KCG **d.** WY

The N-terminus (or N-terminal amino acid) has an unreacted (free) α-amino group. The C-terminus (or C-terminal amino acid) has an unreacted α-carboxylate group. By convention, peptides are always written from the N-terminus to the C-terminus (left to right) whether using full structures or the one- or three-letter abbreviations.

a. Phe–Glu: N-terminus is phenylalanine and C-terminus is glutamate. The two amino acids are linked with one peptide bond.

a.

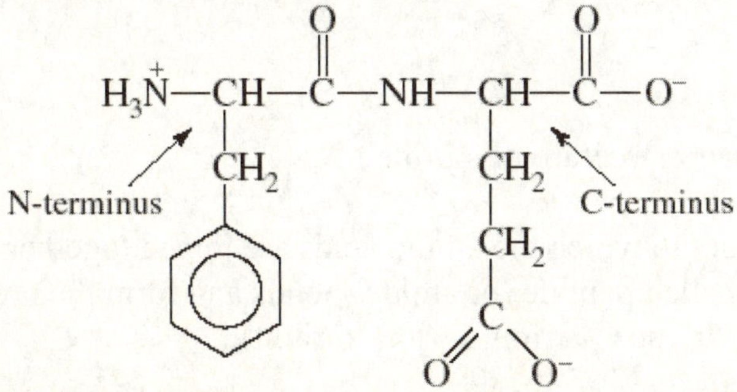

b. KCG: N-terminus is lysine and C-terminus is glycine. The interior amino acid residue is cysteine. The three amino acids are linked with two peptide bonds.

b.

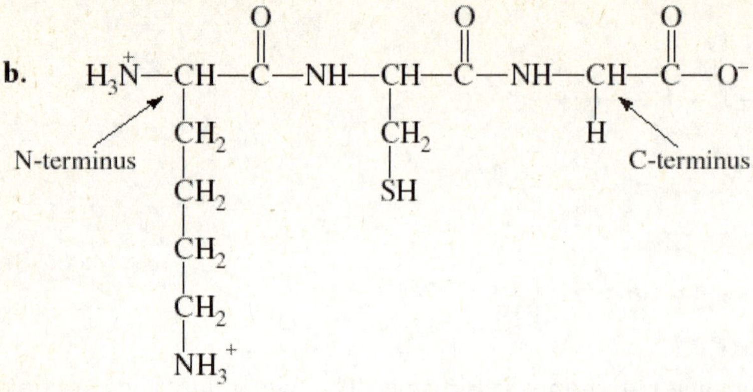

c. His–Met–Gln: N-terminus is histidine and C-terminus is glutamine. The interior amino acid residue is methionine. The three amino acids are linked with two peptide bonds.

c.

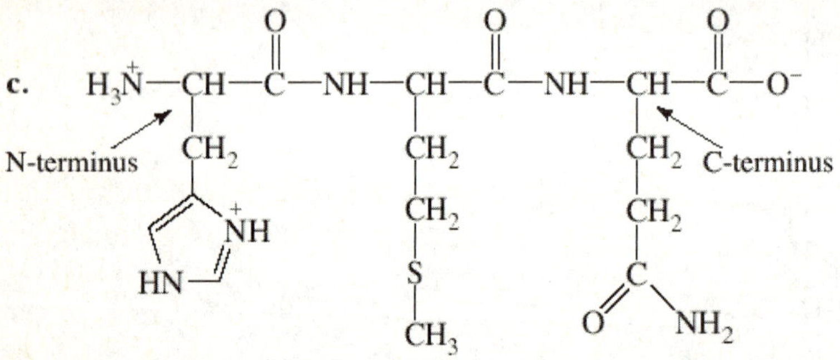

d. WY: N-terminus is tryptophan and C-terminus is tyrosine. The two amino acids are linked with one peptide bond.

d.

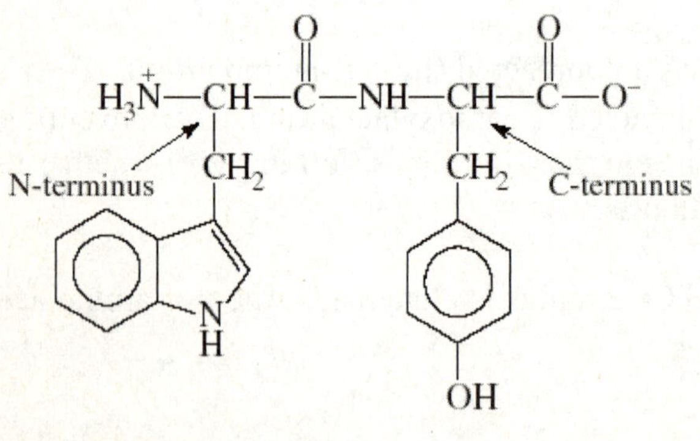

9.17 What type of bonding is present in the primary structure of a protein?

The primary structure of a protein is the order in which the amino acids are joined together by covalent bonds. These covalent bonds are called peptide (or amide) bonds and form the protein backbone that lines up from the N-terminus to the C-terminus (left to right).

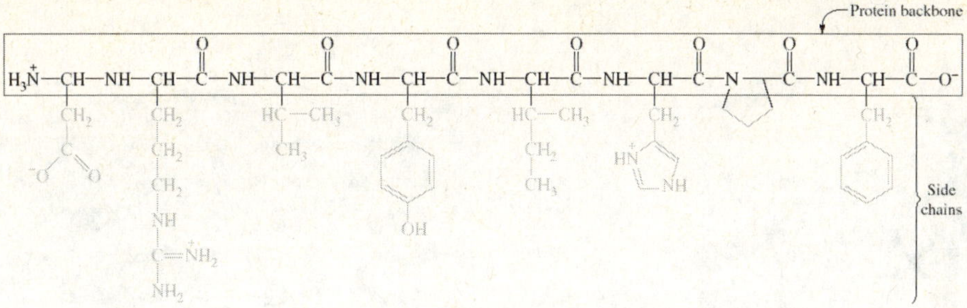

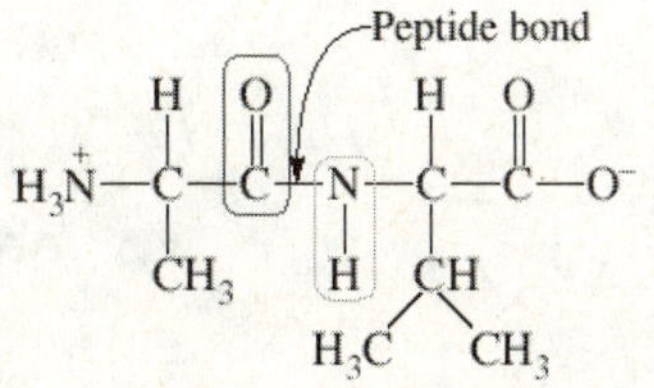

9.19 Name the stabilizing intermolecular force in the secondary structure of a protein.

The secondary structure of a protein describes an area of regular or repeating patterns of structure within the three-dimensional shape of a protein. The secondary structures in a protein are stabilized by hydrogen bonding along the protein backbone and involve amino acids that are near each other in the primary structure. For example, consider the following α-helix.

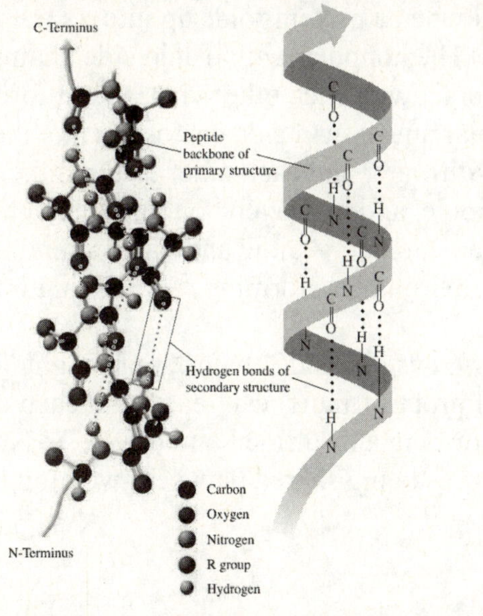

9.21 Describe the differences in the shape of an α helix and a β-pleated sheet.

An α helix is a coiled structure with the amino acid side chains protruding outward from the helix. The β-pleated sheet is an open, extended, zigzag structure with the side chains of the amino acids oriented above and below the sheet. See below: α helix on the left, β-pleated sheet on the right.

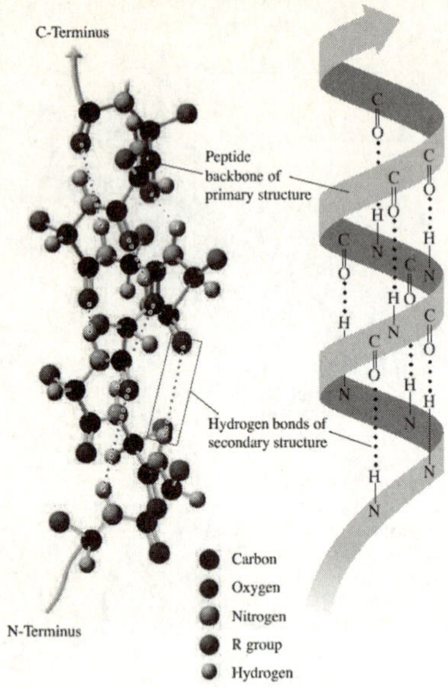

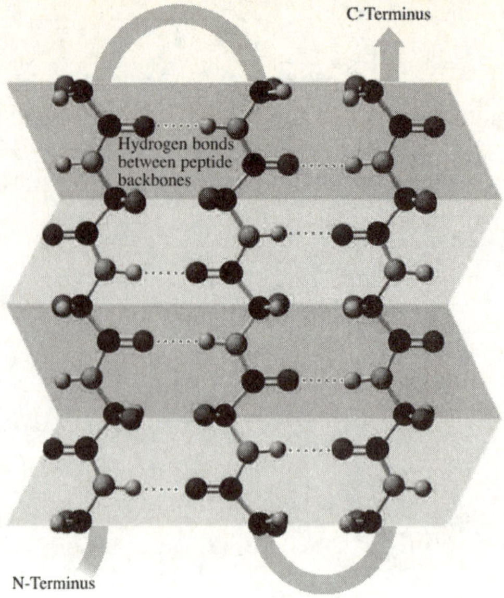

9.23 What type of interaction would you expect between the side chains of each of the following pairs of amino acids in the tertiary structure of a protein?

a. lysine and glutamate
b. leucine and isoleucine
c. threonine and tyrosine
d. glutamine and arginine

The interactions involved when a protein folds up into its tertiary structure include:

1. Nonpolar interactions –The nonpolar amino acid side chains are *unlike* the aqueous (polar) environment and interact with each other via London forces.
2. Polar interactions – Polar amino acid side chains are *like* the surrounding aqueous environment and easily interact with water and each other through the following hydrophilic interactions: dipole-dipole, ion-dipole, and hydrogen bonding interactions.
3. Salt bridges (ionic interactions) – Acidic and basic amino acid side chains exist in their ionized form in an aqueous environment, forming a stabilizing interaction called a salt bridge or ionic interaction.
4. Disulfide bonds – If two cysteine side chains are brought close together during the folding of a protein, the two –SH groups (thiols) can react with each other through an oxidation reaction (losing hydrogens) forming a disulfide bond (–S–S–). Note that the disulfide bond is a covalent bond (electrons being shared between two atoms) in contrast to the other stabilizing attractive forces.

a. lysine and glutamate: glutamate contains the ionized carboxylate group on the side chain and from a salt bridge (ionic interactions) with the ionized amino group on the side chain of lysine.
b. leucine and isoleucine: both contain a nonpolar hydrocarbon side chain, nonpolar interactions (London forces)
c. threonine and tyrosine: threonine and tyrosine both contain —OH groups on their side chains, polar interactions, hydrogen bonds
d. glutamine and arginine: arginine contains an ionized carboxylate group on the side chain, glutamine contains an amide group on the side chain, polar interactions, hydrogen bonds, and ion-dipole

9.25 Determine whether each of the following statements describes primary, secondary, tertiary, or quaternary structure of a protein.

a. Three polypeptide chains interact to form a biologically active protein.
b. Hydrogen bonds form between adjacent segments of the backbone of the same protein to form a "folded fan"-type structure.
c. Nonpolar side chains are repelled by water and move to the interior of the protein.
d. Amino acids react in a condensation reaction to form a peptide bond.

The primary structure of a protein is the order in which the amino acids are joined. The secondary structure of a protein describes an area of regular or repeating patterns of structure within the three-dimensional shape of a protein. The overall three-dimensional shape is the tertiary structure. The highest level of protein structure that involves the association of two or more peptide chains to form a biologically active protein is called quaternary.

a. Three polypeptide chains interact to form a biologically active protein (quaternary structure).
b. Hydrogen bonds form between adjacent segments of the backbone of the same protein to form a "folded fan"-type structure (secondary structure).
c. Nonpolar side chains are repelled by water and move to the interior of the protein (tertiary structure).
d. Amino acids react in a condensation reaction to form a peptide bond (primary structure).

9.27 Match the protein in Column A with its function from Column B.

Column A	Column B
Collagen	Ion transporter
Hemoglobin	Structural connector
Antibody	Oxygen transporter
Na^+/K^+ ATPase	Bind foreign substances in body

Review the discussion of biologically important proteins in Section 9.2

Collagen, Structural connector
Hemoglobin, Oxygen transporter
Antibody, Bind foreign substances in body
Na^+/K^+ ATPase, Ion transporter

9.29 How does the structure of collagen uniquely suit it for its function as a structural protein found in tendons?

Each of the individual helices of the triple helix is strong and flexible. Winding three helices together gives a much stronger structure, but maintains the flexibility. This creates the necessary strength for the tendons to maintain attachments, but provides the flexibility required for motion. See structural representation below.

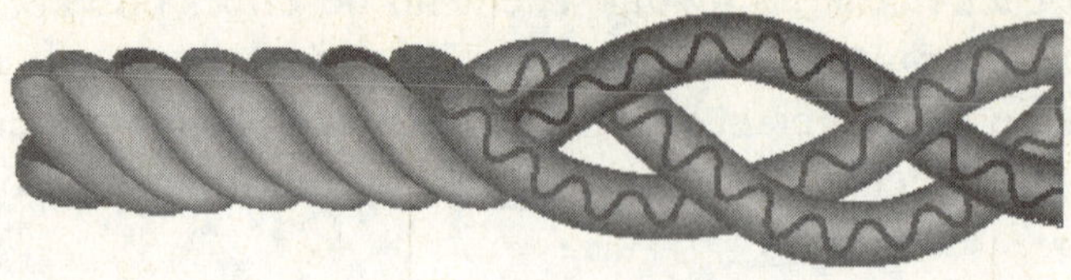

Additional Problems

9.31 What functional groups are common to all α-amino acids?

General structure of an amino acid:

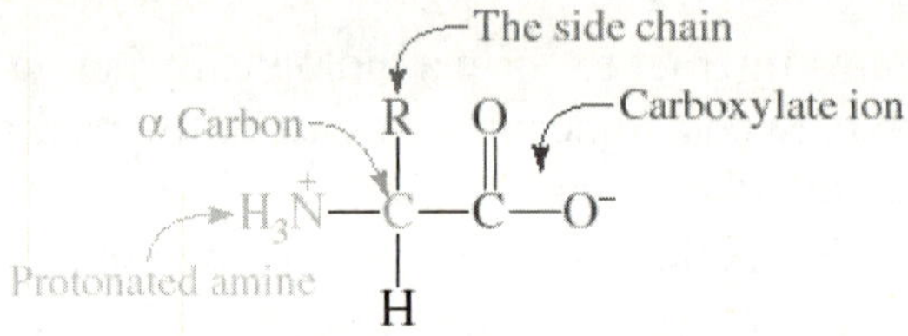

The two common groups are the carboxylate and protonated amine.

9.33 Give the name and three-letter abbreviation for the amino acid described by each of the following:
a. the polar amino acid with a benzene ring in its side chain.
b. the nonpolar amino acid whose side chain forms a ring with its α-amino group.
c. the polar amino acid with a sulfur atom in its side chain.

Refer to Table 9.1 for the structures of the amino acids. Compare the side chains with the descriptions.

a. the polar amino acid with a benzene ring in its side chain.: tyrosine (Tyr)
b. the nonpolar amino acid whose side chain forms a ring with its α-amino group.: proline (Pro)
c. the polar amino acid with a sulfur atom in its side chain.: cysteine (Cys)

9.35 Two of the amino acids have two chiral centers. Give the name and three-letter abbreviation of each.

Refer to Table 9.1 for the structures of the amino acids.

To determine the chiral carbon atoms:

Locate the tetrahedral carbons in the molecule. (Any carbons with double or triple bonds cannot be chiral centers since they do not have four groups attached.) Determine if four different groups are attached to the tetrahedral carbons. If they are, the carbon atom is a chiral center.

threonine (Thr) and isoleucine (Ile)

9.37 Vegetables, nuts, and seeds are often deficient in one or more of the essential amino acids. From the protein source listing in the table below, choose combinations of two or more foods that you could eat to get a complete protein.

Amino Acid Deficiency in Some Foods

Protein Source	**Amino Acid Missing**
Rice	Lys
Oatmeal	Lys
Peas	Met
Beans	Trp, Met
Almonds	Lys, Trp
Corn	Lys, Trp

Refer to Table 9.1 for the structures of the amino acids and the 10 essential amino acids. Proteins that contain all of the essential amino acids are called complete proteins.

A complete protein meal can be obtained by combining foods like rice and beans, peas and corn or peanut butter on whole grain bread.

9.39 Consider the amino acids glycine, proline, and lysine.
a. How many tripeptides can be formed from these three amino acids if each is used only once in the structure?
b. Using three-letter abbreviations for the amino acids, give the sequence for each of the possible tripeptides.
c. Draw the structure of the tripeptide that has proline as its N-terminal amino acid and glycine as its C-terminal amino acid. Circle each peptide bond.

The order of the amino acids is critical to both the structure and the function of the new compound.

a. How many tripeptides can be formed from these three amino acids if each is used only once in the structure?

6 different orders of amino acids are possible

b. Using three-letter abbreviations for the amino acids, give the sequence for each of the possible tripeptides.

Gly-Pro-Lys

Gly-Lys-Pro

Pro-Gly-Lys

Pro-Lys-Gly

Lys-Pro-Gly

Lys-Gly-Pro

c. Draw the structure of the tripeptide that has proline as its N-terminal amino acid and glycine as its C-terminal amino acid. Circle each peptide bond.

Peptides form from the condensation reaction between amino acids. The carboxylate ion (—COO^-) of one amino acid molecule reacts with the protonated amine (—NH_3^+) of a second amino acid. A water molecule is removed, and an amide functional group (peptide bond) is formed.

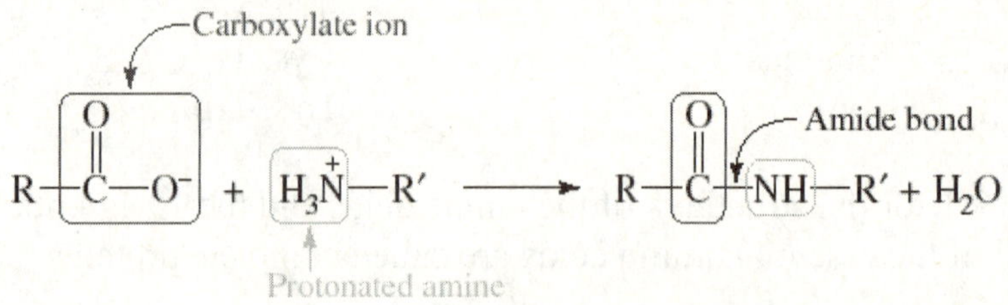

If two amino acids react, a dipeptide is formed with one peptide bond. If three amino acids react, a tripeptide is formed with two peptide bonds.

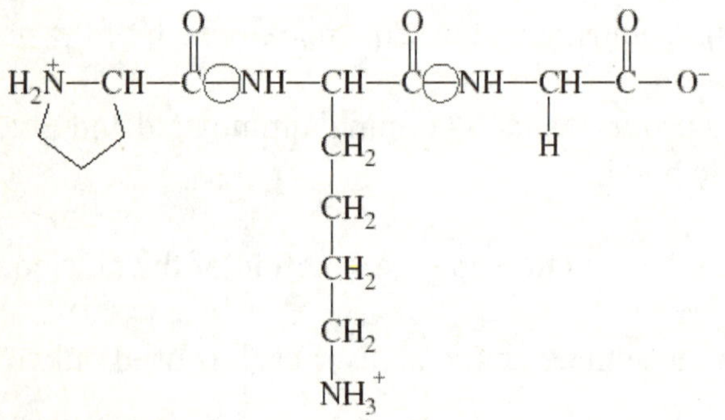

9.41 a. Draw the structure of Ser-Lys-Asp.

b. Would you expect to find this segment at the center or at the surface of a globular protein? Why?

Refer to Table 9.1 for the structures of the amino acids.

Peptides form from the condensation reaction between amino acids. The carboxylate ion (—COO^-) of one amino acid molecule reacts with the protonated amine (—NH_3^+) of a second amino acid. A water molecule is removed, and an amide functional group (peptide bond) is formed.

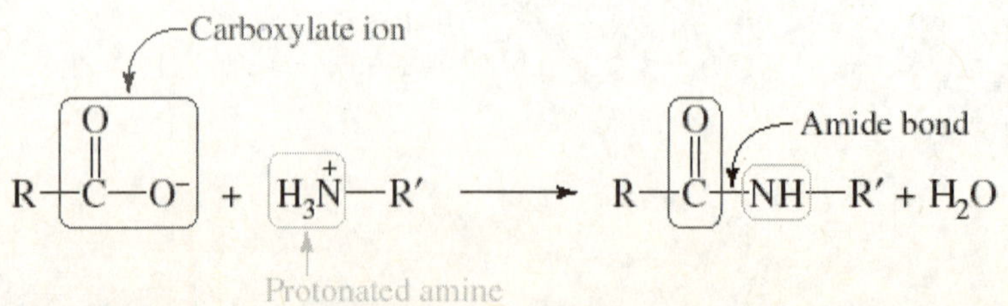

If two amino acids react, a dipeptide is formed with one peptide bond. If three amino acids react, a tripeptide is formed with two peptide bonds.

a. Draw the structure of Ser-Lys-Asp.

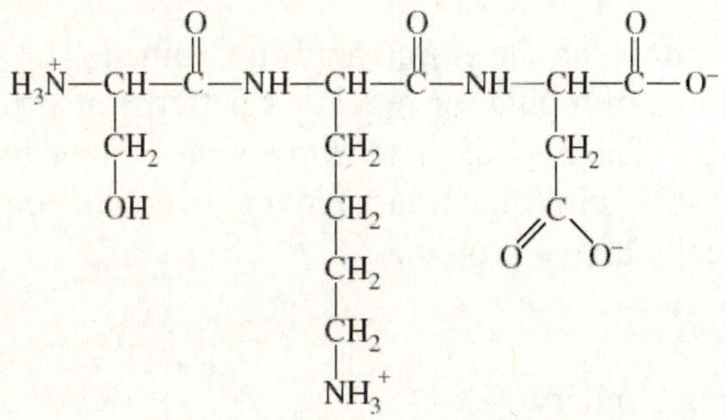

b. Would you expect to find this segment at the center or at the surface of a globular protein? Why?

This segment would be on the surface in the aqueous environment. The side chains are hydrophilic (polar) and would interact with the water in the surrounding environment.

9.43 Identify some differences between the following pairs:
a. primary and secondary protein structures
b. complete and incomplete proteins
c. fibrous and globular proteins

Compare the stabilizing forces in the levels of protein structure. See Table 9.2.

a. primary and secondary protein structures

Primary structures are held together by covalent bonding, and secondary structures are held together by intermolecular forces.

Complete and incomplete refer to the complement of essential amino acids provided by a protein source.

b. complete and incomplete proteins

Complete proteins are those that contain all amino acids and incomplete proteins do not.

Review the tertiary structure of fibrous and globular proteins.

c. fibrous and globular proteins

Fibrous proteins have elongated structures and globular proteins have roughly spherical structures.

9.45 Identify the level of protein structure associated with each of the following:
a. more than one polypeptide
b. hydrogen bonding between backbone atoms
c. the sequence of amino acids
d. intermolecular forces between R groups

The primary structure of a protein is the order in which the amino acids are joined. The secondary structure of a protein describes an area of regular or repeating patterns of structure within the three-dimensional shape of a protein. The overall three-dimensional shape is the tertiary structure. The highest level of protein structure (quaternary) involves the association of two or more peptide chains to form a biologically active protein.

a. more than one polypeptide: quaternary
b. hydrogen bonding between backbone atoms: secondary
c. the sequence of amino acids: primary
d. intermolecular forces between R groups: tertiary and/or quaternary

9.47 Briefly describe the structure of collagen and discuss how it is different from other helix-containing proteins.

Collagen's structure consists of three left-handed helices wrapped around each other to form a right-handed helix called a triple helix. Most of the single helices found in proteins are right-handed, not left-handed, as in collagen. See image below.

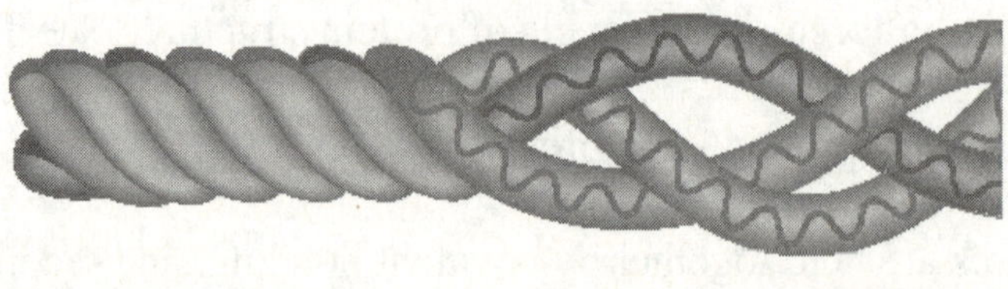

Triple helix 3 α Helix peptide chains

9.49 Indicate what type(s) of intermolecular forces are disrupted and what level of protein structure is changed by the following denaturing treatments.
a. An egg is placed in water at 100°C and boiled for 10 minutes.
b. Acid is added to milk during the preparation of yogurt.
c. Egg whites are whipped in a mixing bowl to make meringue.

Refer to Table 9.3.

TABLE 9.3 EXAMPLES OF PROTEIN DENATURATION

Denaturing Agent	Disrupted Forces	Examples
Heat above 50 °C	Hydrogen bonds and hydrophobic interactions	Cooking food
Acids and bases	Salt bridges and hydrogen bonds	Lactic acid from bacteria, which denatures milk proteins in the preparation of yogurt and cheese
Organic compounds	Disulfide bonds	Thiols, which are used in hairstyling for hair straightening or permanent waves
Heavy metal ions Ag^+, Pb^{2+}, Hg^{2+}	Disulfide bonds and salt bridges	Mercury and lead poisoning
Mechanical agitation	Hydrogen bonds and London forces	Whipped cream and meringue made from egg whites

a. An egg is placed in water at 100°C and boiled for 10 minutes.: hydrogen bonds and nonpolar attractions; secondary and tertiary structure
b. Acid is added to milk during the preparation of yogurt.: hydrogen bonds and salt bridges; secondary and tertiary
c. Egg whites are whipped in a mixing bowl to make meringue.: hydrogen bonds and nonpolar attractions; secondary and tertiary

9.51 What types of covalent bonds can be disrupted when a protein is denatured? Name a denaturing agent that could accomplish this.

The only type of covalent bond involved in tertiary structure is the disulfide bond. One chemical that can disrupt the disulfide bond is ammonium thioglycolate, which can reduce the —S—S— to —SH and —SH.

9.53 Collagen contains an amino acid that is a modified form of the naturally occurring amino acid.
a. Name the natural amino acid.
b. How is the structure of the side chain of this amino acid modified?
c. What is the purpose of the modification in terms of the structure and function of the collagen?

Review the discussion of biologically important proteins.
a. Name the natural amino acid. proline
b. How is the structure of the side chain of this amino acid modified?: An OH group is added to the side chain to form hydroxyproline.
c. What is the purpose of the modification in terms of the structure and function of the collagen?: The OH on the side chain allows for the formation of more hydrogen bonds between side chains, which increases the strength of the collagen.

9.55 Integral membrane proteins are located across a cell membrane and have nonpolar amino acids on their surface and polar amino acids in the interior. This is in contrast to proteins like hemoglobin that are found in the bloodstream. Explain why the surface of the integral membrane proteins is nonpolar.

The middle portion of these proteins spans the nonpolar portion of the membrane. This surface interacts with the nonpolar environment and therefore must be nonpolar itself.

Challenge Problems

9.57 How is the structure of a soap micelle (Chapter 6) similar to the structure of a globular protein? (Hint: Both of these are found in aqueous solution.)

Both have nonpolar interiors and polar surfaces. On a micelle the polar heads of the fatty acid salts face outwards into the aqueous environment, similar to the polar amino acid side chains on the globular protein. The nonpolar tails of the fatty acid salts gather together in the interior of the micelle just as the nonpolar amino acid side chains gather in the interior of the globular protein.

Chapter 10

Enzymes – Nature's Chemists

10.1 Enzymes and Their Substrates

Learning Objectives

Upon completion of this material, a student should be able to do the following:

A. define the following key terms:

thermodynamics	active site	substrate
substrate specificity	cofactor	coenzyme
enzyme-substrate complex	lock-and-key model	induced-fit model

B. describe how thermodynamics relates to enzyme catalyzed reactions
C. distinguish active site, substrate, cofactor, and coenzyme
D. identify the level of protein structure involved in the formation of an enzyme's active site
E. describe what is meant by an enzyme-substrate (ES) complex
F. distinguish the models for enzyme-substrate interaction: lock-and-key versus the induced-fit model

10.2 Thermodynamics of Chemical Reactions

Learning Objectives

Upon completion of this material, a student should be able to do the following:

A. define the following key terms:

activation energy	heat of reaction	rate of reaction

B. describe the differences between an endothermic and an exothermic reaction
C. establish whether a reaction is exothermic or endothermic by using a reaction-energy diagram
D. describe (or use a reaction energy diagram to illustrate) the effect of a catalyst on the:
1.) heat of reaction
2.) activation energy
3.) rate of reaction

10.3 Enzymes and Catalysis

Learning Objectives

Upon completion of this material, a student should be able to do the following:

A. describe the factors that result in the lowering of activation energy by an enzyme
   1.) proximity
   2.) orientation
   3.) bond energy

B. apply the factors that affect enzyme activity to given situations

C. relate enzyme substrate interactions to intermolecular attractive forces

D. apply the general principles of catalysis to the action of hexokinase

10.4 Factors That Affect Enzyme Activity

Learning Objectives

Upon completion of this material, a student should be able to do the following:

A. define the following key terms:

activity	steady state	optimum pH
optimum temperature	inhibitor	reversible inhibition
competitive inhibitor	noncompetitive inhibitor	irreversible inhibition

B. describe the factors affecting enzyme activity

C. predict the effect of changes in conditions on the rate of a reaction

D. distinguish competitive, noncompetitive, and irreversible inhibition

Practice Test for Chapter 10

1. Which of the following terms designate the reactant in an enzyme catalyzed reaction?
   A. enzyme (E)
   B. substrate (S)
   C. enzyme-substrate complex (ES)
   D. active site

2. Enzymes are effective only for one stereoisomer of a pair of enantiomers. This is an example of:
   A. substrate specificity.
   B. competitive inhibition.
   C. enzyme-substrate complex (ES).
   D. enzyme activity.

3. Small organic molecules necessary for the functioning of an enzyme are referred to as:
   A. substrates.
   B. cofactors.
   C. coenzymes.
   D. substrate specificity.

4. Consider the following depiction of the reaction between glucose and adenosine triphosphate.

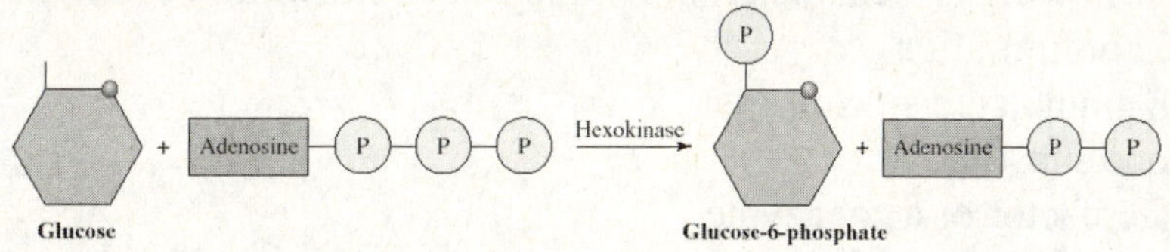

   In this reaction:
   A. glucose-6-phosphate is the ES.
   B. hexokinase is the enzyme.
   C. adenosine triphosphate is a cofactor.
   D. all of the above

5. What level of protein structure is involved in the formation of an enzyme's active site?
   A. primary
   B. secondary
   C. tertiary
   D. quaternary

6. What term is applied to the initial interaction of an enzyme with its substrate?
   A. lock-and-key model
   B. induced-fit model
   C. enzyme-substrate complex
   D. quaternary structure

7. Which of the following models of enzyme activity could be used to explain an enzyme that has more than one substrate?

A.

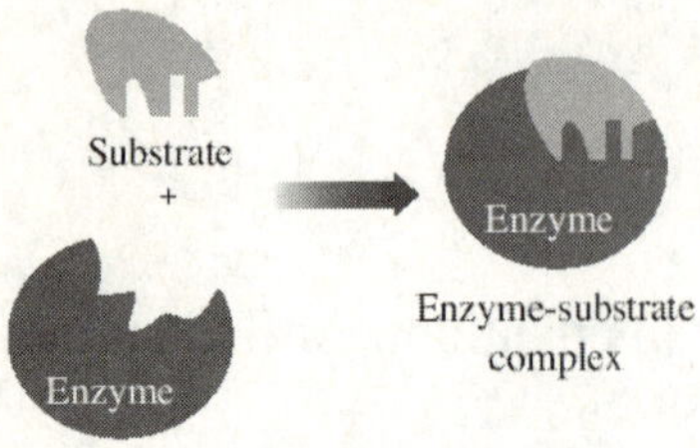

B.

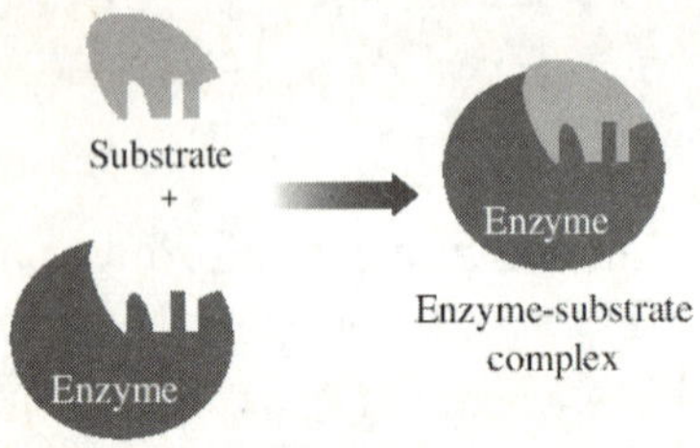

C. Either one could be used.
D. Neither one since enzymes are specific.

8. Which of the following is a characteristic of the active site of **any** enzyme?
A. inflexible configuration
B. lined with amino acid side chains
C. fits only one substrate
D. requires a cofactor or a coenzyme

9. Carboxypeptidase A catalyzes the hydrolysis of peptide bonds. Zn^{2+} is needed in this hydrolysis. In this reaction:
A. carboxypeptidase A is a coenzyme and Zn^{2+} is a cofactor.
B. carboxypeptidase A is the enzyme and Zn^{2+} is a coenzyme.
C. carboxypeptidase A is a cofactor and Zn^{2+} is a coenzyme.
D. carboxypeptidase A is the enzyme and Zn^{2+} is a cofactor.

10. Which of the following is responsible for the formation of the ES?
A. covalent bond formation
B. ionic bond formation
C. intermolecular attractions
D. All of the above can be involved.

11. The following image shows glucose and the enzyme hexokinase (the large molecule). Based on this diagram:

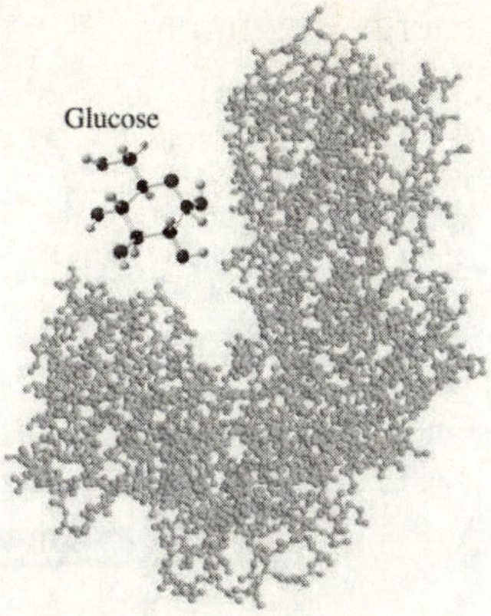

A. the crevice or pocket is the active site of hexokinase.
B. the ES has not yet formed.
C. the crevice or pocket might be lined with polar amino acids.
D. all of the above

12. The energy required to align reactants to form the ES complex is termed:
A. bond formation energy.
B. activation energy.
C. complex energy.
D. heat of reaction.

13. Which of the following correctly characterizes the effect of a catalyst?
A. lowers the activation energy
B. increases the energy of the reactants
C. lowers the energy of the products
D. increases the heat of reaction
E. all of the above

14. The rate of a catalyzed reaction increases with:
A. increasing heat of reaction.
B. increasing activation energy.
C. decreasing heat of reaction.
D. decreasing activation energy.

15. For an exothermic reaction:
A. the energy of the products is less than the reactant.
B. energy is absorbed during the reaction.
C. the temperature will decrease as the reaction proceeds.
D. All of the above describe an exothermic reaction.

16. Consider the following reaction energy diagram.

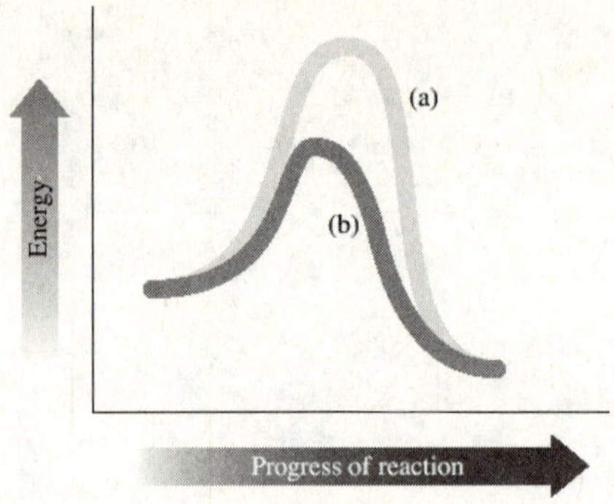

Based on this diagram, which of the following statements is(are) correct?

A. Curve b represents the uncatalyzed pathway.
B. Curve a represents the catalyzed pathway.
C. An exothermic reaction is represented.
D. Curve b has the higher activation energy.
E. None of the statements are correct.

17. Consider the following two reaction energy diagrams.

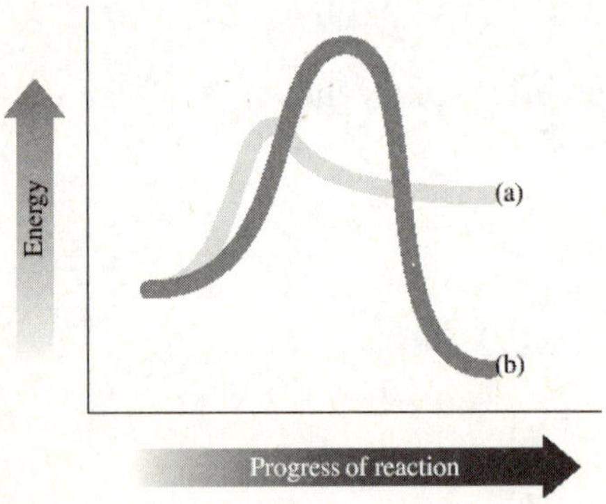

Reaction 1

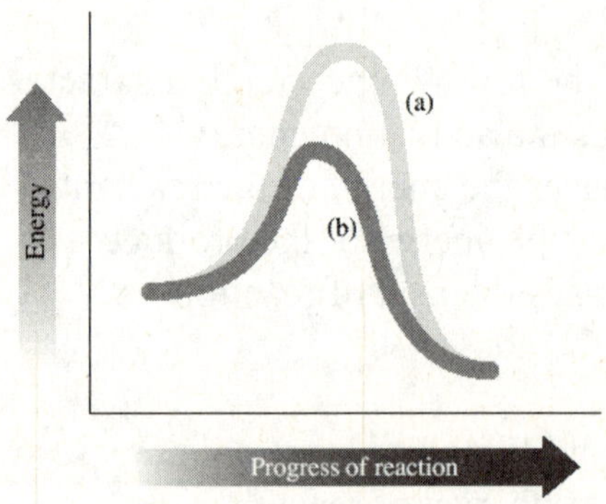

Reaction 2

If in both diagrams the energy scale is the **same**, which is the fastest exothermic reaction?

A. Reaction 1, pathway a
B. Reaction 1, pathway b
C. Reaction 2, pathway a
D. Reaction 2, pathway b

18. A solid is mixed with water in a beaker. The beaker begins to feel cold, this indicates that:
   A. the reaction occurring in the beaker is endothermic.
   B. the energy content of the products is less than the reactants.
   C. the reaction is occurring at a very slow rate.
   D. the following diagram could be used to represent the reaction.

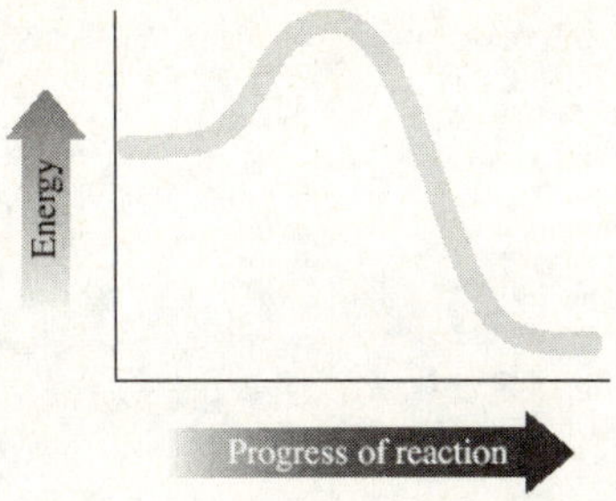

19. Which of the following is **not** a factor in lowering the activation energy in the formation of ES?
   A. heat of reaction
   B. bond energy
   C. orientation
   D. proximity

20. What type of interaction might be involved in the side chain of valine (shown below) and a substrate in the active site?

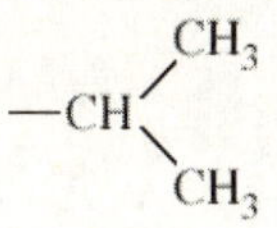

   A. hydrogen bonds
   B. hydrophobic interactions
   C. hydrophilic interactions
   D. ionic interactions

21. The following represents a substrate in the active site.

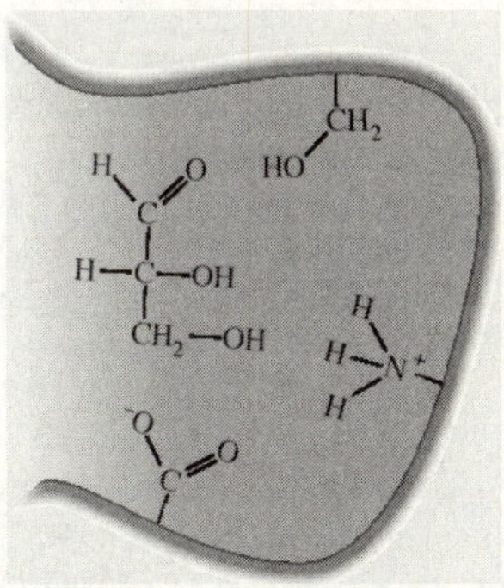

   What type of interaction will develop between the enzyme and the substrate?
   A. hydrogen bonds
   B. hydrophobic interactions
   C. nonpolar interactions
   D. ionic interactions

22. The following diagrams represent an uncatalyzed and a catalyzed reaction, respectively. This diagram represents the lowering of activation energy by the enzyme (E) due to:

A. orientation.
B. temperature.
C. pH.
D. bond energy.

23. The measurement of how fast an enzyme reaction occurs is termed:
A. inhibition.
B. optimum conditions.
C. activity.
D. thermodynamics.

24. When a substrate is being converted to product with the highest efficiency, this condition is referred to as:
A. optimum pH.
B. optimum temperature.
C. activity.
D. steady state.

25. Sucrase catalyzes the breakdown of sucrose to glucose and fructose in the small intestine with an optimum pH of 6.3. Which of the following will decrease the rate of catalysis?
A. increase the pH
B. decrease the H_3O^+ concentration
C. increase the temperature to 65 °C
D. all of the above

26. Consider the following table of data.

TABLE 10.1 OPTIMUM pH FOR SELECTED ENZYMES

Enzyme	Location	Substrate	Optimum pH
Pepsin	Stomach	Peptide bonds	2
Sucrase	Small intestine	Sucrose	6.2
Urease	Liver	Urea	7.4
Hexokinase	All tissues	Glucose	7.5
Trypsin	Small intestine	Peptide bonds	8
Arginase	Liver	Arginine	9.7

If an antacid is taken, what happens to the activity of pepsin?
A. increase
B. decrease
C. remains constant
D. cannot be predicted from the given data

27. The structure on the left has been shown to inhibit the reaction of the substrate on the right.

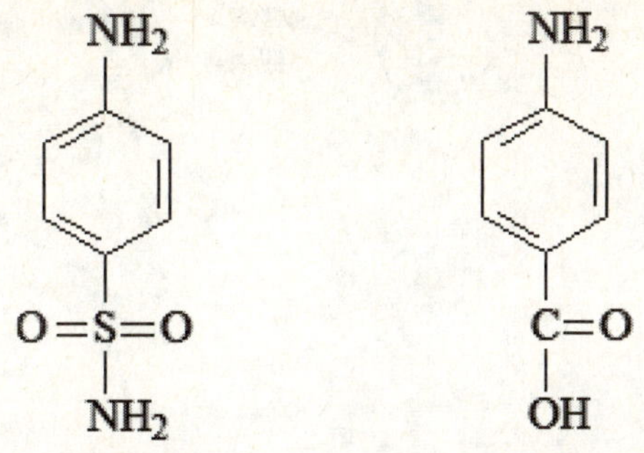

inhibitor substrate

This is most likely an example of:
A. reversible inhibition.
B. irreversible inhibition.
C. competitive inhibition.
D. noncompetitive inhibition.
E. A and C
F. A and D

28. In the presence of an inhibitor, increasing the concentration of the substrate has no effect on enzyme activity. When the inhibitor is removed, activity returns. This is an example of:
A. reversible inhibition.
B. irreversible inhibition.
C. competitive inhibition.
D. noncompetitive inhibition.
E. A and C
F. A and D

29. The image represents the effect of an inhibitor (I) on an enzyme.

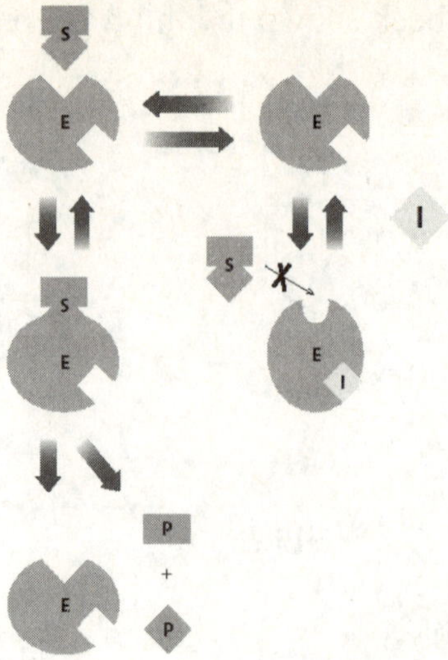

This is an example of:
A. reversible inhibition.
B. irreversible inhibition.
C. competitive inhibition.
D. noncompetitive inhibition.
E. A and D
F. B and C

30. The image represents the effect of an inhibitor (I) on an enzyme.

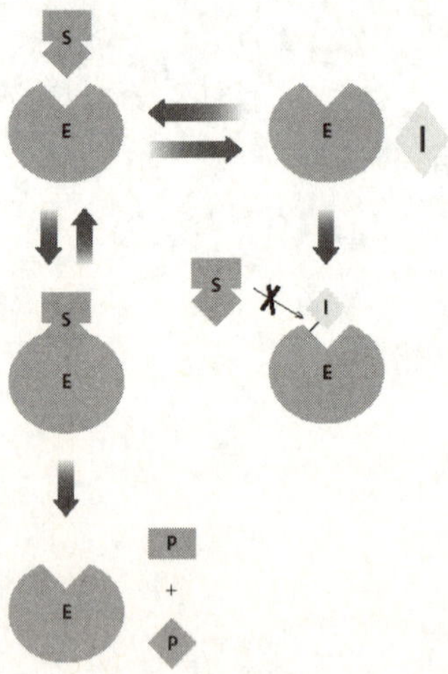

This is an example of:
A. reversible inhibition.
B. irreversible inhibition.
C. competitive inhibition.
D. noncompetitive inhibition.
E. A and C
F. B and C

Answers

1. B	**2.** A	**3.** C	**4.** B	**5.** C	**6.** C	**7.** A	**8.** B	**9.** D	**10.** C
11. D	**12.** B	**13.** A	**14.** D	**15.** A	**16.** C	**17.** D	**18.** A	**19.** A	**20.** B
21. A	**22.** A	**23.** C	**24.** D	**25.** D	**26.** B	**27.** E	**28.** F	**29.** E	**30.** F

Chapter 10 – Solutions to Odd-Numbered Problems

Practice Problems

10.1 What level of protein structure is involved in the formation of an enzyme's active site?

When folding into its tertiary structure, an indentation forms on one part of the protein's surface. This pocket is known as the active site, and it is lined with amino acid side chains. As its name implies, the active site is the functional part of the enzyme where catalysis occurs.

10.3 Distinguish between a cofactor and a coenzyme.

Both are non-protein portions of an enzyme. A coenzyme is small organic molecule, many of which are derived from vitamins. A cofactor is an inorganic substance such as a Mg^{2+} that can be obtained from minerals in the diet.

10.5 Which enzyme-substrate model describes the action of hexokinase? Describe how the enzyme changes when the enzyme and glucose form ES.

In the induced-fit model, the enzyme's active site is flexible and has a shape that is roughly complementary to the shape of its substrate. As the substrate interacts with the enzyme, the active site undergoes a conformational change, adjusting to fit the shape of the substrate. The shape of the substrate may change slightly as well.

The induced-fit model: When glucose is fit into the active site, the enzyme undergoes a conformational change closing around the substrate.

10.7 When water and solid ammonium nitrate (NH_4NO_3) are combined in a test tube, the solid dissolves in an endothermic process. Would the test tube feel hot or cold? Explain your reasoning.

Cold: The reaction is endothermic (heat is a reactant). The reaction absorbs heat from its surroundings (the water and the test tube) and the surroundings become cooler.

10.9 **a.** Describe activation energy for a chemical reaction.
b. How does a catalyst increase the rate of a chemical reaction?
c. In an exothermic reaction, are the products or the reactants lower in energy?

a. The activation energy is the energy necessary to align the reactant molecules and to cause them to collide with enough energy to form products. See diagram below. The activation energy is proportional to the "height of the hill" between reactants and products.

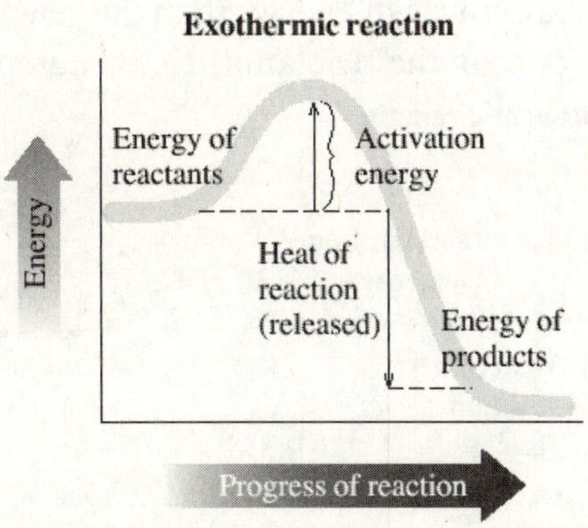

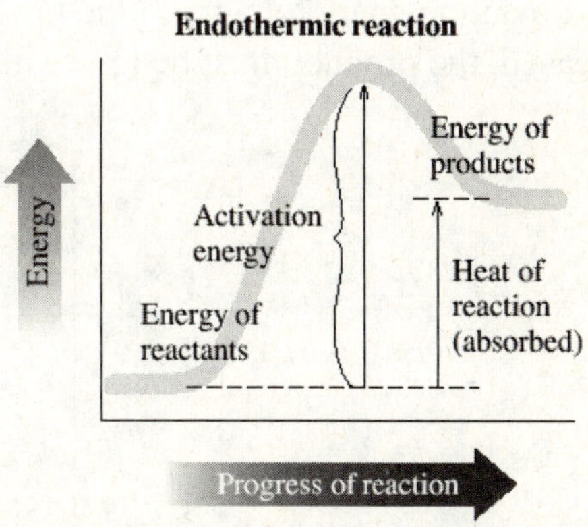

b. A catalyst increases the rate of a chemical reaction by lowering the activation energy. See diagram below. The "height of the hill" between reactants and product is lower in the catalyzed pathway.

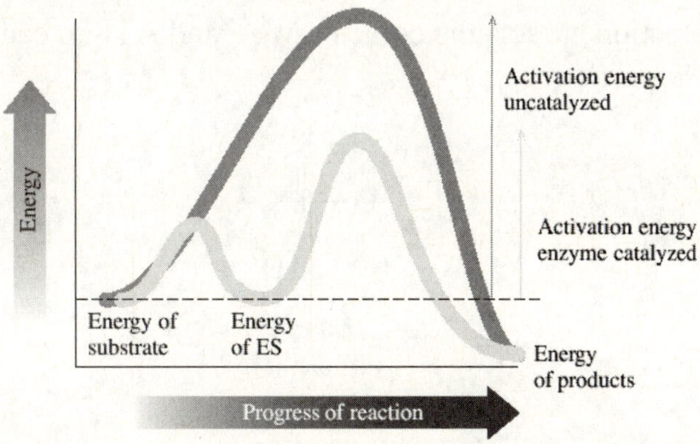

c. Since heat (energy) is released in the reaction, the products are lower in energy than the reactants. See diagram below.

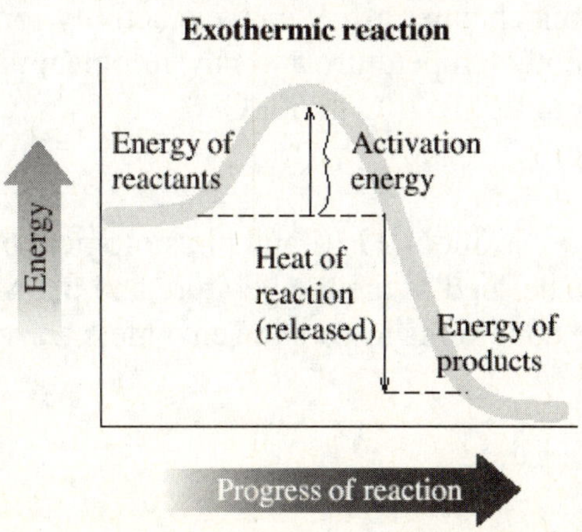

10.11 Classify the following as exothermic or endothermic reactions:
a. The reaction of hexokinase produces energy.
b. On a reaction energy diagram, the products are lower in energy than the reactants.

a. The reaction of hexokinase produces energy. If energy is produced (released), the reaction is exothermic.
b. On a reaction energy diagram, the products are lower in energy than the reactants. If energy is released, the product must be lower in energy than the reactants. See diagram below.

Exothermic reaction

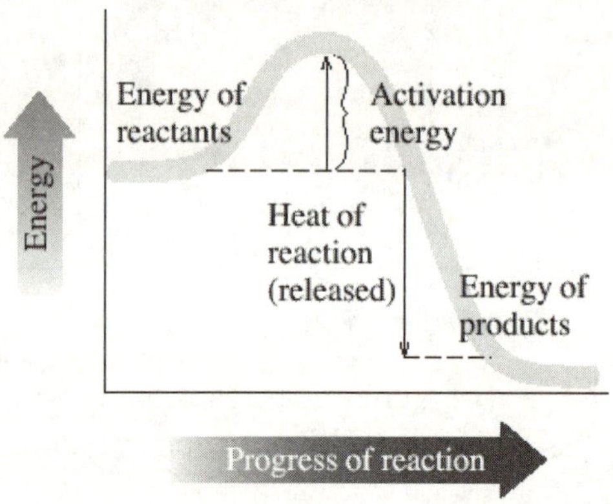

10.13 What kind of interaction attracts the cofactor Mg^{2+} and ATP to each other? (*Hint: look at the structure of the phosphate group.*)

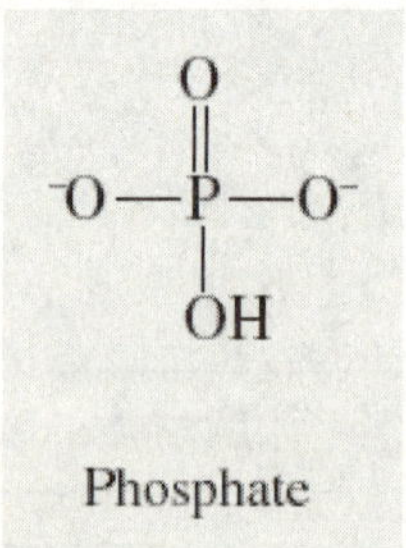

Because the phosphates on ATP have negative charges and Mg^{2+} has positive charges, the attraction is ionic.

10.15 How would the following changes affect enzyme activity for an enzyme whose optimal conditions are normal body temperature and physiological pH?
a. raising the temperature from 37 °C to 60 °C
b. lowering the pH from 7.5 to 4.0

Normal body temperature is about 37 °C and physiological pH is about 7.5. The optimal temperature and pH are defined as the temperature and pH where the enzyme is the most active. Any change from these conditions lowers the enzyme's activity.

10.17 The enzyme urease catalyzes the formation of ammonia and carbon dioxide from urea as shown below:

$$H_2N-\overset{\overset{\displaystyle O}{\|}}{C}-NH_2 + H_2O \xrightarrow{\text{Urease}} 2NH_3 + CO_2$$

Describe what effect the following changes would have on the rate of this reaction:
a. adding excess urea
b. lowering the temperature to 0 °C

a. adding excess urea: For many reactions, increasing the concentration of the reactant (substrate) will cause an increase in the rate of the reaction. This assumes that the concentration of the substrate (urea) is below the saturation level. If the enzyme is saturated with substrate, an increase in substrate concentration will not change the reaction rate (steady state).
b. lowering the temperature to 0 °C: Decreasing the temperature decreases the rate of a reaction

10.19 Penicillin is used to treat bacterial infections by destroying bacteria cells. Why does the antibiotic kill bacteria, but not humans?

Penicillin is an example of an irreversible inhibitor. Penicillin binds to the active site of an enzyme that bacteria use in the synthesis of their cell walls, structures present in bacterial cells but not in human cells.

Additional Problems

10.21 What occurs at the active site of an enzyme?

The active site is the location on an enzyme where catalysis occurs. This is the place where the enzyme binds to the substrate to form the ES complex.

10.23 Match the terms, (1) active site, (2) lock-and-key model, and (3) induced-fit model, with the following descriptions:
a. the portion of an enzyme where catalytic activity occurs
b. the active site adapts to the shape of a substrate
c. the active site has a rigid shape

Definitions:
active site – A pocket on the surface of an enzyme formed during the folding of the tertiary structure and lined with amino acid side chains; portion of the enzyme involved in the catalysis of chemical reactions.
induced-fit model – A model describing the initial interaction of an enzyme to its substrate. An enzyme's active site and its substrate have flexible shapes that are roughly complementary and adjust to allow the formation of the enzyme-substrate complex.
lock-and-key model – A model describing the initial interaction of an enzyme with its substrate. Each enzyme has only one substrate whose shape is complementary to the shape of the active site of the enzyme.

a. the portion of an enzyme where catalytic activity occurs: active site
b. the active site adapts to the shape of a substrate: induced-fit model
c. the active site has a rigid shape: lock-and-key model

10.25 The enzyme trypsin catalyzes the breakdown of many structurally diverse proteins in foods. Does the induced-fit or lock-and-key model explain the action of trypsin best? Explain.

induced-fit model – A model describing the initial interaction of an enzyme to its substrate. An enzyme's active site and its substrate have flexible shapes that are roughly complementary and adjust to allow the formation of the enzyme-substrate complex.
lock-and-key model – A model describing the initial interaction of an enzyme with its substrate. Each enzyme has only one substrate whose shape is complementary to the shape of the active site of the enzyme.

Because trypsin can have more than one substrate, trypsin's action is better described by the induced-fit model.

10.27 A diet deficient in thiamine results in a condition called beriberi, which is characterized by fatigue, weight loss, and poor appetite among several other symptoms. Patients with beriberi do not properly metabolize glucose and other sugars. If thiamine is not an enzyme, what kind of compound do you suppose it is? Explain.

Since thiamine is involved in enzymatic reactions, but is not an enzyme, it is either a cofactor or a coenzyme. Thiamine is organic in nature and is therefore a coenzyme.

10.29 Do each of the following statements describe a simple enzyme, one that requires a cofactor, or one that requires a coenzyme?
a. contains Ca^{2+} in the active site
b. consists of one polypeptide chain in its active form
c. contains vitamin B_6 in its active site

A coenzyme is a small organic molecule, many of which are derived from vitamins. A cofactor is an inorganic substance such as a Mg^{2+} that can be obtained from minerals in the diet.

a. contains Ca^{2+} in the active site: Ca^{2+} is inorganic, cofactor
b. consists of one polypeptide chain in its active form: does not require either an inorganic or organic component beyond the protein: simple enzyme
c. contains vitamin B_6 in its active site: vitamin B_6 is organic, coenzyme

10.31 Which reaction occurs at a faster rate, an exothermic reaction with a low activation energy or an endothermic reaction with a high activation energy? Explain.

An exothermic reaction with a low activation energy will occur at a faster rate. Reactions with lower activation energy occur more quickly regardless of whether they are exothermic or endothermic. The rate of reaction is determined only by the height of the energy barrier and is not influenced by the difference in energy between reactants and products.

10.33 Use the following reaction energy diagram shown to answer the questions:

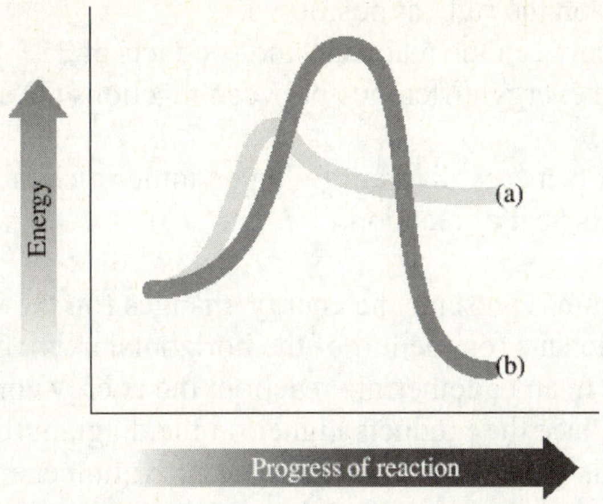

a. Which curve represents the faster reaction, and which represents the slower?
b. Which curve represents an endothermic reaction, and which curve represents an exothermic reaction?

a. Which curve represents the faster reaction, and which represents the slower? The "height of the hill" on each curve represents the activation energy. The lower the hill, the easier it is for the reactants to form products. Reaction energy diagram "a" represents the faster reaction.
b. Which curve represents an endothermic reaction, and which curve represents an exothermic reaction? When the height of the starting materials (reactants) is higher than the ending materials (products), the reaction is exothermic. When the height of the starting materials (reactants) is lower than the ending materials (products), the reaction is endothermic. In reaction energy diagram "b" the starting materials (on the left) are higher on the energy scale than the products (on the right). Reaction energy diagram is exothermic.

10.35 In the following reaction energy diagram, label 1, 2, 3, and 4 as one of the following: (a) energy of reactants, (b) energy of products, (c) activation energy, (d) heat of reaction.

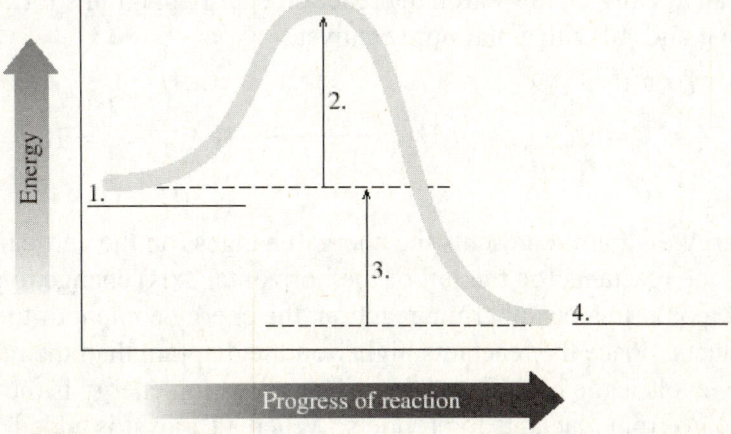

The diagram represents the energy changes (on the vertical axis) that occur along the time scale as the reactants (on the left of the horizontal axis) change to products (on the right of the horizontal axis).

(a) energy of reactants: on the left at position 1
(b) energy of products: on the right at position 4
(c) activation energy: between the reactants and products at 2
(d) heat of reaction: the energy difference between reactions (1) and products (4) or shown by 3

10.37 Draw and label a reaction energy diagram for an endothermic reaction whose activation energy is two times greater than its heat of reaction.

A reaction energy diagram represents the energy changes (on the vertical axis) that occur along the time scale as the reactants (on the left of the horizontal axis) change to products (on the right of the horizontal axis). In an endothermic reaction, the energy content of the products is greater than that of reactants. Place the products higher on the diagram than the reactants. The difference in the levels is the heat of reaction. The activation energy is the "energy hill" that must be scaled to go from reactants to products. Make the height of the "hill" twice as large as the difference between the energy content of reactants and products.

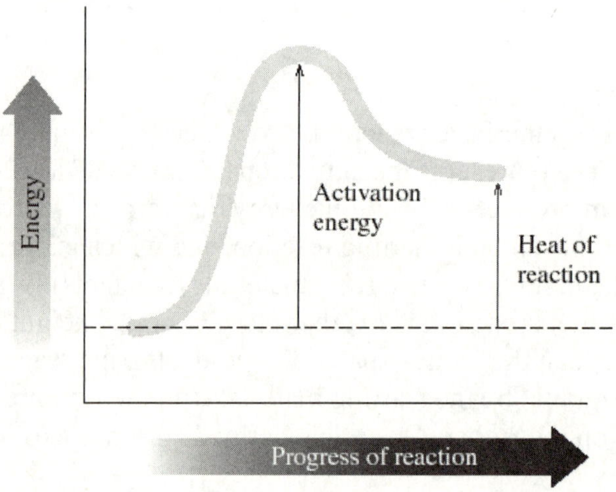

10.39 The hydrogenation reaction shown below is moderately exothermic, but it occurs at a very slow rate in the absence of a catalyst. When a platinum catalyst is added to the reaction, the reaction rate increases dramatically. Draw and label reaction energy diagrams for this reaction with (a) no catalyst present and (b) with a platinum catalyst.

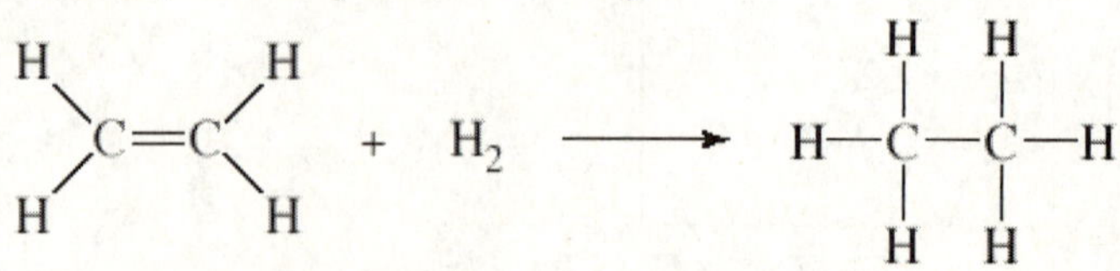

The reaction energy diagram represents the energy changes (on the vertical axis) that occur along the time scale as the reactants (on the left of the horizontal axis) change to products (on the right of the horizontal axis). In an exothermic reaction, the energy content of the reactants is greater than that of products. Place the reactants higher on the diagram than the reactants. The difference in the levels is the heat of reaction. The activation energy is the "energy hill" that must be scaled to go from reactants to products. When a catalyst is added, the energy content of the reactants and products (and the difference between them) is constant but the height of the "energy hill" will be less.

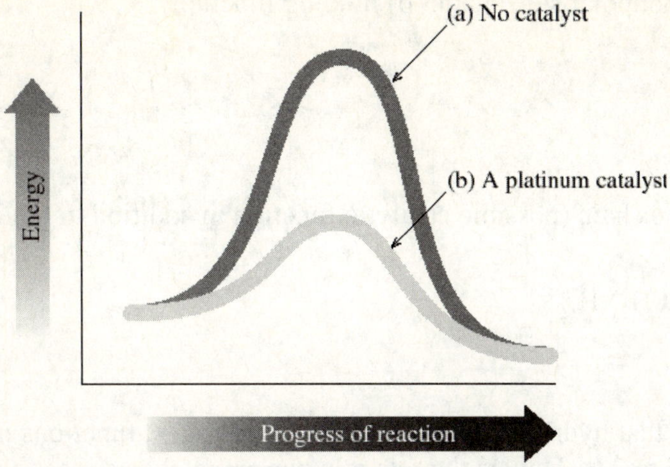

10.41 If each of the following amino acid side chains is present in the active site of an enzyme, indicate whether it would (a) serve a catalytic function, (b) serve to hold the substrate, or (c) both.

a. aspartate

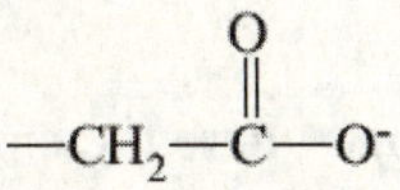

b. phenylalanine

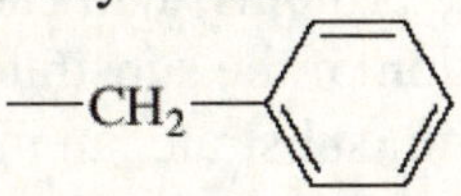

c. valine

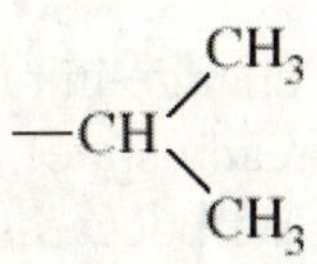

d. lysine

$—CH_2CH_2CH_2CH_2NH_3^+$

The amino acid side chains in the active site of an enzyme are the "tools" used by the enzyme to help facilitate the reaction. These side chains are often the functional groups of the acidic and basic amino acids.

a. aspartate: acidic side chain (possible catalytic function in addition to holding function): c) both

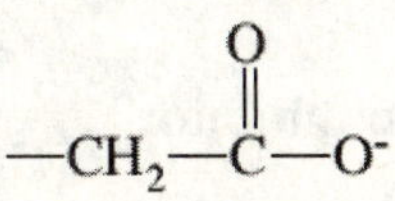

b. phenylalanine: neutral nonpolar side chain: b) holding function

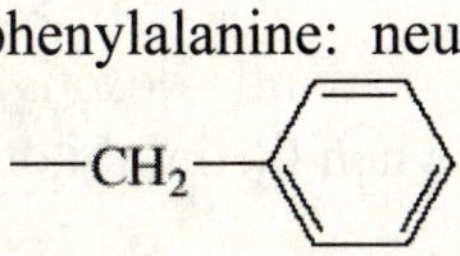

c. valine: neutral nonpolar side chain: b) holding function

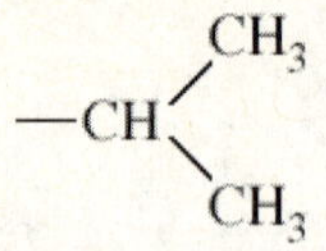

d. lysine: basic side chain (possible catalytic function in addition to holding function), c) both

$—CH_2CH_2CH_2CH_2NH_3^+$

10.43 Pepsin, an enzyme that hydrolyzes peptide bonds in proteins, functions in the stomach at an optimum pH of 1.5 to 2.0. How is the rate of a pepsin-catalyzed reaction affected by each of the following conditions?
a. increasing the concentration of proteins
b. changing the pH to 5.0
c. running the reaction at 0°C

The optimum temperature and pH are defined as the temperature and pH where the enzyme is the most active. Any change from these conditions lowers the enzyme's activity. For many reactions, increasing the concentration of the reactant (substrate, here proteins) will cause an increase in the rate of the reaction. This assumes that the concentration of the substrate (proteins) is below the saturation level. If the enzyme is saturated with substrate, an increase in substrate concentration will not change the reaction rate (steady state).

a. increasing the concentration of proteins: If the enzyme is saturated, no change in rate. If the enzyme is not saturated, the rate will increase until saturation occurs (steady state).
b. changing the pH to 5.0: This pH is outside the optimum range, the rate will decrease.
c. running the reaction at 0°C: Assuming the optimum temperature is body temperature (≈37°C), decreasing the temperature will cause a decrease in reaction rate. This change is fairly drastic and may even denature the proteins.

10.45 After the reaction catalyzed by an enzyme is complete, why do the products of the reaction migrate away from the active site?

The enzyme is less attracted to the products than the substrates. As a result, the products do not fit as snugly into the active site and so are released.

10.47 How does an irreversible inhibitor function differently than a reversible inhibitor?

An irreversible inhibitor forms covalent bonds to the enzyme and permanently deactivates the enzyme. A reversible inhibitor does not form permanent bonds and when the inhibitor is removed the enzyme regains its activity.

10.49 Increasing the substrate concentration of an enzyme-catalyzed reaction increases the rate of the reaction until the enzyme becomes saturated and reaches its maximum activity. If the amount of *enzyme* in the reaction is increased under saturating conditions, would the rate of the reaction increase, decrease, or remain unchanged? Explain.

If the amount of enzyme were increased in a reaction occurring at a steady state, the maximum rate of the reaction would increase since there is now free enzyme to form more ES. More substrate would be converted into products resulting in an increase in reaction rate until saturation again occurs.

10.51 Cadmium is a poisonous metal used in industries that produce batteries and plastics. Cadmium ions (Cd^{2+}) are inhibitors of hexokinase. Increasing the concentration of glucose or ATP, the substrates of hexokinase, or Mg^{2+}, the cofactor of hexokinase, does not change the rate of the cadmium-inhibited reaction. Is cadmium a competitive or noncompetitive inhibitor? Explain.

Cadmium would be a noncompetitive inhibitor because increasing the substrate and cofactor has no effect on the rate. This implies that the cadmium is binding to another site on the enzyme. In competitive inhibition, increasing the substrate and cofactor would increase the rate.

10.53 Many drugs are competitive inhibitors of enzymes. When scientists design inhibitors to serve as drugs, why do you suppose they choose to design competitive inhibitors instead of noncompetitive inhibitors?

When designing an inhibitor, usually the substrate of an enzyme is **known** even if the structure of the enzyme is not. It is easier to design a molecule that resembles the substrate (competitive) than to find an inhibitor that binds to a second site on an enzyme (noncompetitive) when the configuration of other sites on the enzyme are **unknown**. It comes down to working with what you know.

Challenge Problems

10.55 The active site of a hypothetical enzyme contains a pocket of nonpolar amino acid side chains next to a cleft containing acidic and basic amino acid side chains.
a. Which amino acids are likely responsible for catalysis? Explain.
b. Which amino acids are present primarily for holding the substrate in the active site?
c. Would you predict that the substrate is polar or nonpolar based on your answers to a and b? Why?

The amino acid side chains in the active site of an enzyme are the "tools" used by the enzyme to help facilitate the reaction. These side chains are often the functional groups of the acidic and basic amino acids.

a. The acidic and basic amino acids are likely involved in catalysis.
b. The nonpolar amino acids are likely involved in aligning the substrate correctly.
c. The substrate is likely nonpolar with a small polar portion because it, being held in place by nonpolar side chains. (A new way to look at "like dissolves like.")

Chapter 11

Nucleic Acids – Big Molecules with a Big Role

11.1 Components of Nucleic Acids

Learning Objectives

Upon completion of this material, a student should be able to do the following:

A. define the following key terms:

DNA	genome	RNA
nucleotide	nucleoside	

B. distinguish between purines and pyrimidines
C. identify the parts of a nucleoside
D. predict the nucleoside product of the reaction of a sugar and a base
E. identify the parts of a nucleotide
F. provide the products of condensation and hydrolysis of a nucleotide
G. give the names and abbreviations of the four nucleotides in DNA
H. give the names and abbreviations of the four nucleotides in RNA

11.2 Nucleic Acid Formation

Learning Objectives

Upon completion of this material, a student should be able to do the following:

A. define the following key terms:

nucleic acid	phosphodiester bond

B. identify a phosphodiester bond in the structure of a nucleic acid
C. identify the 3' end, 5' end, and the component nucleotides of a nucleic acid
D. formulate standard abbreviations for the structure of a nucleic acid
E. draw structures for RNA dinucleotides
F. draw structures for DNA dinucleotides

11.3 DNA

Learning Objectives

Upon completion of this material, a student should be able to do the following:

A. define the following key terms:

double helix	complementary base pair	supercoiling
chromosome		

B. describe the secondary structure of DNA
C. identify the forces involved in the secondary structure of DNA
D. provide the complementary DNA strand for a given DNA strand
E. distinguish primary, secondary, and tertiary structures of proteins and DNA

11.4 RNA and Protein Synthesis

Learning Objectives

Upon completion of this material, a student should be able to do the following:

A. define the following key terms:

transcription	mRNA	RNA polymerase
rRNA	translation	tRNA
anticodon		

B. describe the differences between DNA and RNA
C. distinguish the types of RNA and how they are used in creating protein from DNA code
D. distinguish the processes of transcription and translation
E. specify the complementary bases found in RNA
F. provide the complementary RNA strand for a given DNA strand
G. describe the structure of tRNA in terms of anticodons

11.5 Putting It Together: The Genetic Code and Protein Synthesis

Learning Objectives

Upon completion of this material, a student should be able to do the following:

A. define the following key terms:

codon	genetic code

B. distinguish codons and anticodons
C. specify the sequence of amino acids coded by a segment of mRNA
D. describe translocation
E. describe termination
F. translate the DNA genetic code into peptides
   1.) write the corresponding segment of mRNA
   2.) identify the triplets
   3.) label the 3' and 5' ends
   4.) write the anticodons
   5.) write the amino acids that will form the peptide chain

11.6 Genetic Mutations

Learning Objectives

Upon completion of this material, a student should be able to do the following:

A. define the following key terms:

mutation	silent mutation	spontaneous mutation
mutagen	carcinogen	somatic cell
germ cell	genetic disease	

B. define genetic mutation
C. describe the effect of the alternation of one nucleotide in a codon on protein synthesis
D. write the amino acid sequence that results from a mutation in a DNA nucleotide
E. distinguish the amino acid sequence resulting from a mutation and the normal sequence based on the structure of mRNA
F. distinguish silent mutations and spontaneous mutations
G. distinguish mutagens and carcinogens
H. distinguish the effects of mutations in somatic and germ cells

11.7 Viruses

Learning Objectives

Upon completion of this material, a student should be able to do the following:

A. define the following key terms:

host	capsid	envelope
retrovirus		

B. describe the general structure of a virus
C. describe how a virus duplicates itself
D. distinguish a virus and a retrovirus
E. specify the purpose of a vaccine
F. distinguish nucleoside analogs and protease inhibitors in HIV-1 virus therapy

11.8 Recombinant DNA Technology

Learning Objectives

Upon completion of this material, a student should be able to do the following:

A. define the following key terms:

recombinant DNA	restriction enzyme	vector
plasmid	expression	resonance hybrid

B. distinguish DNA recombination from organismal cloning
C. describe the steps involved in producing recombinant DNA, including the importance of restriction enzymes, vectors, and expression
D. identify applications of recombinant DNA technology

Practice Test for Chapter 11

1. The following substance is classified as:

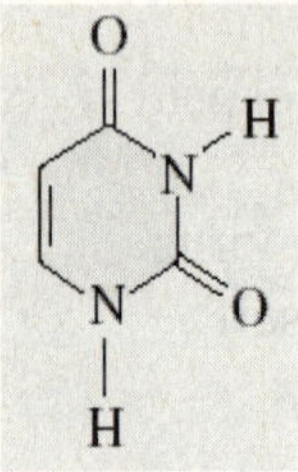

A. a purine.
B. a pyrimidine.
C. ribose.
D. deoxyribose.

2. The following substance is classified as:

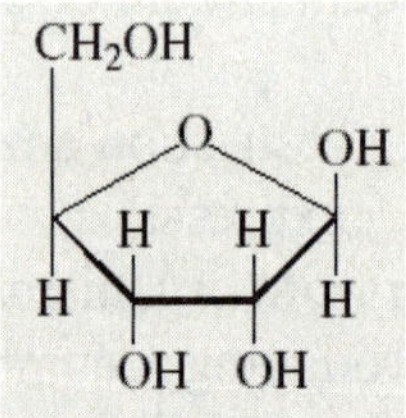

A. a purine.
B. a pyrimidine.
C. ribose.
D. deoxyribose.

3. If a condensation reaction occurred between the following two substances, the type of product formed would be:

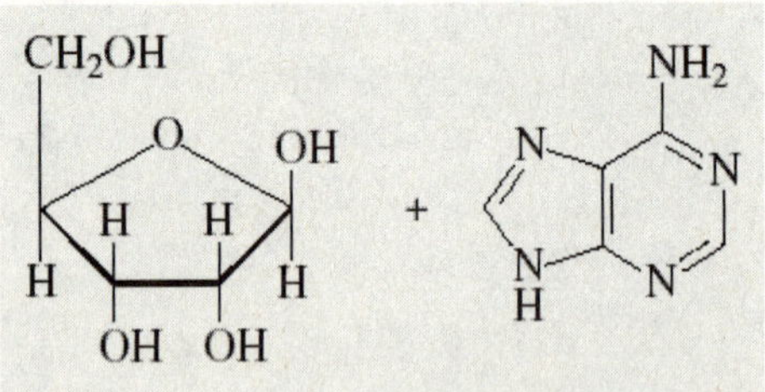

A. DNA.
B. a nucleotide.
C. a nucleoside.
D. RNA.

4. What is the name of the product of the following reaction?

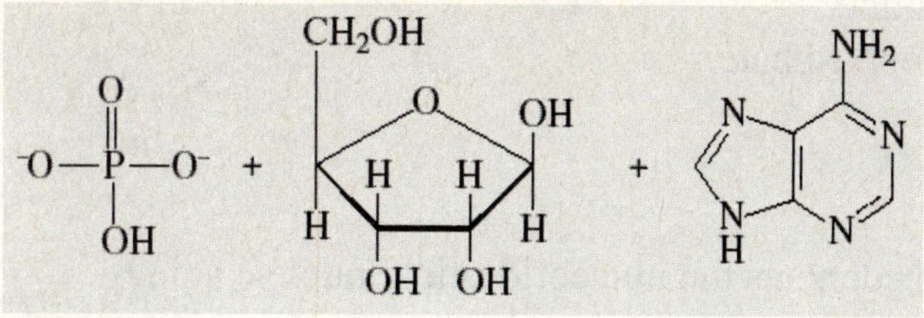

A. adenosine 5'-monophosphate
B. adenosine nucleotide
C. adenosine phosphate
D. AMP
E. A or D

5. Where can the following substance be found?

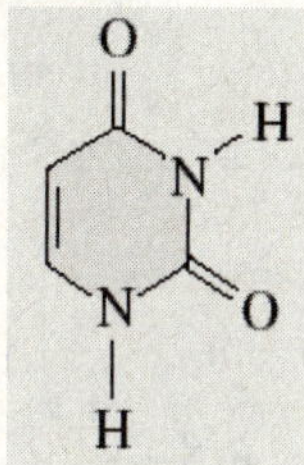

A. RNA
B. DNA
C. both DNA and RNA
D. neither DNA nor RNA

6. The difference between CMP and dCMP is found in which of the components in the make up?
A. sugar
B. phosphate
C. nitrogenous base
D. the bonds linking the components together

7. Consider the following structure.

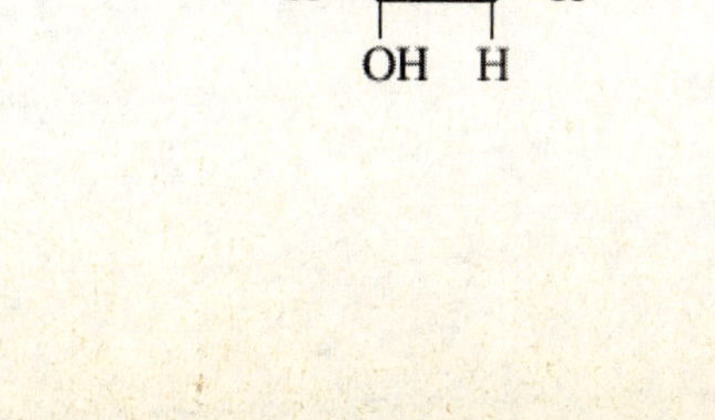

This represents:
A. an RNA nucleotide.
B. uridine 5'-monophosphate.
C. TMP.
D. dTMP.

8. What type of bond connects the nucleotides in a nucleic acid?
A. amide
B. phosphodiester
C. peptide
D. ester

9. Consider the following structure.

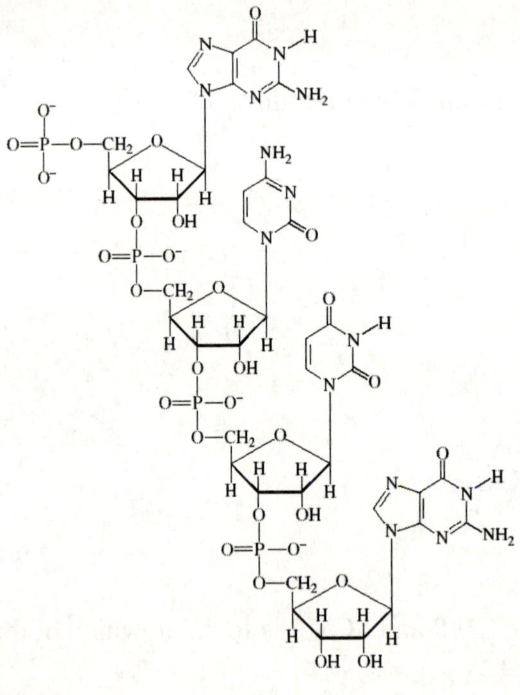

Which of the following is correct?
A. has pyrimidine bases at the 5' and 3' ends
B. contains 4 phosphodiester bonds
C. contains 4 nucleotides
D. could be found in DNA
E. All of the above are correct.

10. What is the abbreviation for the following structure?

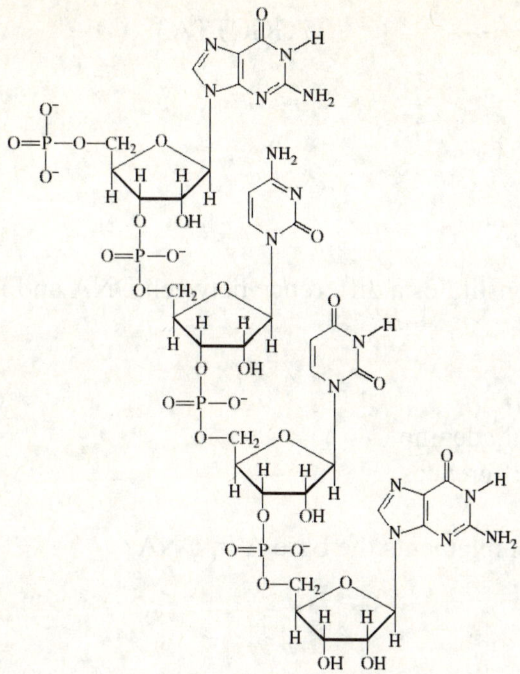

A. 5'GCUG3'
B. 5'GUCG3'
C. 5'GUTG3'
D. 5'GAUG3'

11. The nucleic acid analog of the N-terminus of a protein is:
A. nucleotide.
B. 5' end.
C. 3' end.
D. phosphodiester bond.

12. The double helix structure of DNA is stabilized by:
A. phosphodiester bonds.
B. covalent bonds.
C. hydrogen bonds.
D. disulfide bridges.

13. Which of the following is **not** a complementary base pair found in DNA?
A. A and T
B. G and C
C. A and U
D. All are complementary DNA base pairs.

14. What is the complementary DNA segment 3' to 5' for the segment shown below?

5'GGCTTA3'

A. 3'ATTCGG5'
B. 3'GGCTTA5'
C. 3'TAAGCC5'
D. 3'CCGAAT5'

15. Which of the following constitutes a difference between DNA and RNA?
A. sugar present
B. strand structure
C. size
D. complementary base of adenine
E. All of the above are differences.

16. Which base in mRNA complements the base A in DNA?
A. T
B. U
C. G
D. C

17. The sequence of bases in a DNA template is:

5'ATCGAT3'

What is the corresponding mRNA that is produced from this DNA?
A. 3'TAGCTA5'
B. 3'TAGCTA5'
C. 3'UAGCUA5'
D. 3'AUCGAU5'

18. Which type of RNA provides the amino acids for the growing peptide chain?
A. tRNA
B. mRNA
C. rRNA
D. none of the above

19. The synthesis of a protein based on a DNA template is called:
A. translocation.
B. transcription.
C. translation.
D. transmutation.

20. What do the following mRNA codons have in common?

UAA UAG UGA

A. stop protein synthesis
B. start protein synthesis
C. code for the same amino acid
D. have nothing in common

21. What is the sequence of amino acids coded by the following codons in mRNA?

5' CCA|GUC|AAA|GCC 3'

A. Ala—Lys—Val—Pro
B. Pro—Lys—Val—Ala
C. Pro—Val—Lys—Ala
D. Ala—Val—Lys—Pro

22. The conversion of the information on the codons in mRNA to a peptide chain occurs:
A. in the nucleus of the cell.
B. on the nuclear membrane.
C. in the cytoplasm.
D. in the ribosome.

23. Consider the following representation of tRNA.

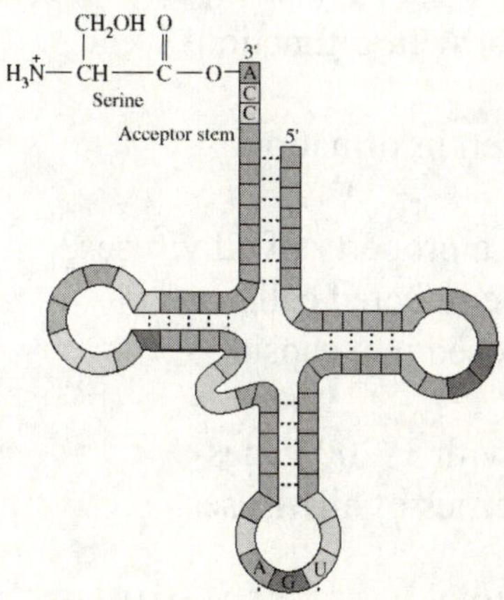

This tRNA is a complement to what codon on mRNA?
A. 5'AGU3'
B. 5'UCA3'
C. 5'TCA3'
D. 5'ACU3'

24. Which of the following represents the correct sequence of events in protein synthesis?
A. transcription, activation, translation, termination
B. transcription, activation, termination, translation
C. translation, activation, transcription, termination
D. activation, transcription, translation, termination

25. Consider the following segment of DNA.

3' GGG|TTA|CAC|ATT 5'

What amino acids will be placed in the peptide chain from this segment?
A. Val—Asn—Pro
B. Pro—Asn—Val—Tyr
C. Pro—Asn—Val
D. Val—Leu—His—Ile

26. If a codon in mRNA was changed from UGU to UGC, this change could most specifically be called a:
A. mutagen.
B. mutation.
C. silent mutation.
D. carcinogen.

27. A normal DNA template produces the following mRNA sequence:

5'ACU|UAC|CGG3'

If a change in the DNA template changes the mRNA sequence to:

5'ACU|UAG|CGG3'

How would this change affect the protein produced?
A. silent mutation
B. mutation affecting protein structure/function
C. no effect at all
D. cannot predict with the given information

28. Which of the following is **not** a property of all viruses?
A. utilize the ribosomes of the infected cells
B. contain nucleic acids enclosed in a capsid
C. can infect any cell type
D. consist of small particles with 3–200 genes
E. All of the above are properties of all viruses.

29. Which of the following correctly pertains to the HIV-1 virus and current therapies?
A. HIV-1 is a retrovirus.
B. Nucleoside analogs are competitive inhibitors.
C. Protease inhibitors halt transcription.
D. Current research involves the use of mutagens.

30. In recombinant DNA technology, the synthesis of a protein cloned into a second organism is called a(n):
A. vector.
B. somatic cell.
C. expression.
D. restriction enzyme.

Answers

1. B	**2.** C	**3.** C	**4.** E	**5.** A	**6.** A	**7.** D	**8.** B	**9.** C	**10.** A
11. B	**12.** C	**13.** C	**14.** D	**15.** E	**16.** C	**17.** A	**18.** B	**19.** C	**20.** A
21. C	**22.** D	**23.** B	**24.** A	**25.** C	**26.** C	**27.** B	**28.** E	**29.** A	**30.** C

Chapter 11 – Solutions to Odd-Numbered Problems

Practice Problems

11.1 Identify each of the following as being a purine or a pyrimidine:

a. thymine **b.**

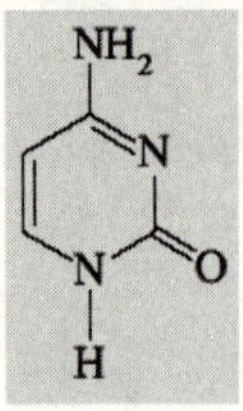

There are five different nitrogenous bases in nucleic acids, and each of the bases contains one of two nitrogen-containing aromatic rings, either the purine ring or the pyrimidine ring.

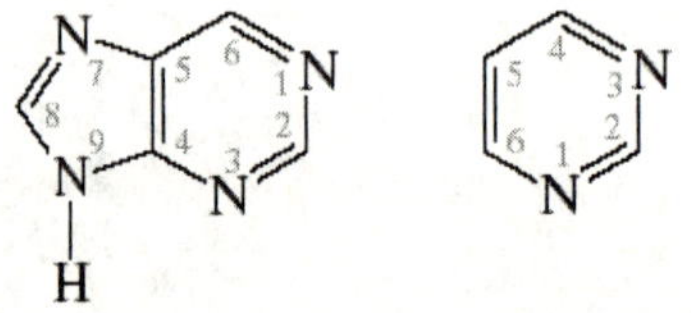

Purine *Pyrimidine*

Compare the ring structures to determine the type of nitrogenous base.

a. pyrimidine: ring structure matches that of the base shown above on the right.

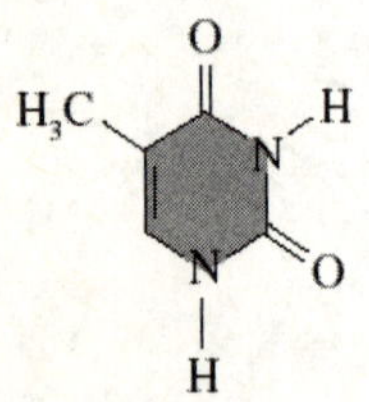

b. pyrimidine: ring structure matches that of the base shown above on the right.

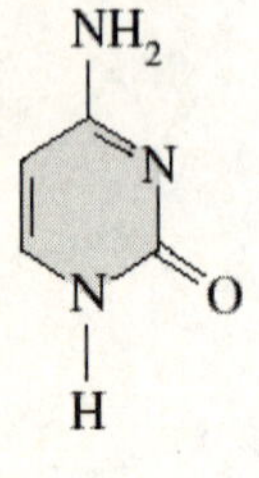

11.3 Identify the bases in problem 11.1 as present in RNA, DNA, or both.

DNA contains two purines, adenine (A) and guanine (G), and two pyrimidines, thymine (T) and cytosine (C). RNA contains the same bases, except thymine (5-methyluracil) is replaced with uracil (U). See structures below.

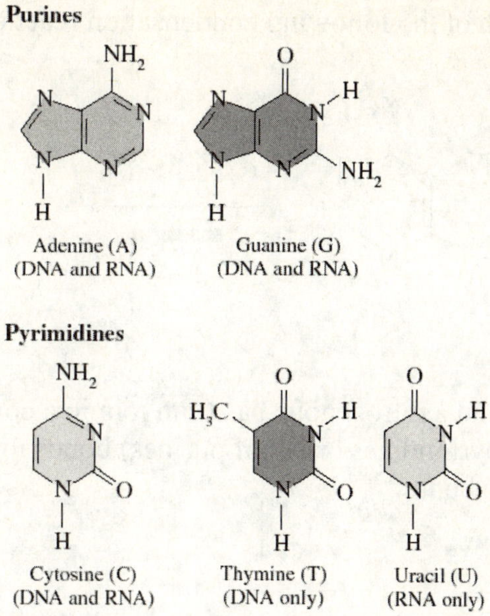

a. thymine: DNA
b. cytosine: DNA and RNA (both)

11.5 List the names and abbreviations of the four nucleotides in DNA.

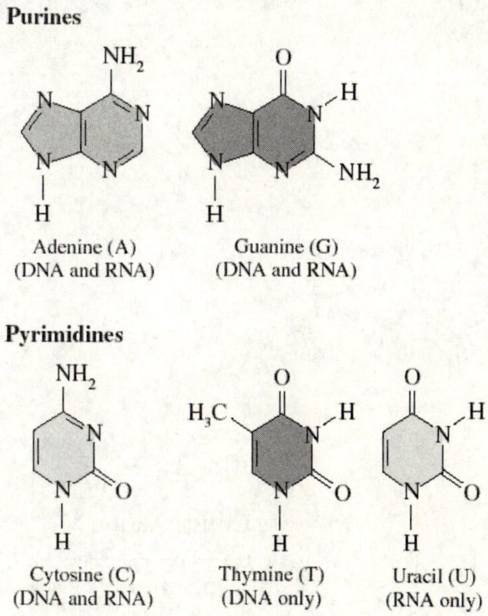

adenine: A
guanine: G
cytosine: C
thymine: T

11.7 Provide the products for each of the following condensation reactions:

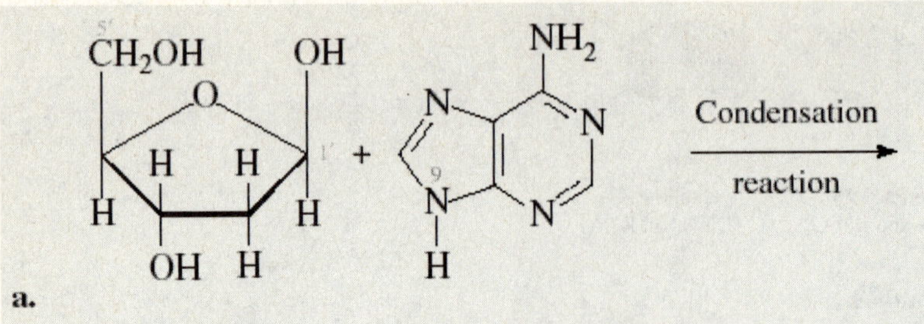

a.

A pentose monosaccharide and a nitrogenous base can join in a condensation reaction when a nitrogen in the base (N1 of pyrimidines or N9 of purines) bonds to C1' of the pentose, forming a carbon-to-nitrogen glycoside bond.

a.

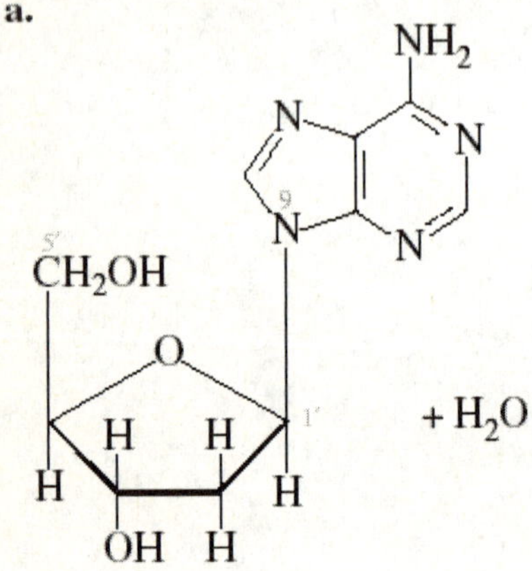

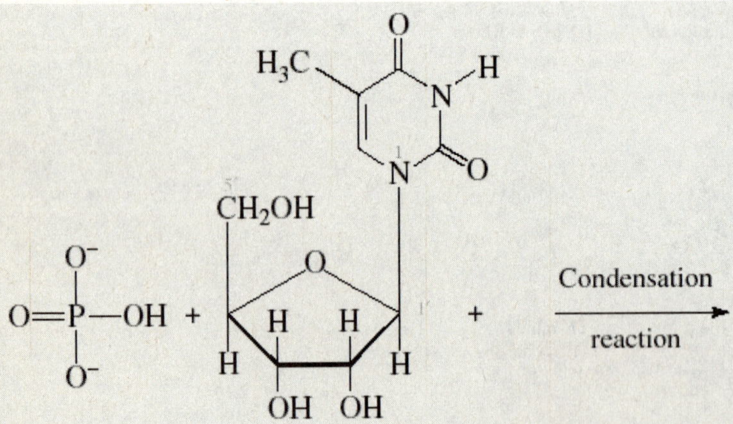

b.

A nucleotide forms in a condensation reaction between a hydrogen phosphate (HPO_4^{2-}) with the C5' -OH of the pentose in the nucleoside molecule. In this reaction a molecule of water is eliminated and the other product of the reaction is the nucleotide.

b.

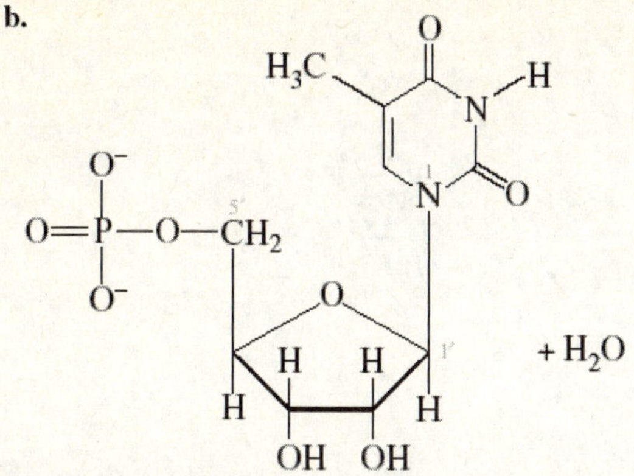

11.9 How are nucleotides bonded together in a nucleic acid chain?

The nucleotides in nucleic acids are held together by phosphodiester bonds between the 3'-OH of a sugar (ribose or deoxyribose) and a phosphate group on the 5'-carbon of another sugar. See example below.

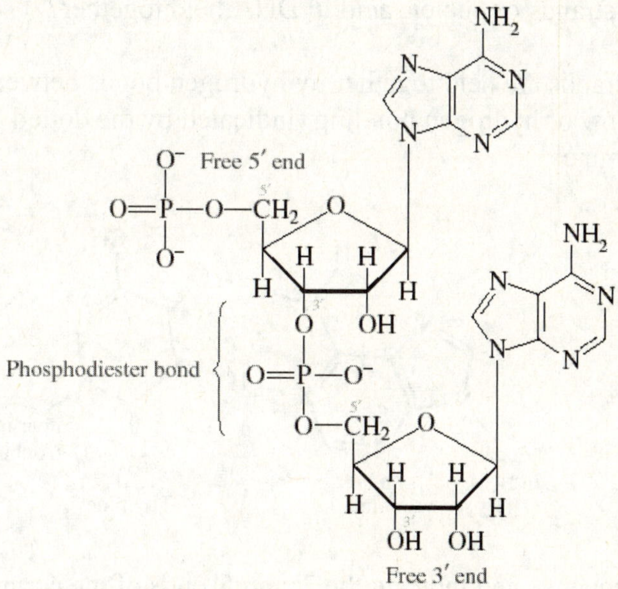

11.11 Draw the dinucleotide GC that would be found in RNA. Label the 5' and 3' end of your structure. Identify the phosphodiester bond.

The G (guanine) nucleotide is at the 5' end and the C (cytosine) nucleotide is at the 3' end. The two nucleotides are linked together with a phosphodiester bond.

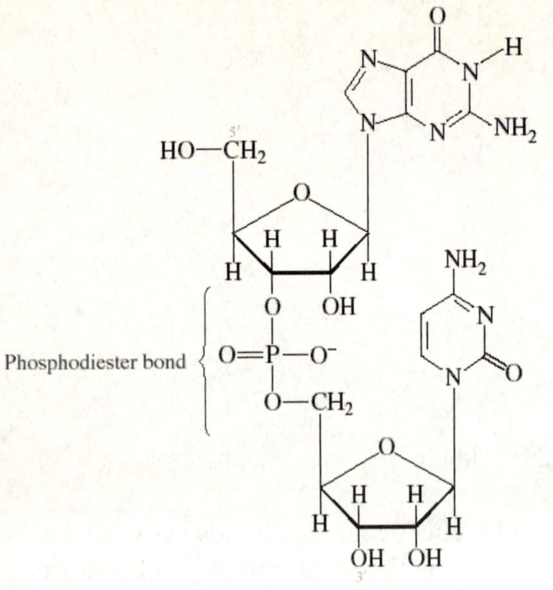

11.13 How are the two strands of nucleic acid in DNA held together?

The two DNA strands are held together by hydrogen bonds between the bases in each strand. See example below of hydrogen bonding (indicated by the dotted lines and arrow) between adenine and thymine

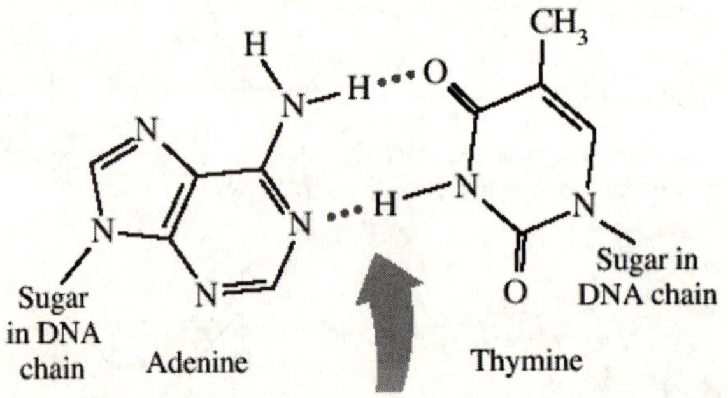

11.15 Write the base sequence and indicate the 3' and 5' ends of the complementary strand for a segment of DNA with the following base sequences:

a. 5'AAAA3'
b. 5'CCCCTTTT3'
c. 5'ACATTGG3'
d. 5'TGTGAACC3'

The pairs A–T and G–C are called complementary base pairs. The two strands of DNA run antiparallel to each other. One strand goes in the 5' to 3' direction and the other strand goes in the 3' to 5' direction.

a. 3'TTTT5'
b. 3'GGGGAAAA5'
c. 3'TGTAACC5'
d. 3'ACACTTGG5'

11.17 Fill in the following table with the analogous nucleic acid structures:

	Protein	Nucleic Acid
Primary structure	Sequence of amino acids	
Secondary structure	Local hydrogen bonding between backbone atoms	
Tertiary structure	Folding of backbone associating amino acids far away in sequence	

	Protein	Nucleic Acid
Primary structure	Sequence of amino acids	Sequence of nucleotides
Secondary structure	Local hydrogen bonding between backbone atoms	Hydrogen bonding between bases in nucleic acid strands
Tertiary structure	Folding of backbone associating amino acids far away in sequence	Supercoiling of DNA into tight, compact structures like chromosomes

11.19 Name the three types of RNA.

Messenger RNA (mRNA), ribosomal RNA (rRNA), and transfer RNA (tRNA).

11.21 In your own words, define the term transcription.

In transcription, the sequence of nucleotides on a DNA template (one strand) is used to produce a complementary mRNA copy using the enzyme DNA polymerase.

11.23 List the mRNA bases that complement the bases A, T, G, and C in DNA.

A in DNA complements with U in mRNA, T in DNA complements with A in mRNA, G in DNA complements with C in mRNA, and C in DNA complements with G in mRNA.

11.25 The sequence of bases in a DNA template strand is 5'TACGGCAAGCTA3'. What is the corresponding mRNA produced?

Use the following to determine the pairing of bases:

A in DNA complements with U in mRNA, T in DNA complements with A in mRNA, G in DNA complements with C in mRNA, and C in DNA complements with G in mRNA.

3'AUGCCGUUCGAU5'

11.27 Why are there at least 20 tRNAs?

There are 20 amino acids that are brought into the ribosome, each bonded to a different tRNA.

11.29 Provide the three-letter amino acid sequence expected from each of the following mRNA segments:

a. 5' AAA | CCC | UUG | GCC 3'
b. 5' CCU | CGA | AGC | CCA | UGA 3'
c. 5' AUG | CAC | AAA | GAA | GUA | CUU 3'

Reference the following table to determine the amino acid to which each mRNA sequence corresponds.

TABLE 11.3 mRNA CODONS: THE GENETIC CODE FOR AMINO ACIDS

First Letter from 5' End	Second Letter: U	C	A	G	Third Letter
U	UUU Phe	UCU Ser	UAU Tyr	UGU Cys	U
	UUC	UCC	UAC	UGC	C
	UUA Leu	UCA	UAA STOP	UGA STOP	A
	UUG	UCG	UAG STOP	UGG Trp	G
C	CUU Leu	CCU Pro	CAU His	CGU Arg	U
	CUC	CCC	CAC	CGC	C
	CUA	CCA	CAA Gln	CGA	A
	CUG	CCG	CAG	CGG	G
A	AUU Ile	ACU Thr	AAU Asn	AGU Ser	U
	AUC	ACC	AAC	AGC	C
	AUA	ACA	AAA Lys	AGA Arg	A
	[a]AUG Met/start	ACG	AAG	AGG	G
G	GUU Val	GCU Ala	GAU Asp	GGU Gly	U
	GUC	GCC	GAC	GGC	C
	GUA	GCA	GAA Glu	GGA	A
	GUG	GCG	GAG	GGG	G

a. Lys—Pro—Leu—Ala
b. Pro—Arg—Ser—Pro—Stop
c. Met—His—Lys—Glu—Val—Leu

11.31 In your own words define translocation.

Translocation occurs when the ribosome moves along the mRNA reading the codons.

11.33 The following portion of DNA is in the template DNA strand:

3' TGT | GGG | GTT | ATT 5'

a. Write the corresponding mRNA section. Show the nucleic acid sequence as triplets and label the 5' and the 3' end.
b. Write the anticodons corresponding to the codons on the mRNA.
c. Write the amino acids that will be placed in a peptide chain.

Use the following to determine the pairing of bases:

A in DNA complements with U in mRNA, T in DNA complements with A in mRNA, G in DNA complements with C in mRNA, and C in DNA complements with G in mRNA.

a. 5' ACA | CCC | CAA | UAA 3'
b. 3' UGU | GGG | GUU | AUU 5'

Reference the following table to determine the amino acid to which each mRNA sequence corresponds.

TABLE 11.3 mRNA CODONS: THE GENETIC CODE FOR AMINO ACIDS

First Letter from 5′ End	Second Letter U	Second Letter C	Second Letter A	Second Letter G	Third Letter
U	UUU Phe	UCU Ser	UAU Tyr	UGU Cys	U
	UUC	UCC	UAC	UGC	C
	UUA Leu	UCA	UAA STOP	UGA STOP	A
	UUG	UCG	UAG STOP	UGG Trp	G
C	CUU Leu	CCU Pro	CAU His	CGU Arg	U
	CUC	CCC	CAC	CGC	C
	CUA	CCA	CAA Gln	CGA	A
	CUG	CCG	CAG	CGG	G
A	AUU Ile	ACU Thr	AAU Asn	AGU Ser	U
	AUC	ACC	AAC	AGC	C
	AUA	ACA	AAA Lys	AGA Arg	A
	[a]AUG Met/start	ACG	AAG	AGG	G
G	GUU Val	GCU Ala	GAU Asp	GGU Gly	U
	GUC	GCC	GAC	GGC	C
	GUA	GCA	GAA Glu	GGA	A
	GUG	GCG	GAG	GGG	G

c. three-letter code: Thr—Pro—Gln—Stop ; one-letter code: TPQ-Stop

11.35 In your own words define mutagen.

A mutagen is an environmental agent that produces a mutation (change in base sequence) in DNA.

11.37 Consider the following portion of mRNA produced by the normal order of DNA nucleotides

5' CUU | AAA | CGA | GUU 3'

a. Write the amino acid sequence that would be produced from this mRNA.

b. Write the amino acid sequence if a mutation changes CUU to AUU. Is this likely to affect protein function?

c. Write the amino acid sequence if a mutation changes CGA to AGA. Is this likely to affect protein function?

d. What happens to protein synthesis if a mutation changes AAA to UAA?

Reference the following table to determine the amino acid to which each mRNA sequence corresponds.

TABLE 11.3 mRNA CODONS: THE GENETIC CODE FOR AMINO ACIDS

First Letter from 5' End	Second Letter: U	C	A	G	Third Letter
U	UUU Phe	UCU Ser	UAU Tyr	UGU Cys	U
	UUC Phe	UCC Ser	UAC Tyr	UGC Cys	C
	UUA Leu	UCA Ser	UAA STOP	UGA STOP	A
	UUG Leu	UCG Ser	UAG STOP	UGG Trp	G
C	CUU Leu	CCU Pro	CAU His	CGU Arg	U
	CUC Leu	CCC Pro	CAC His	CGC Arg	C
	CUA Leu	CCA Pro	CAA Gln	CGA Arg	A
	CUG Leu	CCG Pro	CAG Gln	CGG Arg	G
A	AUU Ile	ACU Thr	AAU Asn	AGU Ser	U
	AUC Ile	ACC Thr	AAC Asn	AGC Ser	C
	AUA Ile	ACA Thr	AAA Lys	AGA Arg	A
	[a]AUG Met/start	ACG Thr	AAG Lys	AGG Arg	G
G	GUU Val	GCU Ala	GAU Asp	GGU Gly	U
	GUC Val	GCC Ala	GAC Asp	GGC Gly	C
	GUA Val	GCA Ala	GAA Glu	GGA Gly	A
	GUG Val	GCG Ala	GAG Glu	GGG Gly	G

a. Write the amino acid sequence that would be produced from this mRNA.

three-letter code: Leu—Lys—Arg—Val; one-letter code: LKRV

b. Write the amino acid sequence if a mutation changes CUU to AUU. Is this likely to affect protein function?

Leucine will be changed to isoleucine, which are both nonpolar, and will not have an effect on the protein function.

c. Write the amino acid sequence if a mutation changes CGA to AGA. Is this likely to affect protein function?

Nothing, both codons code for arginine.

d. What happens to protein synthesis if a mutation changes AAA to UAA?

Leucine will be the last amino acid incorporated into the growing protein chain because UAA is a stop codon.

11.39 Why do viruses need to enter a host cell?

Viruses cannot replicate themselves without a host cell. They do not have the cell machinery (enzymes, organelles, etc.).

11.41 What is the purpose of a vaccine?

A vaccine allows the body to mount an immune response to an inactive form of a virus so that if an active form is encountered later, the body can fight more effectively against it.

11.43 How do protease inhibitors disrupt the life cycle of the HIV-1 virus?

Protease inhibitors inactivate an HIV-1 specific protease that is needed to process viral proteins.

11.45 In your own words define gene expression.

Gene expression is producing a protein from a DNA gene sequence.

11.47 Describe the function of a vector.

A vector is a transporter for donor DNA. It enables incorporation of the donor DNA into the organism DNA.

Additional Problems

11.49 Identify each of the following bases as a pyrimidine or a purine:
a. cytosine **b.** adenine **c.** uracil **d.** thymine **e.** guanine

There are five different nitrogenous bases in nucleic acids, and each of the bases contains one of two nitrogen-containing aromatic rings, either the purine ring or the pyrimidine ring.

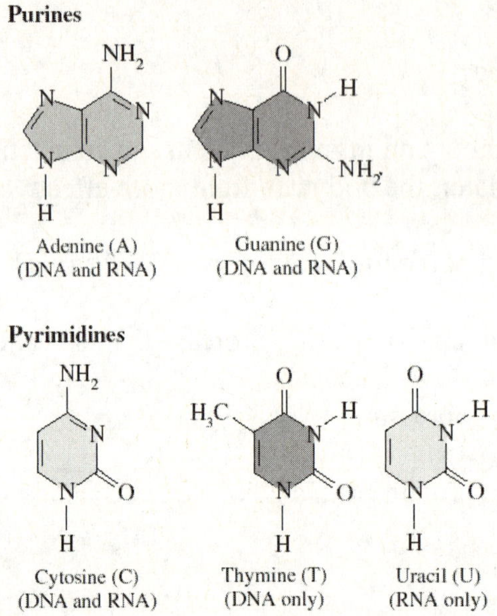

a. cytosine: pyrimidine
b. adenine: purine
c. uracil: pyrimidine
d. thymine: pyrimidine
e. guanine: purine

11.51 Identify the base and sugar in each of the following nucleotides.
a. CMP **b.** dAMP **c.** dGMP **d.** UMP

Nucleotides include the nucleoside name and the number of phosphates present. The base in the nucleoside is indicated with the usual one-letter designations (A, T, C, G, U) and the number of phosphates with the letter M (mono for 1 phosphate), D (di for 2 phosphates), and T (tri for 3 phosphates). If deoxyribose is found in the nucleotide, a lower case *d* is inserted at the beginning of the abbreviation.

a. CMP: cytosine, ribose
b. dAMP: adenine, deoxyribose
c. dGMP: guanine, deoxyribose
d. UMP: uracil, ribose

11.53 How do the bases thymine and uracil differ?

Thymine contains a methyl ($-CH_3$) group that uracil lacks.

11.55 Fill in the following table

	Protein	Nucleic Acid
Repeating unit	Amino acid	
Backbone repeat	N–C_α–C–N–C_α–C, etc.	
One-letter abbreviation	Name of amino acid	
Free left end	Amino or N terminus	
Free right end	Carboxy or C terminus	

	Protein	Nucleic Acid
Repeating unit	Amino acid	Nucleotide
Backbone repeat	N–C_α–C–N–C_α–C, etc.	sugar–phosphate–sugar–phosphate, etc.
One-letter abbreviation	Name of amino acid	Name of base
Free left end	Amino or N terminus	5′ end
Free right end	Carboxy or C terminus	3′ end

11.57 Write the complementary base sequence for each of the following DNA segments. Indicate the 5' and the 3' end.

a. 5'TTACGGACCGC3'
b. 5'ATAGCCCTTACTGG3'
c. 5'GGCCTACCTTAACGACG3'

The pairs A–T and G–C are called complementary base pairs. The two strands of DNA run antiparallel to each other. One strand goes in the 5' to 3' direction and the other strand goes in the 3' to 5' direction.

a. 3'AATGCCTGGCG5'
b. 3'TATCGGGAATGACC5'
c. 3'CCGGATGGAATTGCTGC5'

11.59 Match the following statements with mRNA, rRNA, or tRNA:

a. contains codons for protein synthesis
b. contains anticodons
c. found in the small subunit and the large subunit of the ribosome

Synopsis:
In transcription, DNA's double helix temporarily unwinds so that a complementary copy can be made from one of the strands. This complementary copy is the messenger RNA (mRNA). Once copying is completed, the mRNA travels from the nucleus to an organelle called the ribosome where the mRNA sequence is processed into protein. The ribosome is made up of ribosomal RNA (rRNA) and protein. The second step in the process of making a protein from DNA occurs in the ribosome and is called translation. The mRNA sequence must be translated into a protein sequence. The facilitator for this process is the transfer RNA (tRNA).

a. contains codons for protein synthesis: mRNA
b. contains anticodons: tRNA
c. found in the small subunit and the large subunit of the ribosome: rRNA

11.61 List the possible codons for each of the following amino acids:

a. valine
b. proline
c. histidine

Reference the following table to determine the amino acid to which each mRNA sequence corresponds.

TABLE 11.3 mRNA CODONS: THE GENETIC CODE FOR AMINO ACIDS

First Letter from 5′ End	Second Letter U	Second Letter C	Second Letter A	Second Letter G	Third Letter
U	UUU Phe	UCU Ser	UAU Tyr	UGU Cys	U
	UUC Phe	UCC Ser	UAC Tyr	UGC Cys	C
	UUA Leu	UCA Ser	UAA STOP	UGA STOP	A
	UUG Leu	UCG Ser	UAG STOP	UGG Trp	G
C	CUU Leu	CCU Pro	CAU His	CGU Arg	U
	CUC Leu	CCC Pro	CAC His	CGC Arg	C
	CUA Leu	CCA Pro	CAA Gln	CGA Arg	A
	CUG Leu	CCG Pro	CAG Gln	CGG Arg	G
A	AUU Ile	ACU Thr	AAU Asn	AGU Ser	U
	AUC Ile	ACC Thr	AAC Asn	AGC Ser	C
	AUA Ile	ACA Thr	AAA Lys	AGA Arg	A
	[a]AUG Met/start	ACG Thr	AAG Lys	AGG Arg	G
G	GUU Val	GCU Ala	GAU Asp	GGU Gly	U
	GUC Val	GCC Ala	GAC Asp	GGC Gly	C
	GUA Val	GCA Ala	GAA Glu	GGA Gly	A
	GUG Val	GCG Ala	GAG Glu	GGG Gly	G

a. GUU, GUC, GUA, GUG
b. CCU, CCC, CCA, CCG
c. CAU, CAC

11.63 Provide the amino acid corresponding to each of the following codons:
a. UUG **b.** CGG **c.** AUC

Reference the following table to determine the amino acid to which each mRNA sequence corresponds.

TABLE 11.3 mRNA CODONS: THE GENETIC CODE FOR AMINO ACIDS

First Letter from 5′ End	Second Letter: U	C	A	G	Third Letter
U	UUU Phe	UCU Ser	UAU Tyr	UGU Cys	U
	UUC Phe	UCC Ser	UAC Tyr	UGC Cys	C
	UUA Leu	UCA Ser	UAA STOP	UGA STOP	A
	UUG Leu	UCG Ser	UAG STOP	UGG Trp	G
C	CUU Leu	CCU Pro	CAU His	CGU Arg	U
	CUC Leu	CCC Pro	CAC His	CGC Arg	C
	CUA Leu	CCA Pro	CAA Gln	CGA Arg	A
	CUG Leu	CCG Pro	CAG Gln	CGG Arg	G
A	AUU Ile	ACU Thr	AAU Asn	AGU Ser	U
	AUC Ile	ACC Thr	AAC Asn	AGC Ser	C
	AUA Ile	ACA Thr	AAA Lys	AGA Arg	A
	[a]AUG Met/start	ACG Thr	AAG Lys	AGG Arg	G
G	GUU Val	GCU Ala	GAU Asp	GGU Gly	U
	GUC Val	GCC Ala	GAC Asp	GGC Gly	C
	GUA Val	GCA Ala	GAA Glu	GGA Gly	A
	GUG Val	GCG Ala	GAG Glu	GGG Gly	G

a. leucine
b. arginine
c. isoleucine

11.65 What is the one-letter amino acid sequence formed from the following mRNA that codes for a pentapeptide that is an endorphin called Met-enkephalin?

5' AUG | UAC | GGU | GGA | UUU | AUG | UAA 3'

Reference the following table to determine the amino acid to which each mRNA sequence corresponds.

TABLE 11.3 mRNA CODONS: THE GENETIC CODE FOR AMINO ACIDS

First Letter from 5′ End	Second Letter				Third Letter
	U	C	A	G	
U	UUU Phe	UCU Ser	UAU Tyr	UGU Cys	U
	UUC	UCC	UAC	UGC	C
	UUA Leu	UCA	UAA STOP	UGA STOP	A
	UUG	UCG	UAG STOP	UGG Trp	G
C	CUU Leu	CCU Pro	CAU His	CGU Arg	U
	CUC	CCC	CAC	CGC	C
	CUA	CCA	CAA Gln	CGA	A
	CUG	CCG	CAG	CGG	G
A	AUU Ile	ACU Thr	AAU Asn	AGU Ser	U
	AUC	ACC	AAC	AGC	C
	AUA	ACA	AAA Lys	AGA Arg	A
	[a]AUG Met/start	ACG	AAG	AGG	G
G	GUU Val	GCU Ala	GAU Asp	GGU Gly	U
	GUC	GCC	GAC	GGC	C
	GUA	GCA	GAA Glu	GGA	A
	GUG	GCG	GAG	GGG	G

Reference Table 9.1 for the single-letter codes of the amino acid.

NOTE: The first codon (AUG) is a start codon and the last codon (UAA) is a stop codon.

Amino acid names: tyrosine, glycine, glycine, phenylalanine, methionine

Three-letter codes: Tyr—Gly—Gly—Phe—Met

One-letter codes: YGGFM

11.67 What is the anticodon on tRNA for each of the following codons in a mRNA?

a. GUG b. CCC c. GAA

Use the following to determine the pairing of bases:

A in DNA complements with U in mRNA, T in DNA complements with A in mRNA, G in DNA complements with C in mRNA, and C in DNA complements with G in mRNA.

a. CAC b. GGG c. CUU

11.69 **a.** A base substitution for an enzyme replaces leucine (a nonpolar amino acid) with alanine. Why does this change in amino acids have little effect on the biological activity of the enzyme?

b. A base substitution replaces cytosine in the codon UCA with adenine. How might this substitution affect the amino acids in the protein?

a. Both alanine and leucine have small, nonpolar R-groups. Both amino acids are found in similar environments in proteins so this substitution will unlikely affect protein structure or function.

b. If a serine codon (UCA) is replaced with a stop codon, any remaining amino acids will not be linked onto the growing polypeptide chain. This can affect both the structure and function of the protein.

11.71 Discuss whether each of the following are viable areas for drug development to inactivate a retrovirus:

a. nucleoside analogs
b. integrase inhibitors
c. RNA polymerase inhibitors

a. Nucleoside analogs that can be incorporated into viral DNA to stop reverse transcription are a viable development area.

b. Integrase is the viral enzyme that incorporates viral DNA into host DNA, so this would be a viable method for inactivating the virus.

c. Because RNA polymerase is used to make all proteins in the cell, this would not be a viable enzyme to inactivate.

11.73 How do entry point inhibitors disrupt the life cycle of the HIV-1 virus?

Entry point inhibitors block sites on the outside of a host cell where a virus would attach and eventually enter. If the virus cannot enter the cell, the virus cannot replicate.

11.75 Distinguish between gene cloning and the cloning of an organism.

In gene cloning, a copy of a single gene is created. In organism cloning, many genes are copied, an exact copy of an organism is created.

Challenge Problems

11.77 A protein contains 35 amino acids. How many nucleotides would be found in the mRNA code for this protein?

Calculating formula: (Recall that there are 3 nucleotides in each codon.)

No. of nucleotides/amino acid + 1 start codon + 1 stop codon = $(35 \times 3) + 3 + 3 = 111$ nucleotides

11.79 The DNA double helix can unwind or denature at temperatures between 90–99 °C. Denaturing occurs when H-bonds are broken. Which of the following strands of DNA would be expected to denature at a higher temperature? Provide an explanation.

DNA strand 1	5′ATTTTCCAAAGTATA3′ 3′TAAAAGGTTTCATAT5′
DNA strand 2	5′CGCGAGGGTCCACGC3′ 3′GCGCTCCCAGGTGCG5′

Recall: Each G—C pair will form three hydrogen bonds while each A—T pair forms two hydrogen bonds.

The more hydrogen bonds present, the more energy (as heat) must be applied to denature the DNA. Because strand 2 has more G—C base pairs with three hydrogen bonds each, it is predicted to have the higher denaturation temperature.

Chapter 12

Food as Fuel – A Metabolic Overview

12.1 Overview of Metabolism

Learning Objectives

Upon completion of this material, a student should be able to do the following:

A. define the following key terms:

metabolism	metabolic pathway	metabolite
catabolism	anabolism	cytoplasm
cytosol	mitochondria	

B. identify some important metabolites

C. describe the structure of mitochondria

D. distinguish catabolic and anabolic pathways in metabolism

12.2 Metabolically Relevant Nucleotides

Learning Objectives

Upon completion of this material, a student should be able to do the following:

A. identify the low- and high-energy forms of metabolically relevant nucleotides and their role in metabolism

B. relate the oxidized and reduced forms of metabolically relevant nucleotides to their abbreviations

12.3 Digestion— From Fuel Metabolism to Hydrolysis Products

Learning Objectives

Upon completion of this material, a student should be able to do the following:

A. define the following key terms:

digestion	emulsification	chylomicrons

B. identify the hydrolysis products and location for the digestion of carbohydrates, fats, and proteins

C. identify some important enzymes involved in digestion of carbohydrates, fats, and proteins

D. describe the function of bile salts

E. describe the role of emulsification and chylomicrons in fat digestion

12.4 Glycolysis— From Hydrolysis Products to Common Metabolites

Learning Objectives

Upon completion of this material, a student should be able to do the following:

A. define the following key terms:
   gluconeogenesis, aerobic, anaerobic, fermentation
B. describe the catabolic oxidation of glucose during glycolysis
C. calculate the number of ATP produced during the catabolic oxidation of glucose
D. distinguish aerobic and anaerobic oxidation of pyruvate
E. distinguish the oxidation of glucose and fructose

12.5 The Citric Acid Cycle — Central Processing

Learning Objectives

Upon completion of this material, a student should be able to do the following:

A. define the following key terms:
   citric acid cycle, oxidative decarboxylation
B. describe the catabolic oxidation of glucose during the citric acid cycle
C. name the reactants and products in each step of the citric acid cycle
D. classify the types of reactions found in the citric acid cycle
E. name the reactions of the citric acid cycle that produce NADH and $FADH_2$

12.6 Electron Transport and Oxidative Phosphorylation

Learning Objectives

Upon completion of this material, a student should be able to do the following:

A. define the following key terms:
   chemiosmotic model, oxidative phosphorylation
B. describe the catabolic oxidation of glucose during electron transport and oxidative phosphorylation
C. name and give the abbreviation of the important electron transporters
D. identify and give the function of the electron carriers and complexes in the electron transport chain
E. describe the proton gradient and its importance

12.7 ATP Production

Learning Objectives

Upon completion of this material, a student should be able to do the following:

A. calculate the number of ATP produced during the catabolic oxidation of glucose
B. distinguish the number of ATP produced during the oxidation of NADH from that produced during the oxidation of $FADH_2$
C. list the ATP production from the reactions of glycolysis, oxidation of pyruvate, and the reactions of the citric acid cycle

12.8 Other Fuel Choices

Learning Objectives

Upon completion of this material, a student should be able to do the following:

A. define the following key terms:
   beta (β) oxidation, fatty acyl CoA, ketone bodies, ketosis, transamination, urea cycle

B. describe the oxidation of a fatty acid via β oxidation
C. calculate the number of cycles and amount of ATP produced for a given saturated fatty acid undergoing β oxidation
D. describe how amino acids can be incorporated into catabolic oxidation at various points via transamination
E. relate ketone bodies to diabetes
F. relate the various catabolic pathways, their intermediates, and the high-energy molecules produced

Practice Test for Chapter 12

1. Which of the following reactions would **not** be classified as catabolic?

A.

$$ADP + P_i + Energy \longrightarrow ATP$$

B.

$$Protein + H_2O \xrightarrow[Enzyme]{H^+} Amino\ acids$$

C.

$$Maltose + H_2O \xrightarrow{Maltase} Glucose + glucose$$

D.

$$C_6H_{12}O_6 + 6O_2 \longrightarrow 6CO_2 + 6H_2O$$

2. Consider the substances listed in the choices. Which would be classified as a metabolite?
   A. lactose
   B. carbon dioxide
   C. pyruvate
   D. ATP
   E. All are metabolites.

3. Which of the following is considered to be the energy currency of the body?
   A. FAD
   B. ATP
   C. NADH
   D. Acetyl CoA

4. How is the oxidized form of nicotinamide adenine dinucleotide represented?
   A. $FADH_2$
   B. NADH
   C. FAD
   D. NAD^+
   E. ADP

5. Which of the following is a characteristic of Acetyl CoA?
   A. contains a thioester functional group
   B. high energy form of coenzyme A
   C. exchanges energy when a C—S bond is hydrolyzed
   D. A and B
   E. A, B, and C

6. What is the abbreviation used to designate the substance given below?

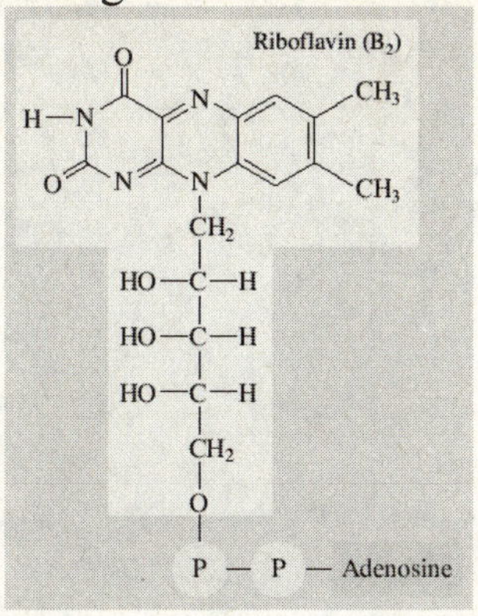

A. ATP
B. FAD
C. NAD^+
D. $FADH_2$
E. Co A
F. Acetyl CoA

7. The following reaction:

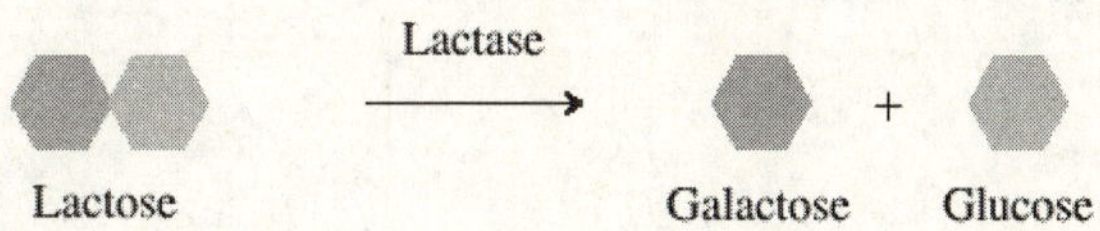

This reaction:
A. occurs in the stomach.
B. is classified as a condensation reaction.
C. produces products that can be transported in the bloodstream.
D. represents emulsification.

8. During the digestion of fats, which of the following transports the triglycerides from the stomach through the bloodstream?
A. chylomicron
B. bile salts
C. cholesterol
D. micelles

9. Of the following reactions, which occurs in the stomach?
A.

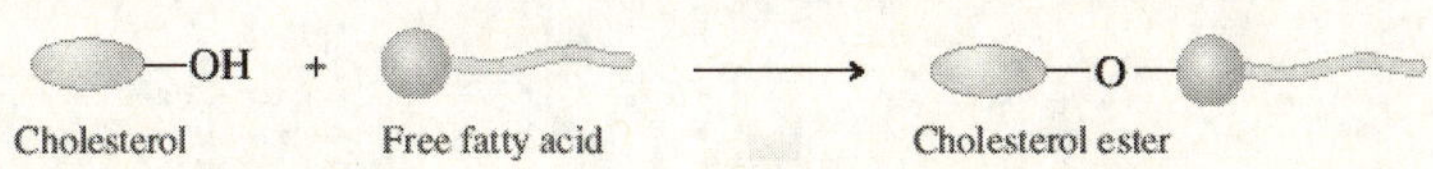

B.

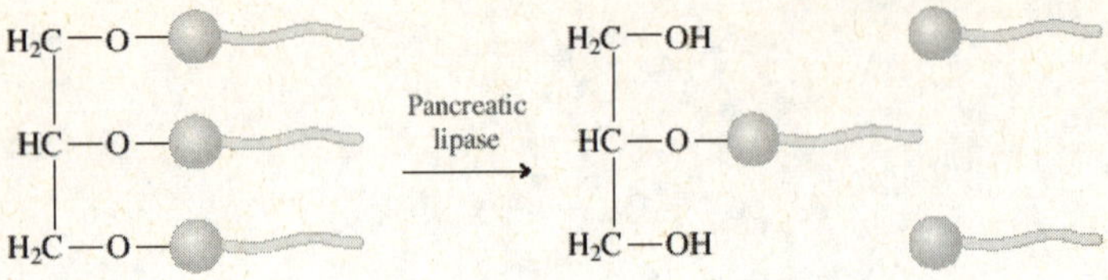

C.

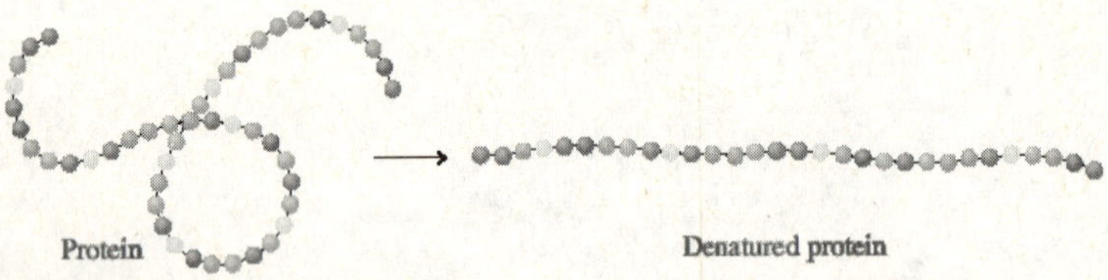

D.

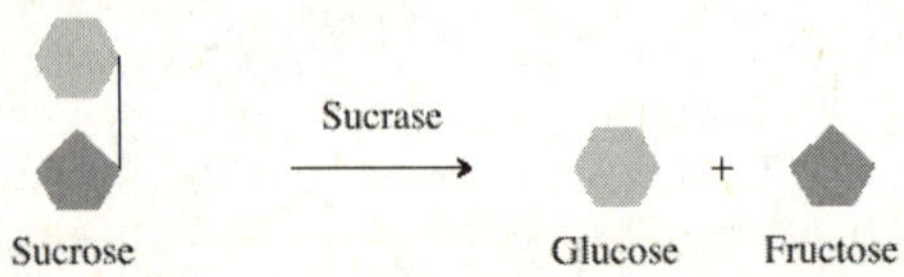

10. The end product of the digestion of starch is
   A. cholesterol.
   B. amino acids.
   C. sucrose.
   D. glucose.

11. The following reaction represents the net chemical reaction of:

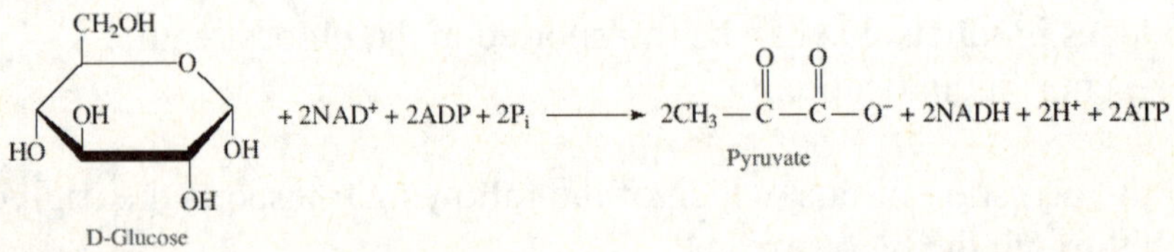

   A. glycolysis.
   B. the citric acid cycle.
   C. electron transport.
   D. oxidative phosphorylation.
   E. all of the above

12. In terms of **high energy molecules**, what is the net output from two molecules of glucose during glycolysis?
   A. 2 ATP
   B. 2 ATP, 2 NADH
   C. 4 ATP, 2 NADH
   D. 4 ATP, 4 NADH
   E. 6 ATP, 6 NADH

13. Which of the following is produced during the aerobic oxidation of pyruvate?
   A. lactate
   B. acetyl CoA
   C. ethanol
   D. All are produced during aerobic oxidation.

14. Fructose can enter into glycolysis:
   A. as fructose-6-phosphate.
   B. as dihydroxyacetone phosphate.
   C. as glyceraldehyde-3-phosphate.
   D. after the regulatory step.
   E. all of the above

15. After 3 molecules of acetyl CoA move through the citric acid cycle:
   A. three molecules of NADH are produced.
   B. six molecules of $FADH_2$ are produced.
   C. six molecules of CO_2 are produced.
   D. six molecules of GTP are produced.

16. In order for the citric acid cycle to run, which of the following are needed?
   A. pyruvate and oxaloacetate
   B. acetyl CoA and pyruvate
   C. acetyl CoA and citrate
   D. oxaloacetate and acetyl CoA

17. What do the following two steps in the citric acid cycle have in common?

   Reaction 3: isocitrate $\longrightarrow$ α-ketoglutarate

   Reaction 8: malate $\longrightarrow$ oxaloacetate

   A. reduce NAD^+
   B. produce CO_2
   C. decarboxylation
   D. isomerization
   E. reduce FAD

18. Which of the following reaction types is not involved in the citric acid cycle?
   A. hydrogenation
   B. hydrolysis
   C. dehydrogenation
   D. isomerization

19. All of the following are involved in electron transport **except**:
   A. cytochrome c.
   B. QH_2.
   C. ADP.
   D. $FADH_2$.

20. The process of producing energy from the oxidation of reduced nucleotides is termed:
    A. electron transport.
    B. oxidative phosphorylation.
    C. oxidative decarboxylation.
    D. dehydrogenation.

21. In the chemiosmotic model, which of the following drives the following reaction?

$$\text{ADP} + \text{P}_i + \text{energy} \longrightarrow \text{ATP}$$

    A. transfer of electrons from complex I to complex II
    B. release of H^+ during the oxidation of NADH and $FADH_2$
    C. reduction of O_2 to form water
    D. movement of H^+ through complex V

22. Electron transport occurs in the:
    A. cytosol.
    B. mitochondrial matrix.
    C. inner membrane of the mitochondria.
    D. cell membrane.

23. NADH enters electron transport at:
    A. complex I.
    B. complex II.
    C. complex III.
    D. complex IV.
    E. complex V.

24. For the oxidation of one molecule of glucose, which of the following produces the largest amount of ATP?
    A. glycolysis
    B. oxidation of pyruvate
    C. citric acid cycle
    D. All produce approximately equal amounts.

25. The net production of ATP from NADH:
    A. is greater than that from $FADH_2$ because it takes two hydrogen atoms to reduce FAD.
    B. is greater than $FADH_2$ because NADH enters the electron transport chain earlier than $FADH_2$.
    C. because NADH pumps 6 protons into the inner mitochondrial membrane while $FADH_2$ pumps 4 protons.
    D. A and C
    E. None of the above, $FADH_2$ produces more ATP per molecule.

26. Which of the following conversions directly produces the highest yield of ATP as a product of the reaction as written?

A. glucose $\longrightarrow$ 2 pyruvate

B. pyruvate $\longrightarrow$ lactate

C. GTP $\longrightarrow$ GDP + P_i

D. pyruvate $\longrightarrow$ acetyl CoA

27. Calculate the number of ATPs produced from the β oxidation of the following fatty acid entering β oxidation as an acyl CoA.

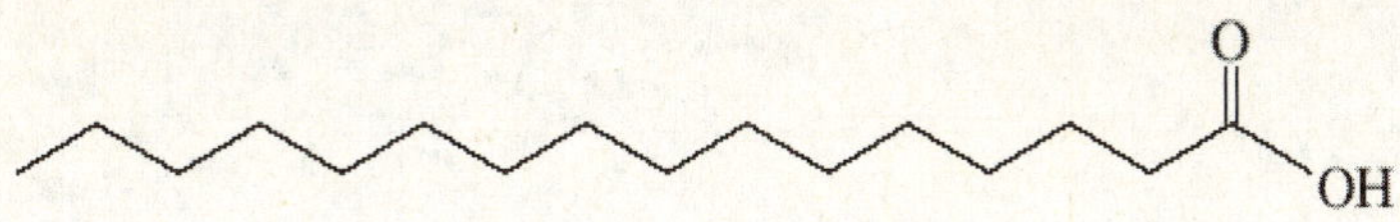

A. 80 ATP
B. 97.5 ATP
C. 108 ATP
D. 112 ATP

28. The following are the four steps involved in β oxidation. In which is the length of the fatty acyl CoA reduced by two carbon atoms?
A. 1 - oxidation
B. 2 - hydration
C. 3 - oxidation
D. 4 - removal of acetyl CoA
E. None of the steps reduce the length by only two carbon atoms.

29. Which of the following characterizes the condition known as ketosis?
A. accumulation of β-hydroxybutyrate, acetoacetate, and acetone in liver
B. occurs in the presence of high concentration of carbohydrates
C. low fat levels in the diet
D. results in high blood pH

30. During metabolism, all but of the following relate to amino acids?
A. undergo transamination before excretion
B. have the ability to directly generate ATP
C. nitrogen content eventually excreted as urea
D. can regenerate intermediates in the citric acid cycle

Answers

1. A	**2.** C	**3.** B	**4.** D	**5.** E	**6.** B	**7.** C	**8.** A	**9.** C	**10.** D
11. A	**12.** D	**13.** B	**14.** E	**15.** C	**16.** D	**17.** A	**18.** A	**19.** C	**20.** B
21. D	**22.** C	**23.** A	**24.** C	**25.** B	**26.** D	**27.** C	**28.** D	**29.** A	**30.** B

Chapter 12 – Solutions to Odd-Numbered Problems

Practice Problems

12.1 How could you identify a catabolic reaction?

Catabolism refers to chemical reactions in which larger molecules are broken down into a few common smaller molecules called metabolites. See Figure 12.3 below.

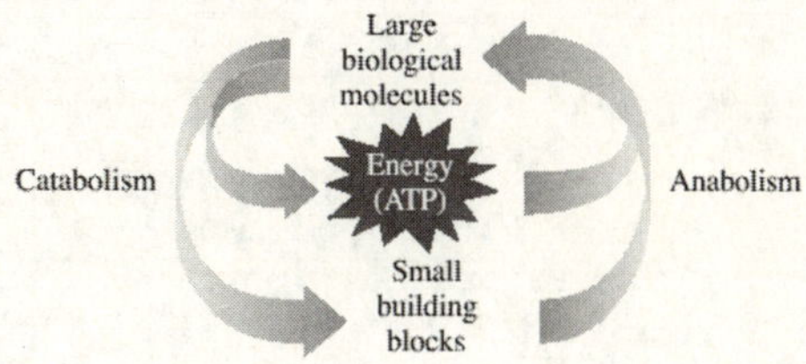

A catabolic reaction will break apart of larger molecule into smaller molecules, usually with the release of energy. Catabolic reactions tend to be oxidative, involving the loss of hydrogen or the addition of oxygen.

12.3 Name a chemical reaction that breaks down biomolecules into their component parts.

A few examples: hydrolysis reactions, digestion of polysaccharides, oxidation of carbohydrates to carbon dioxide and water

12.5 The ultimate end product for the catabolism of glucose (6 carbons) is __________.

The end products of the oxidation of glucose are carbon dioxide and water.

12.7 Identify the cell organelle considered the energy-producing factory for the cell.

Located within the cytoplasm are the mitochondria. The matrix and inner membrane of the mitochondria contain the enzymes that catalyze the energy-producing oxidation of carbohydrates, fats, and amino acids.

12.9 Match one of the following metabolic nucleotides to its description:

NADH	ATP
NAD^+	ADP
$FADH_2$	Acetyl CoA
FAD	CoA

a. the reduced form of nicotinamide adenine dinucleotide
b. exchanges energy when a C—S bond is hydrolyzed
c. the oxidized form of flavin adenine dinucleotide

The abbreviations are derived from the name of the nucleotide and the reduced forms contain more hydrogen atoms than the corresponding oxidized forms. Specific functions of the nucleotides can be reviewed in the text.

a. the reduced form of **n**icotinamide **a**denine **d**inucleotide: From the name NAD and if in the reduced form it will contain more hydrogen atoms, NADH
b. exchanges energy when a C—S bond is hydrolyzed: Review of functions of the nucleotides indicates that Acetyl CoA is the high energy form of CoA, Acetyl CoA
c. the oxidized form of **f**lavin **a**denine **d**inucleotide: From the name FAD and if in the oxidized form contains fewer hydrogen atoms, FAD

12.11 Name a carbohydrate (if any) that undergoes digestion in each of the following sites:
a. mouth **b.** stomach **c.** small intestine

Refer to Figure 12.5.

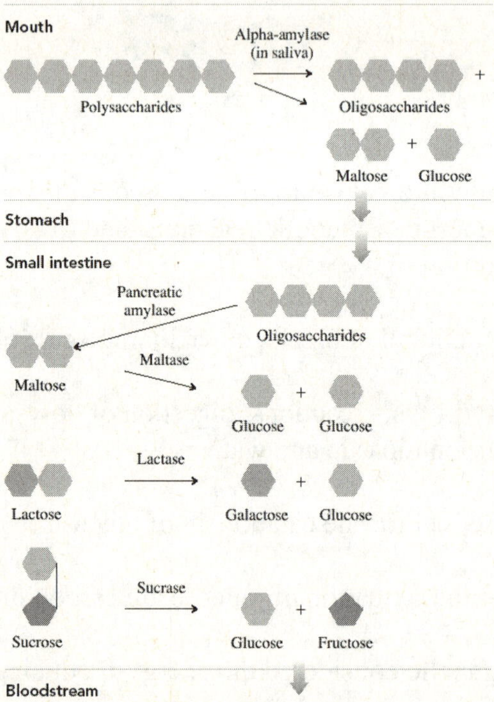

a. mouth: starch
b. stomach: none
c. small intestine: oligosaccharides and disaccharides

12.13 Name the chemical reaction occurring during most digestive processes.

Many digestive processes break down the repeating units that make up large biomolecules through the addition of water to the specific type of bond holding the units together: (peptide bonds are broken to yield amino acids; glycosidic bonds yield simpler saccharides). This type of reaction is referred to as hydrolysis.

12.15 Describe how cholesterol is packaged after absorption in the intestine.

Cholesterol is linked to a free fatty acid forming a cholesterol ester, which is then repackaged with protein as a lipoprotein called a chylomicron.

12.17 Name the end products for digestion of proteins.

Protein digestion begins in the stomach where proteins are denatured (unfolded) by the acidic digestive juices. Digestive enzymes hydrolyze peptide bonds as the denatured proteins are broken down into their constituent amino acids.

12.19 Name the end product of glycolysis.

There are ten chemical reactions in glycolysis that result in the formation of two molecules of pyruvate from one molecule of glucose.

12.21 How many ATP are produced when one molecule of glucose undergoes glycolysis?

Reactions 6 through 10 of glycolysis generate two high-energy NADH molecules during the addition of two more phosphates and four ATP molecules when the four phosphates are removed from the sugar phosphates. The first five reactions require an energy investment of two molecules of ATP, so the net production is two molecules of ATP.

12.23 In terms of high energy molecules, what is the net output for one molecule of glucose undergoing glycolysis?

Through the ten reactions in glycolysis, one molecule of glucose is converted into two molecules of pyruvate. Because two molecules of ATP were originally required in the energy-investment phase and four total are produced in the energy-generation phase, the net energy output for one molecule of glucose is two NADH and two ATP. The net chemical reaction is:

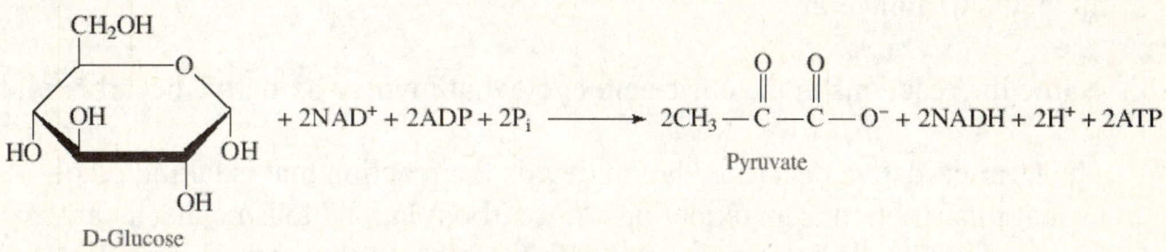

12.25 Name the coenzyme produced during the conversion of pyruvate to lactate.

Under these anaerobic conditions, the middle carbonyl in pyruvate is reduced (hydrogen added) to an alcohol functional group, and lactate is formed. The hydrogen (and energy) required for this reaction is supplied by NADH and H^+, producing NAD^+, the coenzyme.

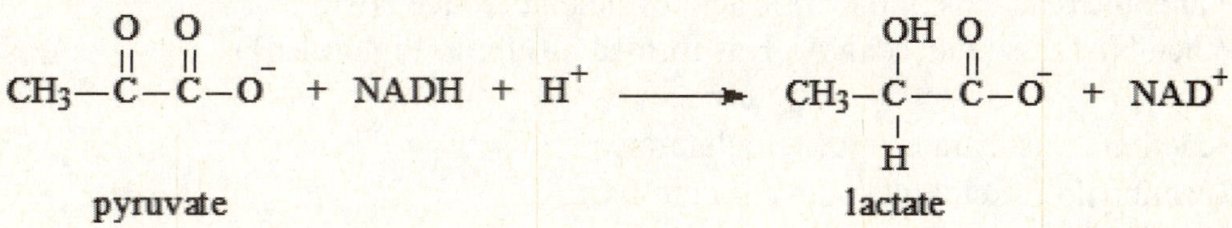

12.27 Name the products of fermentation.

Yeast converts pyruvate to ethanol under anaerobic conditions. This process is called fermentation. Yeast produces pyruvate from glucose under low oxygen conditions and transforms pyruvate into ethanol, CO_2 and NAD^+. The reaction is:

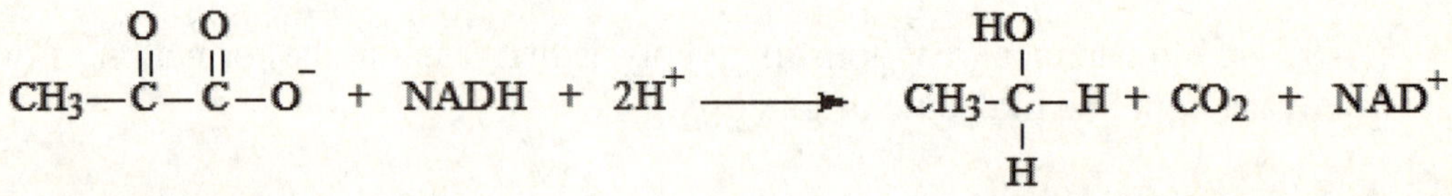

pyruvate ethanol

12.29 What is the main regulation point of glycolysis? How does this explain why fructose is readily converted into fat?

The main regulation point of glycolysis is phosphofructokinase, the enzyme involved in step three. Because the products of the metabolism of fructose enter glycolysis at step five, they bypass the main regulation point, which leads to the production of excess pyruvate and acetyl CoA that is not needed in the cells and is ultimately converted to fat.

12.31 Name the compounds needed to start the citric acid cycle.

The name citric acid cycle originated because the first step in the cycle involves the formation of the six-carbon molecule citrate, the conjugate base of citric acid. This initial reaction is a condensation reaction between an entering **acetyl CoA** molecule and the four-carbon molecule **oxaloacetate**.

12.33 Name the reactions in the citric acid cycle that involve oxidative decarboxylation.

The term oxidative decarboxylation details the reaction that is taking place. “Oxidative” tells us that this reaction is an oxidation. “Decarboxylation” tells us that a carboxylate (COO^-) is removed as CO_2. Look for the production of CO_2 in the citric acid cycle. Refer to Figure 12.11.

Reaction 3: isocitrate → α-ketoglutarate

Reaction 4: α-ketoglutarate → succinyl CoA

12.35 Name the reactions of the citric acid cycle that reduce NAD^+.
When NAD^+ is reduced, NADH is formed. Refer to Figure 12.11.

Reaction 3: isocitrate → α-ketoglutarate
Reaction 4: α-ketoglutarate → succinyl CoA
Reaction 8: malate → oxaloacetate

12.37 Name the electron carrier that transports electrons from complex I to complex III.

Two of the electron carriers, coenzyme Q and cytochrome *c*, not firmly attached to any one complex, serve to shuttle electrons between the complexes. See Figure 12.12 below.

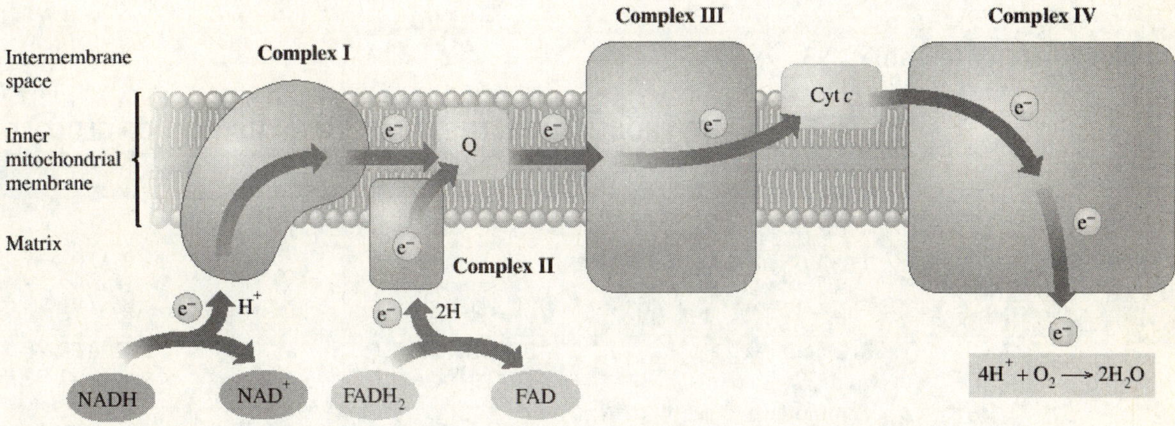

Coenzyme Q or ubiquinone.

12.39 Name the complex where water is produced.

At complex IV, electrons are transferred from cytochrome *c* through another set of electron acceptors to combine with hydrogen ions and oxygen (O_2) to form water. This is the final stop for the electrons. See Figure 12.12 below.

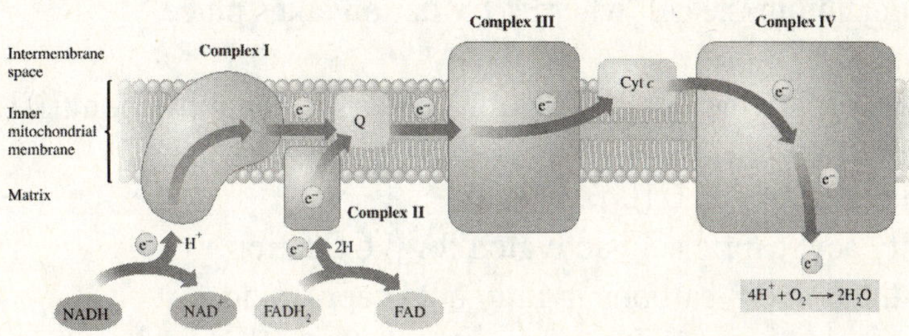

12.41 According to the chemiosmotic theory, how does the proton gradient provide energy to synthesize ATP?

The movement of protons from an area with many protons to an area with fewer protons releases energy. As protons flow back into the matrix through complex V, the resulting release of energy drives the synthesis of ATP.

$$ADP + P_i + energy \longrightarrow ATP$$

12.43 List the energy yield in ATP molecules for each of the following:
a. NADH $\rightarrow$ NAD^+
b. glucose $\rightarrow$ 2 pyruvate
c. 2 pyruvate $\rightarrow$ 2 acetyl CoA + $2CO_2$
d. acetyl CoA $\rightarrow$ $2CO_2$

Refer to Table 12.3:

TABLE 12.3 ATP PRODUCED BY THE COMPLETE OXIDATION OF GLUCOSE

Pathway	Reduced Nucleotides Produced	ATP Yield
Glycolysis	$2NADH_{cytosol} \longrightarrow 2NADH_{matrix}$	5 ATP
(Produced directly in pathway)		2 ATP
2 Pyruvate $\longrightarrow$ 2 Acetyl CoA	2NADH	5 ATP
Citric acid cycle Two turns of the cycle accommodates 2 acetyl CoA	6 NADH $2FADH_2$	15 ATP 3 ATP
(Produced as GTP in pathway)		2 ATP
	TOTAL ATP	32 ATP

a. NADH $\rightarrow$ NAD^+ : 2.5 ATP
b. glucose $\rightarrow$ 2 pyruvate: 7 ATP (directly produced and that from 2 NADH)
c. 2 pyruvate $\rightarrow$ 2 acetyl CoA + $2CO_2$: 5 ATP
d. acetyl CoA $\rightarrow$ $2CO_2$: 10 ATP (one turn of the citric acid cycle)

12.45 Name the location in the cell where β oxidation takes place.

The oxidation of fatty acids (β oxidation) occurs in the mitochondrial matrix.

12.47 Capric acid is a saturated C10 fatty acid.
a. Draw fatty acyl capric acid activated for β oxidation.
b. Identify the α and β carbons in fatty acyl capric acid.
c. Write the overall equation for the complete β oxidation of capric acid.

a. Draw fatty acyl capric acid activated for β oxidation.

An activated fatty acid consists of a fatty acid bonded to a CoA, called a fatty acyl CoA.

b. Identify the α and β carbons in fatty acyl capric acid.

The α carbon is the carbon atom adjacent to the carbonyl carbon, while the β carbon is the second carbon from the carbonyl carbon.

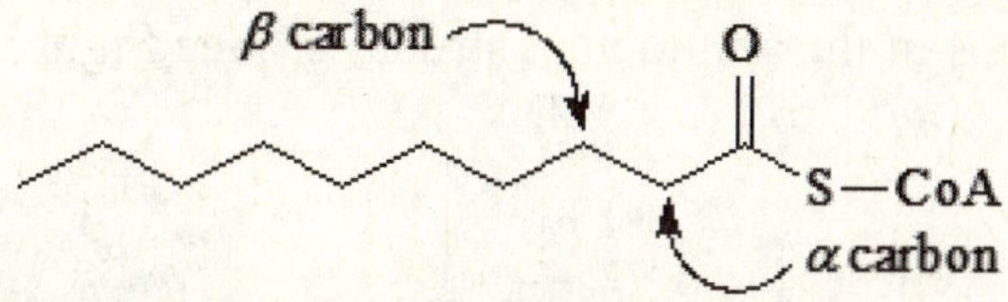

c. Write the overall equation for the complete β oxidation of capric acid.

Fatty acyl CoA is oxidized, thereby producing a new fatty acyl CoA that is two carbons shorter and one molecule of acetyl CoA. Each turn of β oxidation produces one NADH and one $FADH_2$. A C10 acid requires 4 turns of the cycle.

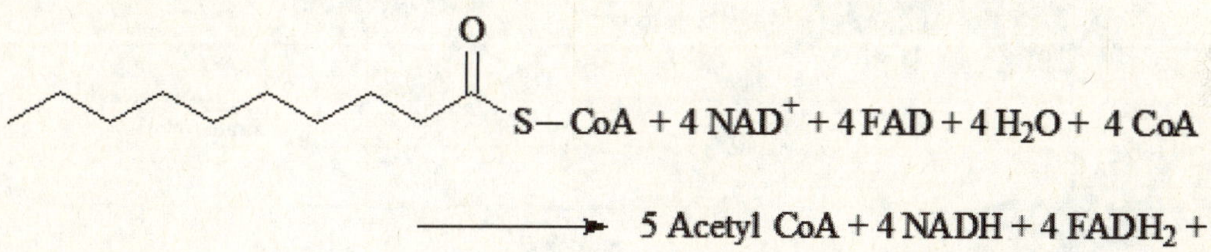

12.49 Under what conditions are ketone bodies produced in the body?

The oxidation of large amounts of fatty acids can cause acetyl CoA molecules to accumulate in the liver. When this occurs, the two carbon acetyl units condense in the liver, forming the four-carbon ketone molecules, β-hydroxybutyrate and acetoacetate, and the molecule acetone. These are collectively referred to as ketone bodies, with their formation termed ketogenesis.

12.51 Why does the body convert NH_4^+ into urea?

The excess ammonium ions produced in the metabolism of proteins must be excreted from the body because they are toxic if allowed to accumulate. A series of reactions called the urea cycle converts ammonium ions (NH_4^+) into urea, which can then be excreted in the urine.

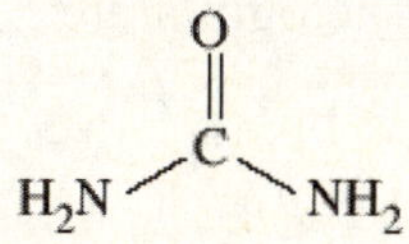

Urea

12.53 Name the metabolic substrate(s) that can be produced from the carbon atoms of each of the following amino acids:
a. alanine
b. aspartate

c. valine
d. glutamine

Various amino acids offer a way to replenish the intermediates in the citric acid cycle and therefore have the ability to generate ATP. Which intermediate they replenish depends on the number of carbons in the amino acid structure. See Figure 12.17.

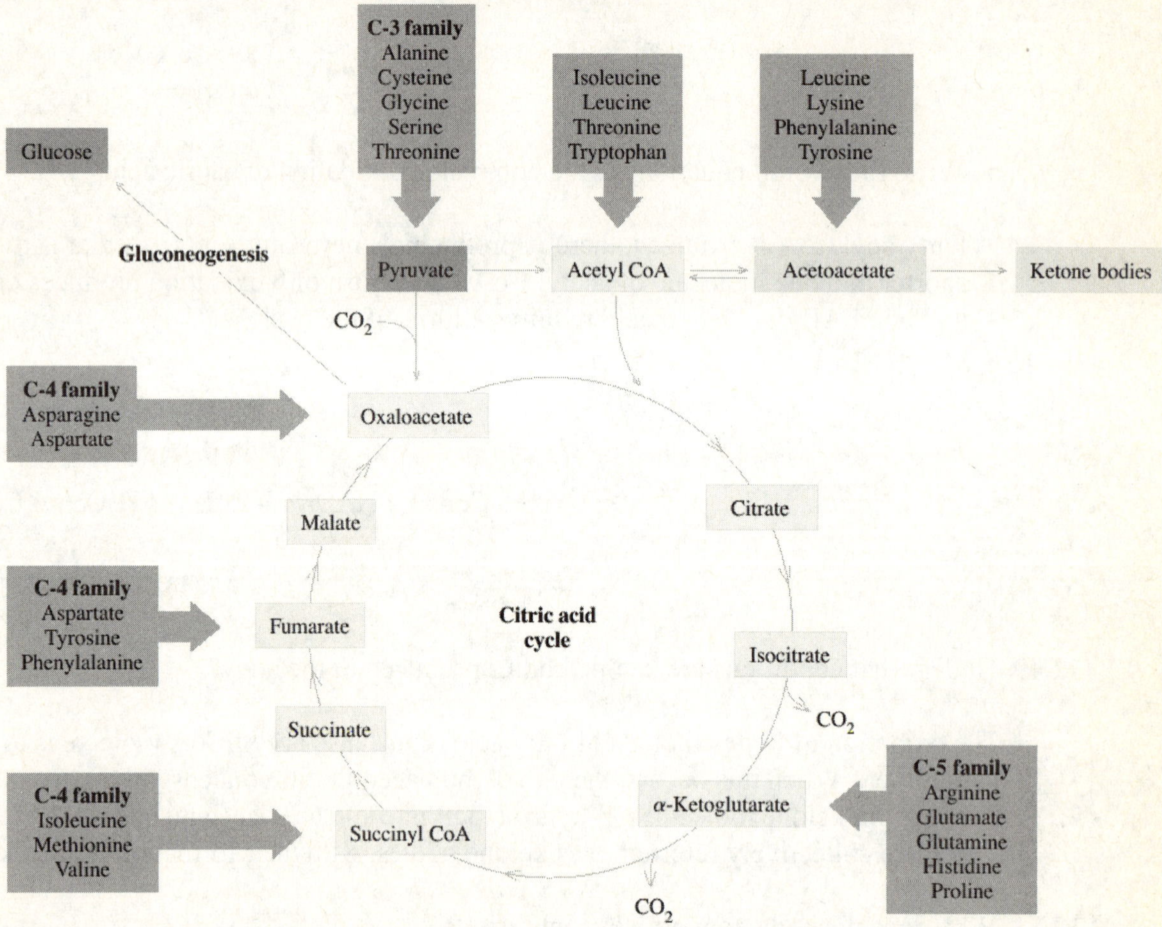

a. alanine: C-3 family - pyruvate
b. aspartate: C-4 family - oxaloacetate and fumarate
c. valine: C-4 family - succinyl CoA
d. glutamine: C-5 family - α-ketoglutarate

Additional Problems

12.55 Name the location in the cell where the following catabolic processes take place:

a. glycolysis

b. citric acid cycle

c. ATP synthesis

Refer to Figure 12.2 and Table 12.1.

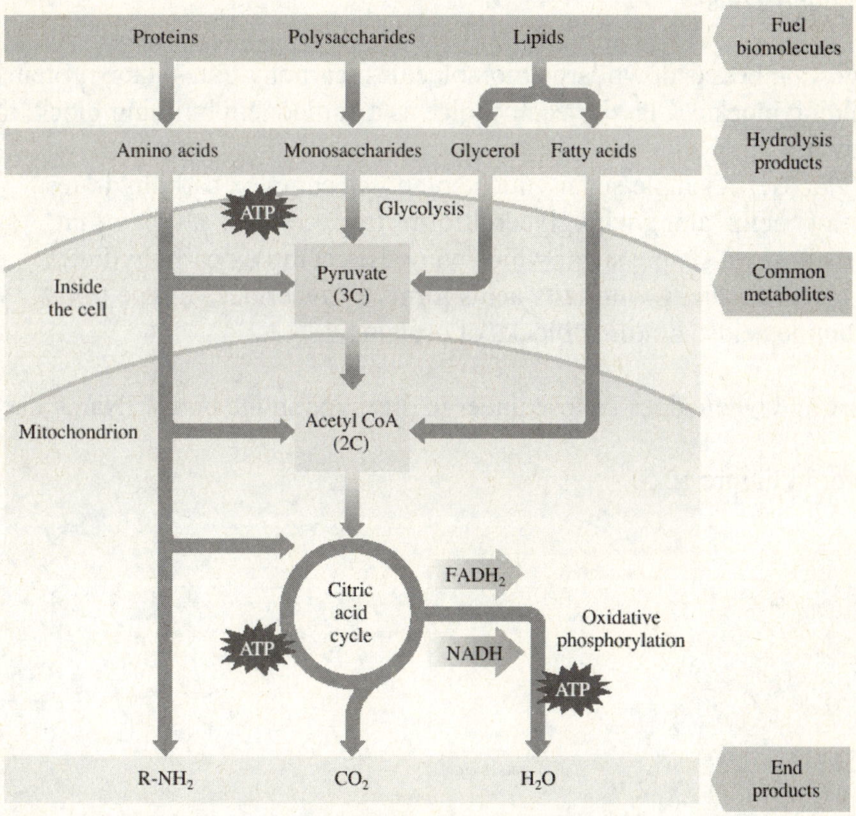

TABLE 12.1 LOCATIONS AND FUNCTIONS OF COMPONENTS IN ANIMAL CELLS

Component	Description and Function
Cell membrane	Separates the contents of a cell from the external environment and contains structures that communicate with other cells
Cytoplasm	Consists of all the cellular contents between the cell membrane and nucleus
Cytosol	Is the fluid part of the cytoplasm that contains enzymes for many of the cell's chemical reactions
Endoplasmic reticulum	Rough type processes proteins for secretion and synthesizes phospholipids; smooth type synthesizes fats and steroids
Golgi complex	Modifies and secretes proteins from the endoplasmic reticulum and synthesizes glycoproteins and cell membranes
Lysosomes	Contain hydrolytic enzymes that digest and recycle old cell structures
Mitochondria	Contain the structures for the synthesis of ATP from energy-producing reactions
Nucleus	Contains genetic information for the replication of DNA and the synthesis of protein
Ribosomes	Are the sites of protein synthesis using mRNA templates

a. glycolysis: cytosol
b. citric acid cycle: mitochondrial matrix
c. ATP synthesis: mitochondrial matrix

12.57 Identify the type of food–carbohydrate, fat, or protein–that gives each of the following digestion products:
a. glucose
b. fatty acid
c. maltose
d. glycerol
e. amino acids

Digestion breaks down large biomolecules (carbohydrates, fats, proteins) into the component building blocks. Classify each choice as a component building block, then the biomolecule is known.
a. glucose: a simple sugar which composes complex carbohydrates
b. fatty acid: along with glycerol forms triglycerides, a type of fat
c. maltose: a simple sugar which composes complex carbohydrates
d. glycerol: along with fatty acids forms triglycerides, a type of fat
e. amino acids: building blocks of proteins

12.59 How and where does lactose undergo digestion in the body? Name the products.

Refer to Figure 12.5.

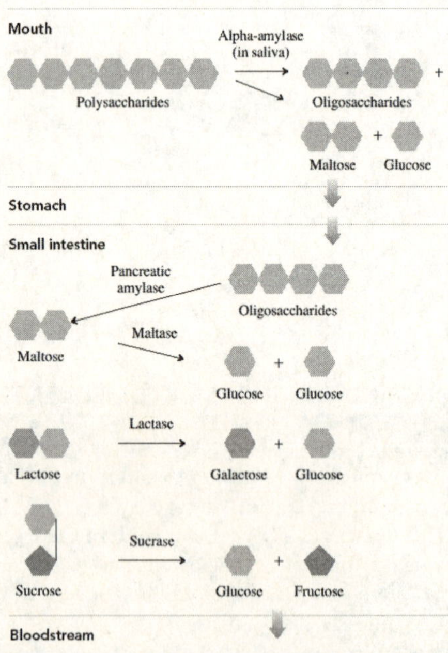

Lactose undergoes digestion in the small intestine to yield galactose and glucose.

12.61 Which of the following reactions in glycolysis produce ATP or NADH?
a. 1,3-bisphosphoglycerate to 3-phosphoglycerate
b. glucose-6-phosphate to fructose-6-phosphate
c. phosphoenolpyruvate to pyruvate

Refer to Figure 12.8.

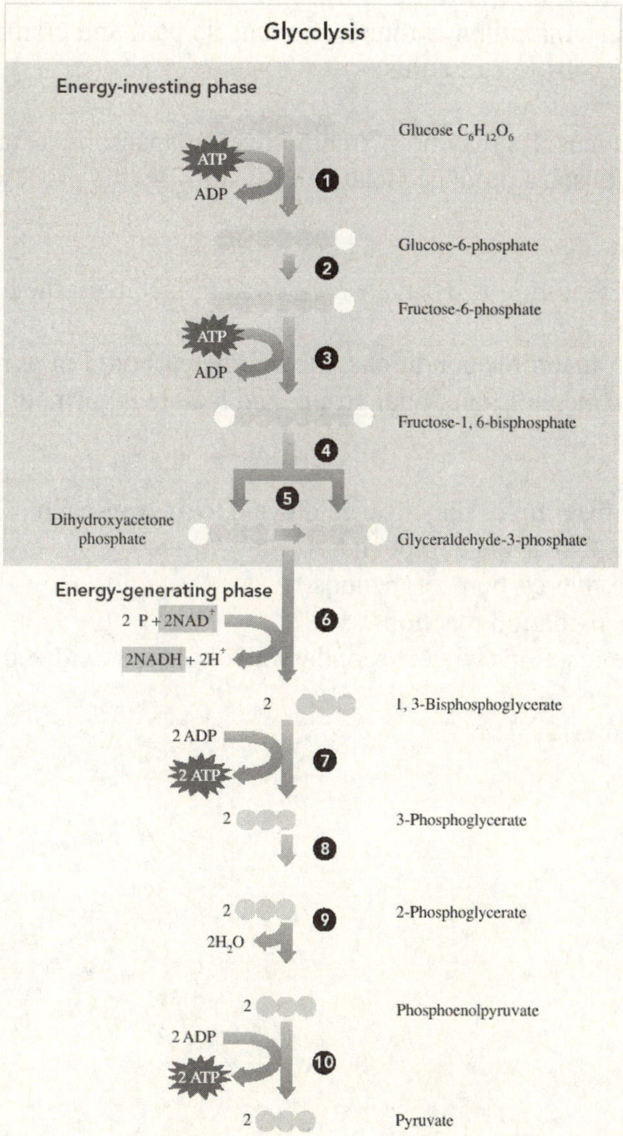

a. 1,3-bisphosphoglycerate to 3-phosphoglycerate: produces ATP
b. glucose-6-phosphate to fructose-6-phosphate: produces neither ATP or NADH
c. phosphoenolpyruvate to pyruvate: produces ATP

12.63 Which of the reactions given in problem 12.61 represent isomerizations where the reactants and products are structural isomers?

Recall that glucose and fructose are isomers. The reaction listed in b, the conversion of glucose-6-phosphate to fructose-6-phosphate, is an isomerization.

12.65 After running a marathon, a runner has muscle pain and cramping. What might have occurred in the muscle cells to cause this?

The runner's muscles may be switching over to anaerobic catabolism to keep ATP production going. The muscles produce lactate and H^+ during this process, causing soreness.

12.67 Under what condition is pyruvate converted to lactate in the body?

Under these anaerobic conditions, the middle carbonyl in pyruvate is reduced (hydrogen added) to an alcohol functional group, and lactate is formed.

12.69 Refer to the diagram of the citric acid cycle to answer each of the following:

a. Name the six-carbon compounds.
b. Name the five-carbon compounds.
c. Name the oxidation reactions.
d. Name the reactions where secondary alcohols are oxidized to ketones.

Refer to Figure 12.11.

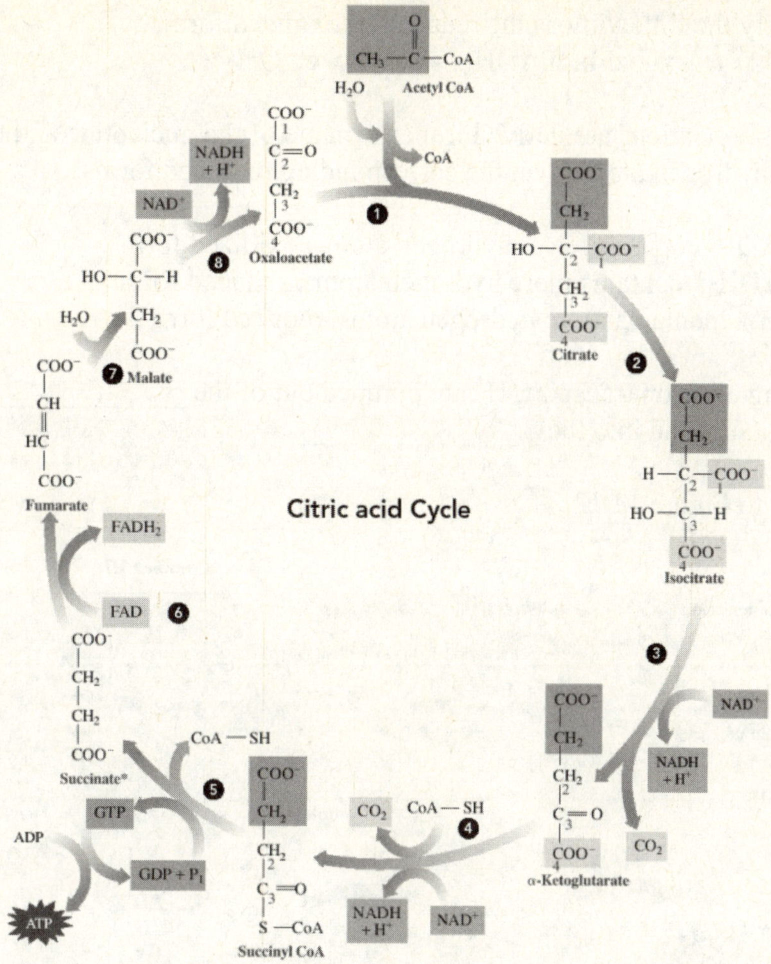

a. Name the six-carbon compounds: citrate and isocitrate

b. Name the five-carbon compounds: α-ketoglutarate

c. Name the oxidation reactions:

Reaction 3: isocitrate → α-ketoglutarate
Reaction 4: α-ketoglutarate → succinyl CoA
Reaction 6: succinate → fumarate
Reaction 8: malate → oxaloacetate

d. Name the reactions where secondary alcohols are oxidized to ketones.

Reaction 3: isocitrate → α-ketoglutarate
Reaction 8: malate → oxaloacetate

12.71 If there are no reactions in the citric acid cycle that use oxygen, O_2, why does the cycle operate only in aerobic conditions?

Aerobic refers to any process occurring in the presence of oxygen. O_2 is used during electron transport. The coenzymes NAD^+ and FAD needed in the citric acid cycle are only available if they are oxidized during electron transport. Thus without electron transport, there is a lack of NAD^+ and FAD, without which the citric acid cycle cannot proceed.

12.73 Identify the following as the reduced or oxidized form:

a. NAD^+ **b.** $FADH_2$ **c.** QH_2

The abbreviations are derived from the name of the nucleotide and the reduced forms contain more hydrogen atoms than the corresponding oxidized forms.

a. NAD^+: contains fewer hydrogen atoms, oxidized form
b. $FADH_2$: contains more hydrogen atoms, reduced form
c. QH_2:: contains more hydrogen atoms, reduced form

12.75 During electron transport, H^+ are pumped out of the ____________ , across the inner membrane, and into the ________________.

Refer to Figure 12.12.

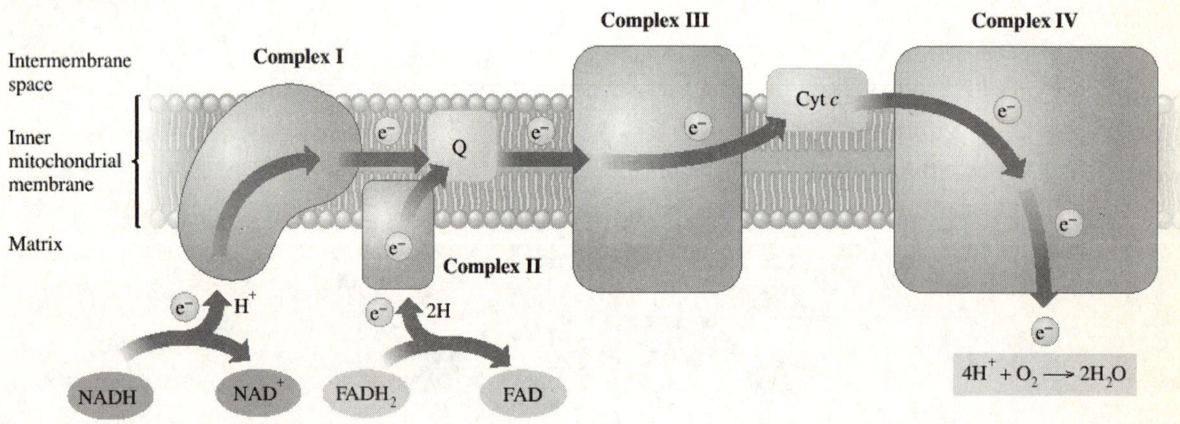

matrix; inner membrane space

12.77 In the chemiosmotic model, how is energy provided to synthesize ATP?

The chemiosmotic model links electron transport to the generation of a proton (H^+) difference (gradient) on either side of the inner membrane of the mitochondria, which results in the production of ATP. In this model, three of the complexes (I, III, and IV) span the inner membrane and pump (relocate) protons out of the matrix and into the intermembrane space as electrons are shuttled through the complexes. A difference in both charge (electrical) and concentration (chemical) of protons on either side of the membrane results in an electrochemical gradient. The formation of the proton gradient across the inner mitochondrial membrane provides the energy for ATP synthesis

12.79 How many ATPs are produced when glucose is oxidized to pyruvate compared to when glucose is oxidized to CO_2 and H_2O?

See Table 12.3.

TABLE 12.3 ATP PRODUCED BY THE COMPLETE OXIDATION OF GLUCOSE

Pathway	Reduced Nucleotides Produced	ATP Yield
Glycolysis	$2NADH_{cytosol} \longrightarrow 2NADH_{matrix}$	5 ATP
(Produced directly in pathway)		2 ATP
2 Pyruvate $\longrightarrow$ 2 Acetyl CoA	2NADH	5 ATP
Citric acid cycle Two turns of the cycle accommodates 2 acetyl CoA	6 NADH $2FADH_2$	15 ATP 3 ATP
(Produced as GTP in pathway)		2 ATP
	TOTAL ATP	32 ATP

The oxidation of glucose to pyruvate produces 5 ATP whereas 32 ATP are produced from the complete oxidation of glucose to CO_2 and H_2O.

12.81 Consider the complete oxidation of capric acid, a saturated C10 fatty acid.
a. How many acetyl CoA units are produced?
b. How many turns of β oxidation are needed?
c. How many ATPs are generated from the complete oxidation of capric acid?

In β oxidation, carbons are removed two at a time from an activated fatty acid. An activated fatty acid consists of a fatty acid bonded to a CoA, called a fatty acyl CoA. The removal of two carbons during β oxidation produces an acetyl CoA and a fatty acyl CoA shortened by two carbons. The cycle repeats until the original fatty acid is completely degraded to two-carbon acetyl CoA units. The acetyl CoA's can then enter the citric acid cycle in the same way as acetyl CoA derived from glucose. This reaction cycle occurs in the mitochondrial matrix.

a. How many acetyl CoA units are produced?: 10/2 = 5, 5 acetyl CoA units
b. How many turns of β oxidation are needed?
The net reaction for one cycle of β oxidation is summarized as

$$\text{Fatty Acyl CoA}_{n\text{ carbons}} + NAD^+ + FAD + H_2O + CoA \longrightarrow$$

$$\text{Fatty Acyl CoA}_{n\text{-2 carbons}} + \text{Acetyl CoA} + NADH + H^+ + FADH_2$$

It takes $(n/2) - 1$ cycles to break a fatty acid with n carbon atoms: $(10/2) - 1 = 4$.

c. How many ATPs are generated from the complete oxidation of capric acid?

Use the following table as a guide but modified for a C10 fatty acid producing 5 Acetyl CoA, 4 turns of β oxidation (4 NADH and 4 $FADH_2$) .

ATP PRODUCTION FROM β OXIDATION FOR A STEARIC ACID (18C) MOLECULE	
9 Acetyl CoA 9 Acetyl CoA × 10 ATP/acetyl CoA	90 ATP
8 turns of β oxidation 8 NADH × 2.5 ATP/NADH	20 ATP
8 $FADH_2$ × 1.5 ATP/$FADH_2$	12 ATP
Total	122 ATP

5 Acetyl CoA × 10 ATP/acetyl CoA = 50 ATP
4 NADH × 2.5 ATP/NADH = 10 ATP
4 $FADH_2$ × 1.5 ATP/$FADH_2$ = 6 ATP
Total: 66 ATP

Challenge Problems

12.83 Identify all steps in the oxidative catabolism of glucose that can be considered oxidative decarboxylations.

Consider both glycolysis and the citric acid cycle.

Oxidation of pyruvate: pyruvate → acetyl CoA
Reaction 3 of citric acid cycle: isocitrate → α-ketoglutarate
Reaction 4 of the citric acid cycle: α-ketoglutarate → succinyl CoA

12.85 Lauric acid, a saturated C12 fatty acid, is found in coconut oil.

a. Draw fatty acyl lauric acid activated for β oxidation.
b. Identify the α and β carbons in fatty acyl lauric acid.
c. Write the overall equation for the complete β oxidation of lauric acid.
d. How many acetyl CoA units are produced?
e. How many cycles of β oxidation are needed?
f. Account for the total ATP yield from β oxidation of lauric acid (C12) by completing the following table:

______Acetyl CoA × 10 ATP/acetyl CoA	______ATP
______NADH × 2.5 ATP/NADH	______ATP
______$FADH_2$ × 1.5 ATP/$FADH_2$	______ATP
Total	94 ATP

a. Draw fatty acyl lauric acid activated for β oxidation.

An activated fatty acid consists of a fatty acid bonded to a CoA, called a fatty acyl CoA. Lauric acid is a 12 carbon fatty acid that when activated is:

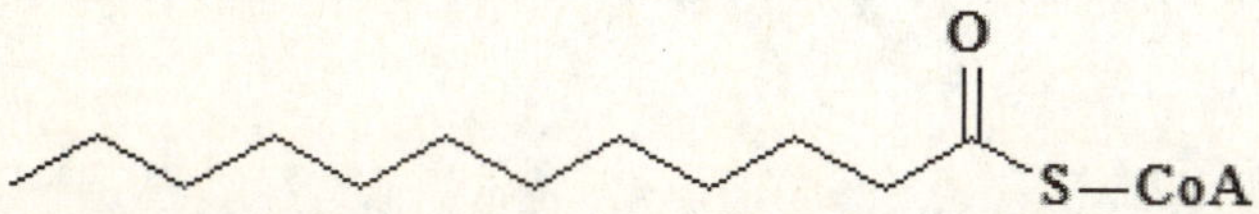

b. Identify the α and β carbons in fatty acyl lauric acid.

The α carbon is the carbon atom adjacent to the carbonyl carbon, while the β carbon is the second carbon from the carbonyl carbon.

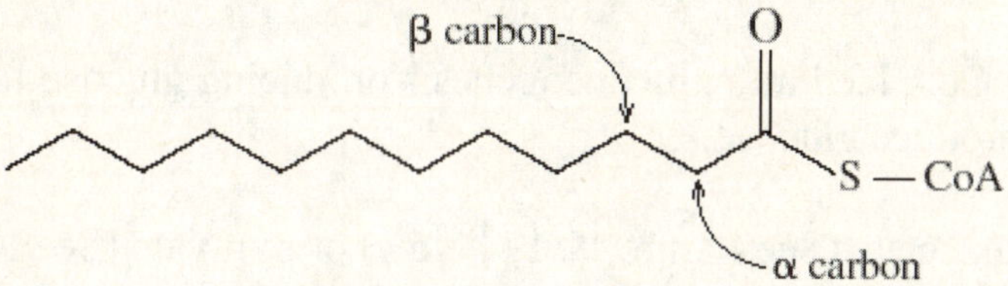

c. Write the overall equation for the complete β oxidation of lauric acid.

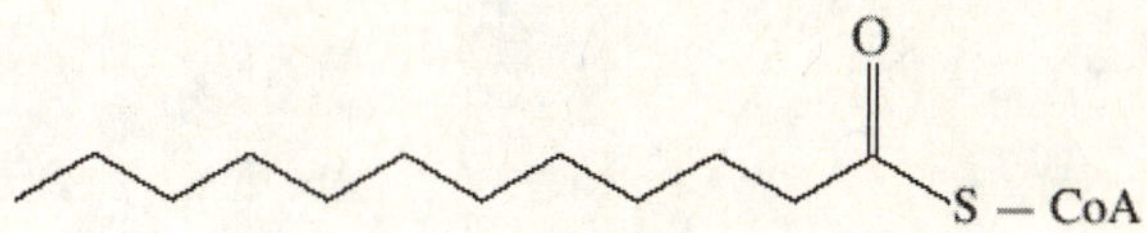

$$+ \ 5\,NAD^+ + 5\,FAD + 5\,H_2O + 5\,CoA \longrightarrow$$

$$6\,\text{Acetyl CoA} + 5\,NADH + 5\,H^+ + 5\,FADH_2$$

d. How many acetyl CoA units are produced?

In β oxidation, carbons are removed two at a time from an activated fatty acid. With 12 carbon atoms in lauric acid, 6 acetyl CoA units are produced.

e. How many cycles of β oxidation are needed?

The net reaction for one cycle of β oxidation is summarized as

$$\text{Fatty Acyl CoA}_{n\text{ carbons}} + NAD^+ + FAD + H_2O + CoA \longrightarrow$$

$$\text{Fatty Acyl CoA}_{n\text{-2 carbons}} + \text{Acetyl CoA} + NADH + H^+ + FADH_2$$

It takes $(n/2) - 1$ cycles to break a fatty acid with n carbon atoms: $(12/2) - 1 = 5$.

f. Account for the total ATP yield from β oxidation of lauric acid (C12) by completing the following table:

Using the production of 6 Acetyl CoA and 5 turns through the spiral:

_____6_____Acetyl CoA × 10 ATP/acetyl CoA	60 ATP
_____5_____NADH × 2.5 ATP/NADH	12.5 ATP
_____5_____$FADH_2$ × 1.5 ATP/$FADH_2$	7.5 ATP
Total	80 ATP

12.87 Can acetyl CoA feed into gluconeogenesis producing glucose for the body? Can fatty acids be used to produce glucose?

Only oxaloacetate (See Figure 12.17 below) or pyruvate (See Figure 12.18 below) are starting points for gluconeogenesis. Acetyl CoA cannot be converted to pyruvate, so fatty acids cannot generate glucose.

Figure 12.17

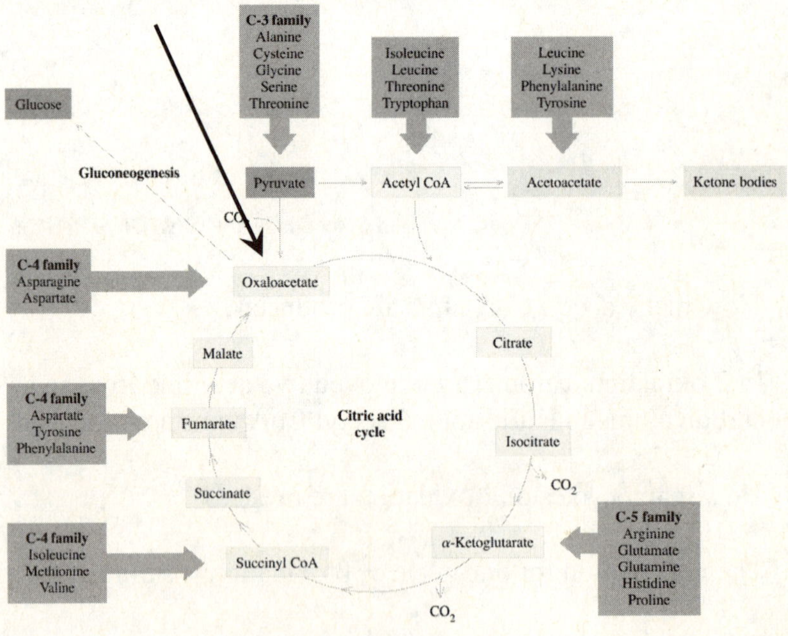

Figure 12.18

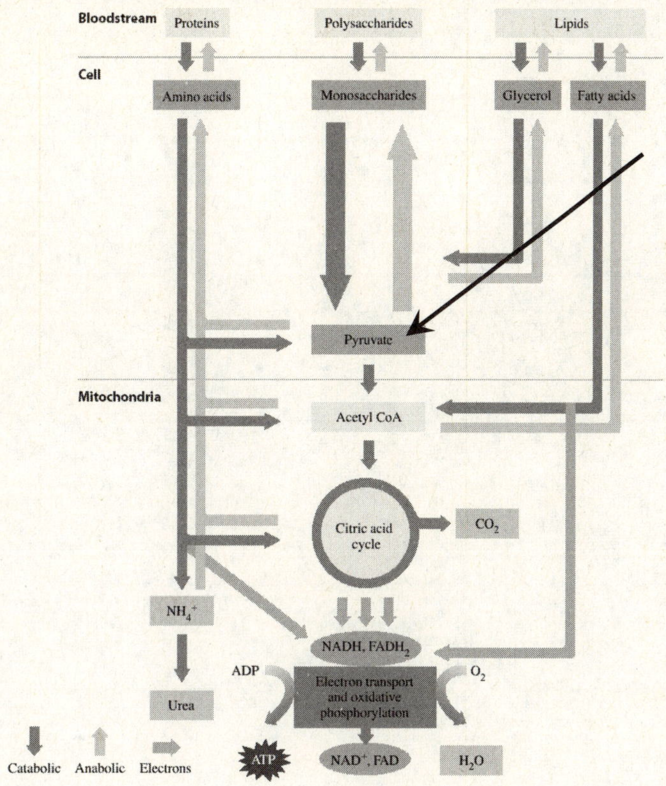
Bloodstream
Proteins
Polysaccharides
Lipids
Cell
Amino acids
Monosaccharides
Glycerol
Fatty acids
Pyruvate
Mitochondria
Acetyl CoA
Citric acid cycle
CO_2
NH_4^+
Urea
NADH, $FADH_2$
ADP
Electron transport and oxidative phosphorylation
O_2
ATP
NAD^+, FAD
H_2O
Catabolic
Anabolic
Electrons